Boundary Element Methods for Soil-Structure Interaction

Boundary Element Methods for Soil-Structure Interaction

Edited by

W.S. HALL
University of Teesside,
Middlesbrough, United Kingdom

and

G. OLIVETO
University of Catania,
Catania, Italy

KLUWER ACADEMIC PUBLISHERS
DORDRECHT / BOSTON / LONDON

A C.I.P. Catalogue record for this book is available from the Library of Congress.

ISBN 1-4020-1300-0

Published by Kluwer Academic Publishers,
P.O. Box 17, 3300 AA Dordrecht, The Netherlands.

Sold and distributed in North, Central and South America
by Kluwer Academic Publishers,
101 Philip Drive, Norwell, MA 02061, U.S.A.

In all other countries, sold and distributed
by Kluwer Academic Publishers,
P.O. Box 322, 3300 AH Dordrecht, The Netherlands.

Printed on acid-free paper

Printed in the Netherlands.

CONTENTS

PART 2. RELATED TOPICS AND APPLICATIONS

7. BEM FOR CRACK DYNAMICS

M H Aliabadi (London)

8. SYMMETRIC GALERKIN BOUNDARY ELEMENT ANALYSIS IN THREE-DIMENSIONAL LINEAR-ELASTIC FRACTURE MECHANICS

A Frangi (Milan), G Maier(Milan), G Novati(Trento), R Springhetti (Trento)

CONTRIBUTORS

M. H. ALIABADI
Department of Engineering,
Queen Mary College,
University of London,
London, E1 4NS, UK.

D. AUBRY
Laboratoire de Mécanique de
Sols-Structures-Matériaux,
CNRS UMR 8370, École Centrale
de Paris, Châtenay Malabry,
France.

D. CLOUTEAU
Laboratoire de Mécanique de
Sols-Structures-Matériaux,
CNRS UMR 8370, École Centrale
de Paris, Châtenay Malabry,
France.

J. DOMÍNGUEZ
Escuela Superior de Ingenieros,
Universidad de Sevilla,
Camino de los Descubrimientos
s/n. 41092 Sevilla, SPAIN.

A. FRANGI
Department of Structural
Engineering,
Politecnico di Milano, P.za L.
da Vinci 32, 20133 Milan, Italy.

C.Z. KARAKOSTAS
Institute of Engineering
Seismology and Earthquake
Engineering,
P.O. Box 53, GR 551 02 Finikas,
Thessaloniki, Greece.

G. MAIER
Department of Structural
Engineering,
Politecnico di Milano, P.za L.
da Vinci 32, 20133 Milan, Italy.

G.D. MANOLIS
Department of Civil Engineering,
Aristotle University,
P.O. Box 502, GR 540 06,
Thessaloniki, Greece.

G. NOVATI
Department of Mechanical and Structural Engineering, University of Trento, via Mesiano 77, 38050 Trento, Italy.

C POLIZZOTTO
Dipartimento di Ingegneria Strutturale e Geotecnica, Università di Palermo, Viale della Scienze, 90128 Palermo, Italy.

H. POWER
Department of Mechanical Engineering, University of Nottingham, University Park, Nottingham, NG7 2RD, UK.

E. REINOSO
Instituto de Ingenieria, UNAM, Ciudad Universitaria, Apartado Postal 70-472, Mexico City, D.F. 04510, Mexico.

R. SPRINGHETTI
Department of Mechanical and Structural Engineering, University of Trento, via Mesiano 77, 38050 Trento, Italy.

C C. SPYRAKOS
Earthquake Engineering Laboratory, Civil Engineering Department, National Technical University Of Athens, Greece.

C. SONG
School of Civil and Environment Engineering, University of New South Wales, Sydney, NSW 2052, Australia.

J. P. WOLF
Institute of Hydraulics and Energy, Department of Civil Engineering, Swiss Federal Institute of Technology, Lausanne, Switzerland.

L. C.WROBEL
Department of Mechanical Engineering, Brunel University, Uxbridge, UB8 3PH, UK.

INTRODUCTION

W S HALL
School of Computing and Mathematics, University of Teesside, Middlesbrough, TS1 3BA UK
G OLIVETO
Division of Structural Engineering, Department of Civil and Environmental Engineering, University of Catania, Viale A. Doria 6, 95125 Catania, Italy

Soil-Structure Interaction is a challenging multidisciplinary subject which covers several areas of Civil Engineering. Virtually every construction is connected to the ground and the interaction between the artefact and the foundation medium may affect considerably both the superstructure and the foundation soil. The Soil-Structure Interaction problem has become an important feature of Structural Engineering with the advent of massive constructions on soft soils such as nuclear power plants, concrete and earth dams. Buildings, bridges, tunnels and underground structures may also require particular attention to be given to the problems of Soil-Structure Interaction. Dynamic Soil-Structure Interaction is prominent in Earthquake Engineering problems.

The complexity of the problem, due also to its multidisciplinary nature and to the fact of having to consider bounded and unbounded media of different mechanical characteristics, requires a numerical treatment for any application of engineering significance. The Boundary Element Method appears to be well suited to solve problems of Soil-Structure Interaction through its ability to discretize only the boundaries of complex and often unbounded geometries. Non-linear problems which often arise in Soil-Structure Interaction may also be treated advantageously by a judicious mix of Boundary and Finite Element discretizations. The purpose of this state of the art book on "Boundary Element Methods for Soil-Structure Interaction" is to review progress made in the applications of the Boundary Element Method in the solution of Soil-Structure Interaction for the scientific communities of Structural and Earthquake Engineering. The object is to provide these communities with a wealth of efficient computational methods for the solution of problems which would otherwise require less accurate and/or computationally more expensive procedures.

The book contains nine chapters from leading European experts on Boundary Element Methods and Soil-Structure Interaction. Its concept originated at the EUROMECH Colloquium 414 on Boundary Element Methods for Soil-Structure Interaction which took place in Catania, Italy from 21 to 23 June, 2000 at which the authors made short presentations on the state-of-the-art in their particular area of expertise. Since that time each author has developed these first ideas into a significant contribution to the subject. Scientific papers also presented at the Colloquium have already appeared as a Special Issue of Meccanica (Advances in Boundary Element Methods in Soil-Structure Interaction and Other Applications, Volume 36, Issue 4, 2001).

The book is organised into two parts. Part 1, containing five of the nine chapters that constitute the book, deals with problems specific to Soil-Structure or Fluid-Structure-Soil Interaction. Part 2, containing the remaining four, is devoted to related topics and applications that nevertheless are of interest to specific aspects of Soil-Structure Interaction.

In Part 1 the first Chapter is by Professor J Dominguez of the University of Seville and contains a review of 25 years of dynamic Soil-Structure Interaction. The material is introduced from an engineering point of view and after a brief introduction of the Soil-Structure Interaction problem deals with the dynamic stiffness of foundations, the seismic response of foundations and with seismic problems related to gravity and arch dams. In particular the situations of empty and full reservoir are covered and the effects of bottom sediments and travelling waves are considered.

The second Chapter by Dr. D Clouteau and Professor D Aubry of the University of Paris is devoted to formulation and computational aspects of the Soil-Structure Interaction problem. An introduction is provided to briefly describe the physical and numerical models used in the treatment of the problem. A physical and mathematical formulation of the problem is provided in the second section. Then the concept of domain decomposition is introduced, together with several techniques useful for the reduction of degrees of freedom to be considered in the analyses. The fourth section is specifically dedicated to boundary integral equations, the Boundary Element Method and to coupling with other numerical techniques, particularly to the FEM-BEM coupling. The fifth section considers unbounded interfaces, invariant operators and invariant domains in connection with their application to specific problems. The next section deals with Green functions in layered half-spaces. The seventh and final section of this Chapter is devoted to applications. The techniques described in the previous sections are therefore applied to Soil-Structure-Fluid Interaction in arch dams and sheet-piles, Soil-Structure Interaction for a nuclear reactor resting on a layered half-space with random heterogeneities and to geophysics boreholes. Topographic effects and the characteristic city effect are also described.

The third Chapter by Dr J P Wolf of the University of Lausanne and Dr C Song of the University of Sydney presents an alternative approximate approach to the solution of the dynamic Soil-Structure Interaction problem which is essentially based on the Boundary Element Method but does not require fundamental solutions. This method is appealing when the fundamental solutions are not known or when they are difficult to evaluate. After an introduction to the literature on the method, the dynamic unbounded Soil-Structure Interaction problem is defined and the unknown quantities are identified. The next section presents the main concept on which the scaled-boundary element method is based. In sections 4 and 5 two derivations of this approximate method are presented with the first being mathematically motivated and the second mechanically oriented. In section 6 the analytical solution in the radial direction is explicitly provided in the frequency domain while in section 7 the corresponding numerical solutions are formulated in the frequency and time domains. Several possible extensions of the procedure are discussed in section 8 while a set of numerical applications is reported in section 9. Section 10 presents some results obtained for bounded media and the final section of this Chapter discusses problems connected with the implementation of the method, its advantages and its limitations.

The fourth Chapter by Professor G D Manolis & Dr. C Z Karakostas of the University of Thessaloniki addresses the problem of Soil-Structure Interaction in random media. The first section presents an introduction to the problem and an outline of the presentation of the material. The second section presents a review of the literature on soil dynamics and dynamic Soil-Structure Interaction when the soil is considered as a random medium. Section 3 presents the general formulation of the problem in the form of a stochastic integral equation, a formal analytical solution and a closure approximation for zero mean forcing function. Section 4 addresses the problem of forced vibration in random soil media. After reporting an analytical solution for a simple problem, a BEM based approximate solution making use of the perturbation theory is developed and illustrated by means of a numerical example. The fifth section is used for the formal development of a BEM formulation based on the perturbation theory. Two examples concerning unlined circular tunnels are used to show the excellent performance of the theory in the presence of small random perturbations. For large random perturbations a different BEM formulation based on orthogonal polynomial expansions is developed in section 6. An example considering the propagation of an SH wave in a random medium shows why small perturbation theory cannot reliably predict results when the medium randomness is large.

The fifth Chapter by Professor C C Spyrakos of the University of Athens considers Soil-Structure Interaction as it is currently used in engineering practice with reference to buildings and bridges. Initially two brief literature reviews are given separately for buildings and bridges. Then the second section is dedicated to the seismic design of buildings including the effects of Soil-Structure Interaction. Two design procedures derived from seismic codes and guidelines are presented and applied to the case of an actual building showing the effects of Soil-Structure Interaction on the structural response and on the resulting design. The third section is devoted to the seismic analysis of bridges including SSI. It starts by providing information on the modelling of the various structural parts, especially soil and abutments, and continues by presenting two analysis procedures: a linear iterative static procedure and a non-linear static procedure. The section concludes with a numerical application illustrating the linear iterative procedure and with a parametric study considering various soil types. Seismic codes and guidelines also mainly inspire this Chapter.

The sixth Chapter, by Professor C Polizzotto of the University of Palermo (which is the first of Part 2 covering related topics and applications of the Boundary Element Method) considers the problem of non-local elasticity. This problem is of interest in Soil-Structure Interaction because some classical soil models are non-local and because the singularity problems of local elasticity vanish if the non-local approach is adopted. After the introduction, the Eringen non-local elastic model is reviewed in section 2. The third section introduces a non-local hyperelastic material through thermodynamically consistent constitutive equations. The fourth section discusses the static boundary value problem for such a material proving that, whenever it exists, the solution is unique. Moreover the problem is formulated as a classic linear elastic one of the local type with an unknown initial strain field accounting for the non-local behaviour. In section 5 some variational principles of local elasticity are extended to the non-local model considered while in section 6 a stationary principle is provided in terms of boundary integrals. The symmetric Galerkin and non-symmetric collocation BEM formulations for non-local elasticity are presented in the next two sections 7 and 8.

In the seventh Chapter by Professor M H Aliabadi of the University of London the application of the Boundary Element Method to crack problems in dynamic fracture mechanics is presented. After a review of the literature on the subject provided in the introductory section, a formulation of the dual Boundary Element Method for three-dimensional crack problems in the time domain is presented in section 2 together with a numerical solution procedure. The third section presents a formulation of the problem in the Laplace transformed domain with the Durbin method used to bring the solution back into the time domain. In the fourth section the dual reciprocity BEM method is presented leading to a system of coupled second order ordinary differential equations which can be solved by direct time integration methods. The fifth section points to the singularity problems that must be addressed in each of the previously presented formulations and provides the necessary lead to the relevant literature on the subject. Finally two examples are presented in the section on numerical applications where the results obtained by the three previously mentioned methods are compared among themselves and against solutions available in the literature.

The eighth Chapter by Professors A Frangi and G Maier of Milan Polytechnic, Professor G Novati and Dr R Springhetti of the University of Trento is devoted to the application of the symmetric Galerkin Boundary Element Method (SGBEM) to the solution of three-dimensional linear elastic problems in Fracture Mechanics. A brief introductory section reviews the literature on the subject and focuses on the particular problem in hand. In the second section two relevant boundary integral equations are formulated for displacements and tractions. A special regularising procedure is then applied to remove high order singularities arriving at a couple of self-adjoint boundary integral equations containing only weakly singular terms. The symmetrical structure of the problem can be maintained also in the discrete boundary element formulation if a Galerkin interpolation scheme is used. The third section deals with the numerical evaluation of the weakly singular integrals. Specific integration formulae, based on appropriate co-ordinate transformations, are provided for the cases of coincident elements, adjacent elements with a common edge and elements having in common only a vertex. In the fourth section several numerical applications are carried out and compared against results available in the literature. The first three examples refer to typical fractures in an infinite medium: a penny shaped crack, an elliptical crack and a spherical-cap crack. The results provided in the form of displacement discontinuities or stress intensity factors compare favourably against analytical or numerical results available in the literature. The other three examples refer to edge cracks in finite domains: an edge crack in a bar, a circular edge crack in a plate, a quarter elliptic corner crack in a plate. Once again the results compare very well with others available in literature.

The ninth and final Chapter of the book by Professor L C Wrobel of Brunel University, Dr E Reinoso of the University of Mexico City and Professor H Power of the University of Nottingham presents an important application of the BEM to a specific and relevant topic in dynamic Soil-Structure Interaction and earthquake engineering, typically the problem of site or local amplification effect. An introductory section explains how local amplification effects are predicted by one-dimensional theories and why more comprehensive two- and three-dimensional theories may be required in many practical applications. The same section also gives a review of the subject showing how the problem has been dealt with in the literature. The second section summarises the main results of the theory of wave propagation in an elastic, homogeneous and isotropic half space in a way suitable for the envisaged applications. In particular P, SH, SV and Rayleigh waves are described. Section 3 presents a two-dimensional BEM formulation for SH incident waves in canyons and valleys. Application of the model to the Mexico City valley situation shows that the one-dimensional theory predicts good results towards the centre of the valley, but is not adequate towards the edges where the response is much more irregular. The fourth section presents a similar formulation for P, SV and Rayleigh incident waves.

The next three sections deal, respectively, with the observed amplifications in the Mexico City valley, with the amplifications predicted by the one-dimensional theory and with those predicted by the two-dimensional BEM model. The conclusion is that, although the one-dimensional theory can often predict the average amplification behaviour, in many cases the two-dimensional theory is more adequate and in some cases only a full three-dimensional model can explain the complete behavioural pattern.

Overall the book provides an authoritative guide to the literature on the subject covered and is expected to be an invaluable tool for practising engineers, students and scholars in the fields of structural, geotechnical and earthquake engineering. Engineers and students may readily locate the material or methods available for the solution of their particular problem while scholars may discover methods previously not considered for the particular application being considered. The book should also be of interest to the larger community of applied mathematicians and software developers in seeing a field where the Boundary Element Method can provide a wealth of relevant and efficient solutions. Finally the book can be used as a starting point for research and for the investigation of unsolved problems in Soil-Structure and Fluid-Structure-Soil Interaction, particularly non-linear coupled problems which could be advantageously approached by means of Boundary Element Methods.

W S Hall, Middlesbrough
G Oliveto, Catania
February 2003

ACKNOWLEDGMENTS

The editors would like to take this opportunity to thank CRUI, the British Council, the European Mechanics Society, GNDT-CNR and MIUR for their support over a number of years, first for a bilateral research project between the University of Catania and the University of Teesside. This eventually lead to the Catania EUROMECH Colloquium 414 in June 2000, at which were laid the foundations for the present volume and related special issue of the journal Meccanica.

PART 1

SOIL-STRUCTURE INTERACTION

CHAPTER 1.
TWENTY FIVE YEARS OF BOUNDARY ELEMENTS FOR DYNAMIC SOIL-STRUCTURE INTERACTION

J. DOMÍNGUEZ
Escuela Superior de Ingenieros, Universidad de Sevilla,
Camino de los Descubrimientos s/n. 41092 Sevilla, SPAIN.

1. Introduction

This chapter is intended to show the applicability of the Boundary Element Method (BEM) to Dynamic Soil Structure Interaction (DSSI) problems. To show this a review of the work carried out by the author and his co-workers during the last twenty five years is presented. Reference is made to the work done by many others, however, the chapter is not a state of the art review of all the work done in the field of numerical dynamic soil-structure interaction.

The behaviour of structures based on compliant soils and subject to dynamic actions may depend to a large extent on the soil properties and on the foundation characteristics. The analysis of this behaviour requires a model which takes into account not only the structure but also the soil and the dynamic interaction forces existing between them. The first DSSI problems, studied during the late thirties, were related to the vibration of large machines mounted on massive foundations. The dynamic behaviour of these machines could only be understood by taking into account the dynamic interaction between the soil and the machine foundation. Tall buildings, or any other structure based on the ground, subject to wind loads are also examples of problems where DSSI effects may be important and where the excitation is directly applied to the structure. The analysis of structures under the effects of earthquakes leads to a second kind of DSSI problem where the excitation is transmitted through the soil.

To show in a simple manner the important effects of DSSI on the dynamic response of ground based structures the following simple problem is analyzed. Consider a single degree of freedom system consisting of a

W.S. Hall and G. Oliveto (eds.), Boundary Element Methods For Soil-Structure Interaction, 1–60.

concentrated mass M which can move horizontally (Figure 1), connected to an elastic foundation through a flexural bar with stiffness K. The foundation may move horizontally with a stiffness K_x and rotate with a stiffness K_ϕ. Any vertical motion is restricted.

The horizontal displacement of the mass under a ground motion excitation can be written as:

$$u = u_g + u_f + \phi_f h + u_e \tag{1}$$

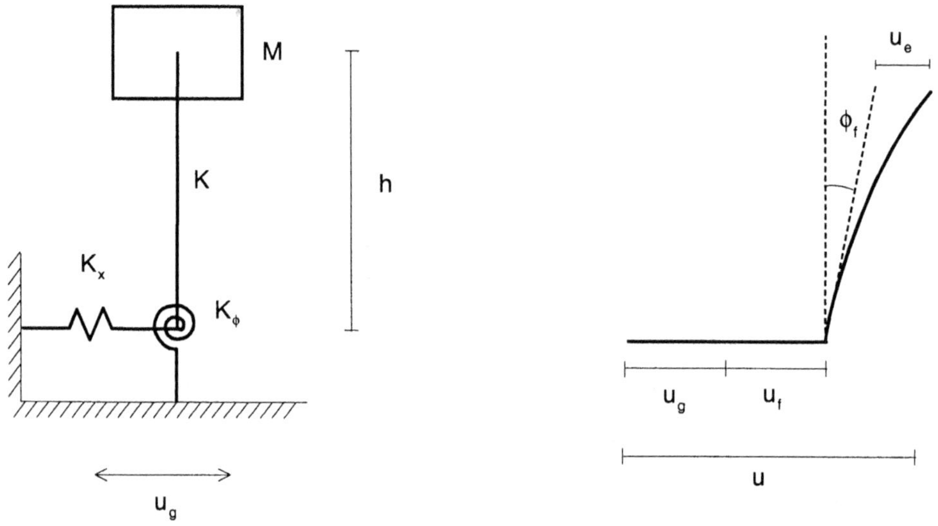

Figure 1. Simple model including dynamic soil-structure interaction effects.

where u_g represents the ground displacement, u_f the horizontal displacement of the foundation, ϕ_f the foundation rotation and u_e the elastic deformation of the flexural member. The mass acceleration ü is written as:

$$\ddot{u} = \ddot{u}_g + \ddot{u}_f + \ddot{\phi}_f h + \ddot{u}_e \tag{2}$$

and the equilibrium equation for the mass M as:

$$M\,\ddot{u} + K\,u_e = 0 \tag{3}$$

The total mass acceleration can be obtained in terms of the ground acceleration and the acceleration due to the elastic deformation $\ddot{u}_e$. From the two equilibrium equations of the flexural member:

$$K_x u_f = K u_e \tag{4}$$

$$K_\phi \phi_f = K u_e h \tag{5}$$

one obtains

$$\ddot{u}_f = \frac{K}{K_x} \ddot{u}_e \tag{6}$$

$$\ddot{\phi}_f = \frac{Kh}{K_\phi} \ddot{u}_e \tag{7}$$

and by substitution into equation (2),

$$\ddot{u} = \ddot{u}_g + \left(\frac{K}{K_x} + \frac{Kh^2}{K_\phi} + 1 \right) \ddot{u}_e \tag{8}$$

The equilibrium equation (3) becomes:

$$M \ddot{u}_g + M \left(\frac{K}{K_x} + \frac{Kh^2}{K_\phi} + 1 \right) \ddot{u}_e + K u_e \tag{9}$$

or

$$M \left(\frac{K}{K_x} + \frac{Kh^2}{K_\phi} + 1 \right) \ddot{u}_e + K u_e = -M \ddot{u}_g \tag{10}$$

A comparison of this equation to that corresponding to the same system on a rigid foundation ($K_x \sim \infty$, $K_\phi \sim \infty$, $\delta = \phi = \delta = \phi = 0$)

$$M \ddot{u}_e + K u_e = -M \ddot{u}_g \tag{11}$$

shows that both systems have natural frequencies that can be very different. Thus, when the foundation compliance is considered

$$\omega_n^2 = \frac{K}{M} \frac{1}{\left(\frac{K}{K_x} + \frac{Kh^2}{K\phi} + 1 \right)} = \frac{\omega_n^2 \; rigidbase}{1 + \frac{K}{K_x} + \frac{Kh^2}{K\phi}} \tag{12}$$

The foundation flexibility can modify the natural frequency of the system to an important extent and therefore its response to dynamic loads of any kind.

As mentioned before, design of machine foundations was the first engineering problem where DSSI effects were considered. The basic goal in this case is to limit the foundation motion amplitudes which allow for a satisfactory operation of the machine and do not disturb the people in the vicinity. The design rules for machine foundations were based on tradition and rules-of-thumb during the first half of the twentieth century. Those methods were often obtained from a Winkler elastic reaction of the soil and an added mass corresponding to part of the soil that would be vibrating in phase with the foundation. A revision of the classical methods may be found in the books by Whitman and Richart (1967) and by Richart et al. (1970).

The basic foundation stiffness problem can be seen in Figure 2. The foundation is assumed to be a rigid block on a half-space representing the soil, under the effects of a vertical force *p(t)*. To compute the foundation stiffness, its mass is considered to be zero. The relation between the dynamic force *p(t)* applied to the foundation and its displacement *x(t)* gives the foundation stiffness.

$$p(t) = K(t)x(t) \tag{13}$$

Assuming a time harmonic excitation with frequency ω,

$$p(t) = P_o \, e^{i\omega t} \tag{14}$$

the foundation displacement is

$$x(t) = X_o \, e^{i\omega t} \tag{15}$$

where X_o is a complex number. The foundation stiffness is also a complex number:

$$K(\omega) = \frac{p(t)}{x(t)} = \frac{P_o}{X_o} \tag{16}$$

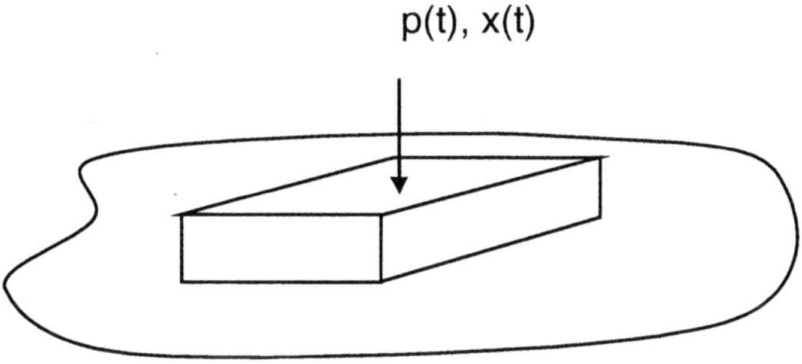

Figure 2. Rigid foundation on uniform half-space.

In order to visualize the meaning of the foundation stiffness one may consider a simple analogy drawn by Roesset (1980). Assume a single degree of freedom system as shown in Figure 3 to represent the soil under the footing.

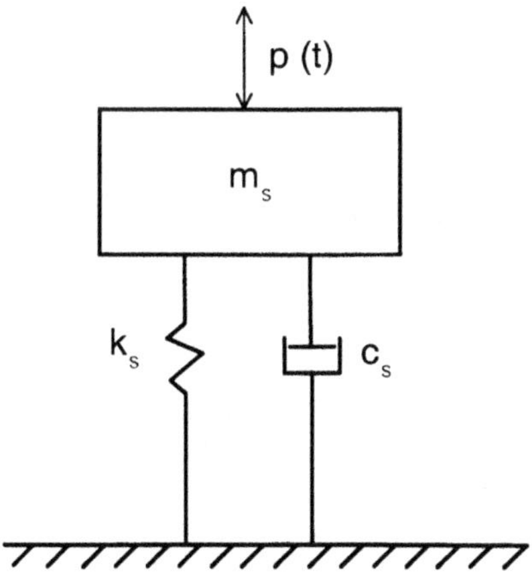

Figure 3. Analogy for soil dynamic response.

Its equilibrium equation

$$p(t) = m_s \ddot{x} + c_s \dot{x} + k_s x \tag{17}$$

under time harmonic loading gives

$$\frac{p(t)}{x(t)} = (k_s - m_s \omega^2) + i\omega c_s \tag{18}$$

By comparison of equations (16) and (18) one obtains:

$$K(\omega) = (k_s - m_s \omega^2) + i\omega c_s = k_1 + ik_2 \tag{19}$$

or

$$K(\omega) = k_s (1 - \frac{\omega^2}{\omega_n^2}) + 2i\beta \frac{\omega}{\omega_n} \tag{20}$$

where $\omega_n = \sqrt{\frac{k_s}{m_s}}$ and $\beta = \frac{c_s}{c_{cr}} = \frac{c_s}{2\sqrt{k_s m_s}}$. This quantity can be written as:

$$K(\omega) = k_s \{k + i\omega c^*\} \tag{21}$$

Thus, the spring constant of the simple model represents the static stiffness of the foundation whereas the two frequency dependent coefficients k and c^* represent the real and the imaginary parts of the dynamic foundation stiffness, respectively. These coefficients can be represented versus the angular frequency ω as shown in Figure 4.

The actual dynamic vertical stiffness coefficients for a rigid circular massless foundation on an elastic half-space as obtained by Veletsos and Wei (1971) are shown in Figure 5. The qualitative agreement between both figures is clear.

DSSI effects when loads act directly on the structure (wind loads, moving machinery, traffic on bridges, etc) are basically due to the foundation compliance. They can be taken into account by using the foundation stiffness matrix (a frequency dependent matrix when the analysis is done in the frequency domain). In many other cases the

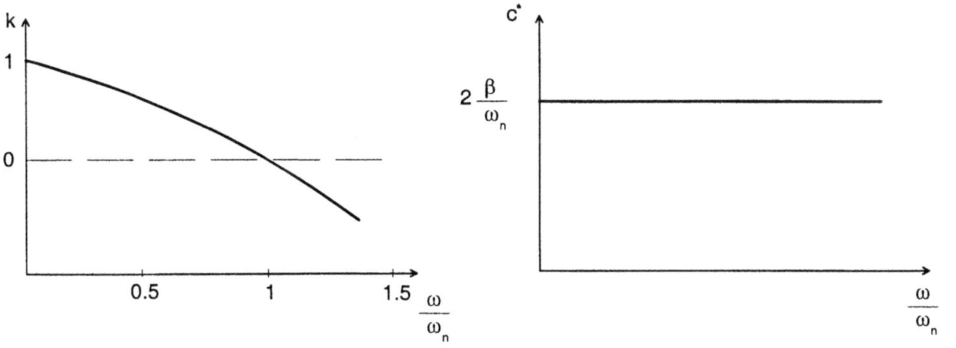

Figure 4. Stiffness coefficients for model of Figure 3.

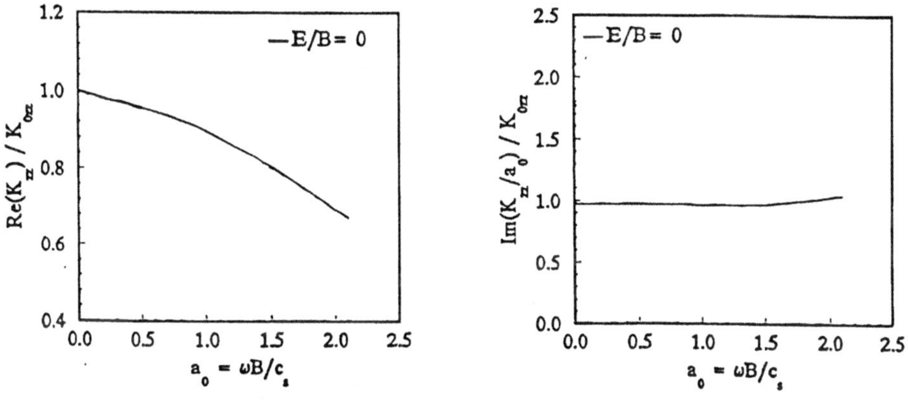

Figure 5. Dynamic vertical stiffness coefficients for a rigid square foundation on an elastic half-space.

dynamic excitation comes from the soil (earthquakes, nearby road or railway traffic, underground explosions, etc). In those cases, the influence of DSSI on the structural response is twofold: on the one hand the excitation due to waves impinging on the structure depends on the soil properties and the foundation characteristics; on the other hand, the response of the structure to the excitation also depends on DSSI effects.

As general statement it can be said that the influence of DSSI on the response of structures to ground motion is important for large and massive structures. Power plants, bridges, dams and large buildings are typical examples where this phenomenon is relevant. When a large structure is excited by waves travelling through the soil, as occurs in the event of an earthquake, two important effects associated to the size of the foundation and the structure are present. The first one is called kinematic interaction and is associated to the size and geometry of the foundation. The existence of a large massless foundation would produce by itself a filtering of the incident waves in such a way that the foundation response is a function of its own geometry. The phenomenon can be illustrated by the image of two very light boats on the surface of the sea; one very small and the other equally light but large. The first one would follow the free surface motion without any change in it; the second one would have its own motion and would change the sea motion in its vicinity.

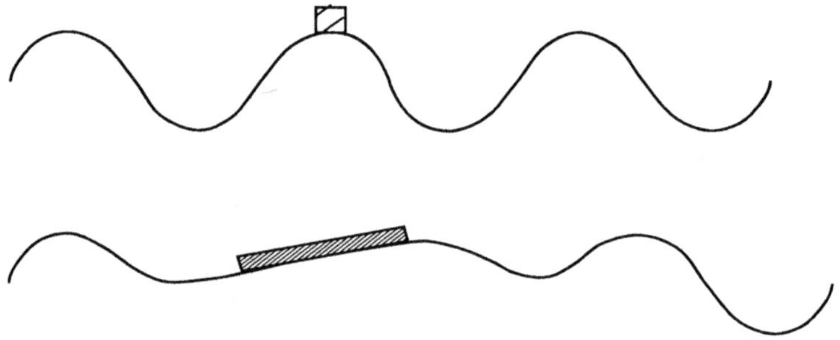

Figure 6. Kinematic interaction.

The second effect is known as the travelling-wave effect. It takes place when the characteristic length of the structure is of the same order as the wavelength of the seismic waves. For instance, harmonic waves with a 0.2 s period in a rock with a shear wave velocity of 2500m/s have a wavelength of 500 m. Over a distance of 125 m., which can be the length of a bridge or a dam, the ground motion changes from its maximum value to zero. The importance of this effect depends on the size of the structure and on the type, frequency, and direction of the waves.

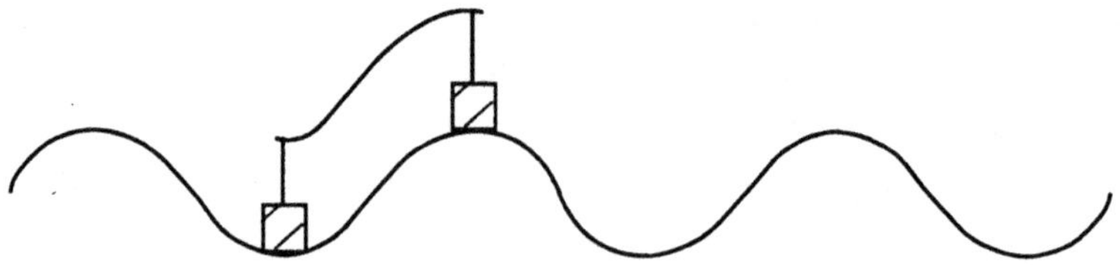

Figure 7. Travelling-wave effect.

Soil-structure interaction is, in most cases, studied assuming linear elastic behaviour. Under this assumption the soil-structure system can be analyzed in the frequency domain using a substructure technique. Foundations are in many cases massive and may be assumed to be rigid. Their dynamic behaviour is characterized by the stiffness (rigidity) matrix, which relates the force vector (forces and moments) applied to the foundation assumed massless with the resulting displacement vector (displacements and rotations). Once the dynamic stiffness of a foundation is known, its response including the mass, or that of any structure supported on it, may be immediately evaluated in those cases where the dynamic excitation is directly applied to the structure. When the system is excited by waves travelling through the soil, prior to the analysis of the structure mounted on the springs defined by the foundation stiffness, the excitation of such a system must be determined. To this end, the forces and moments needed to avoid any motion of the massless foundation impinged by the waves travelling through the soil (kinematic interaction) are computed. Opposite forces and moments are applied to the foundation in the complete soil-foundation-structure model in order to compute the response of the structure to the incoming waves.

The analysis of the seismic response of structures on flexible foundations or large structures where the travelling wave effects are important require the use of a model where soil and structure are studied together as will be seen in Section 4 of this Chapter. Soil-structure interaction problems where non-linear effects are important require a direct time domain analysis. Non-linear contact conditions and non-linear behaviour of the structure are the most frequent situations for which a time domain analysis is required.

2. Dynamic Stiffness of Foundations.

The first study of the stiffness of a foundation representing the soil as a linear elastic half-space was carried out by Reissner (1936). He studied the response of a disc on the surface of the soil subjected to vertical harmonic forces. A uniform distribution of stresses under the disc was assumed. Knowing that the actual stress distribution was far from being uniform, in the mid 1950´s, several authors carried out studies assuming certain stress distributions for circular and rectangular foundations (Arnold et al; 1965; Bycroft, 1956). The mixed boundary value problem, with prescribed displacements under the rigid footing and zero traction over the remaining portion of the surface, was studied during the 1960's and early 1970's (Paul, 1967; Veletsos and Wei, 1971; Luco and Westman, 1971). Relaxed boundary conditions were assumed under the footing. Several studies were also made using viscoelastic soil models (Veletsos and Verbic, 1973).

Wong and Luco (1976) computed dynamic compliances (stiffness inverse) of a surface rigid massless foundation of arbitrary shape on an elastic half-space by dividing the soil-foundation interface into rectangular elements. The tractions were considered to be uniformly distributed within the elements and a relation between the tractions over an element and the displacements on the soil surface was obtained by integration of Lamb's point load solution (1904). This method is, in fact, a Boundary Element Method with a half-space fundamental solution. However, the integration of this fundamental solution is rather involved and only surface foundations may be analyzed.

The first numerical technique widely used for computation of foundation stiffness was the Finite Element Method (FEM). The development of energy absorbing boundaries for 2-D by Waas (1972) and for axisymmetric problems by Kausel (1974) made possible the analysis of foundations resting on, or embedded in, layered soils. The finite element models, however, contain assumptions like the existence of a rigid bedrock at the bottom, or a parallel layered geometry extending to infinity, that are not always realistic. In addition, 3-D dynamic soil-structure interaction problems present important difficulties for finite element models due to the large number of elements involved in the analysis and the lack of infinite elements such as those existing for 2-D problems.

Boundary Element Methods (BEM) based on boundary integral equations are very well suited for dynamic soil-structure interaction problems and they have also become a very widely used approach for the solution of this type of problems. Unbounded regions are naturally represented. The radiation of waves towards infinity is automatically included in the model, which is based on an integral representation valid for internal and external regions.

The first BE application for DSSI problems was presented by Domínguez in 1978(a). The direct formulation of the BEM was applied to compute dynamic stiffness of foundations. The frequency domain formulation was used to obtain stiffness of rectangular foundations resting on, or embedded in, a viscoelastic half-space. Otternstrener and Schmid (1981) and Otternstrener (1982), followed the same approach to study, respectively, dynamic stiffness of foundations and cross-interaction between two foundations. Non-homogeneous soils have been studied by Abascal and Domínguez (1986). Apsel and Luco (1983) used an indirect BEM in combination with semi-explicit Green's functions to compute stiffness of circular foundations embedded in a layered half-space. Dynamic stiffness of circular foundations on the surface or embedded in layered soils have been computed using the direct BEM by Alarcón et al. (1989) and Emperador and Domínguez (1989). Karabalis and Beskos (1984) computed dynamic stiffness of surface foundations excited by non-harmonic forces using the time domain BEM. Also in the time domain, Mohammadi and Karabalis (1990) studied the use of adaptive discretization techniques and compared "relaxed" versus "non-relaxed" boundary conditions. The BEM has also been used to compute dynamic stiffness of foundations when soil-foundation separation exists, by Hillmer and Schmid (1988), and Abascal and Domínguez (1990).

In most problems where the soil-structure interaction effect is important the foundation is massive and may be studied as a rigid body. When the foundation is a strip footing that may be represented by a plane model (Figure 8), it has three degrees of freedom corresponding to the horizontal, vertical and rocking (rotation) co-ordinates. For 3-D foundations (Figure 9) each vector has six components: one vertical, two horizontal, two rocking and one torsional.

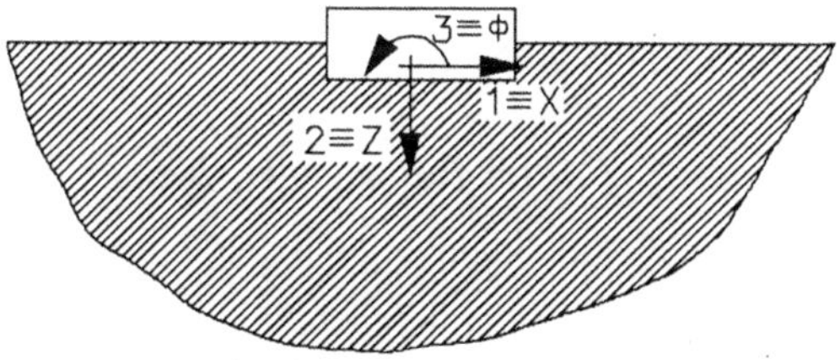

Figure 8. Strip foundation.

For a harmonic excitation with frequency ω, the dynamic stiffness matrix relates the vector of forces (and moments) $\boldsymbol{R}$, applied to the foundation and the resulting vector of displacements (and rotations) $\boldsymbol{u}$, when the foundation is assumed to be massless.

$$\boldsymbol{R} = \boldsymbol{K}\,\boldsymbol{u} \tag{22}$$

The terms of the matrix ***K*** are functions of the frequency *ω*. Properly speaking, the matrix ***K*** should be called stiffness or impedance of the soil for a given shape of the foundation.

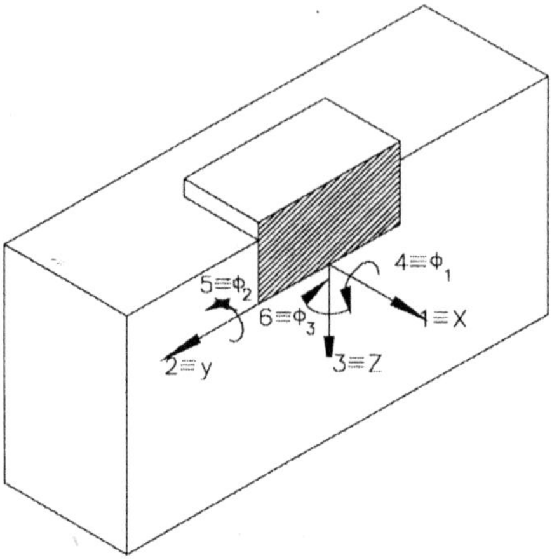

Figure 9. Three-dimensional foundation.

It is worth saying that the dynamic forces and displacements related by equation (22) are generally out of phase. It is convenient, then, to use complex notation to represent forces and displacements. The stiffness components are also written as

$$K_{ij}(\omega) = Re(K_{ij}) + i\, Im(K_{ij}) \tag{23}$$

where $i = (-1)^{1/2}$.

The real part of the stiffnesses is related to the stiffness and inertia properties of the soil. The imaginary part shows the damping of the system. The main damping effect is due to the energy dissipated by the waves propagating away from the foundation (radiation damping). It is obvious that since this kind of damping is associated to the wave radiation, it exists for linear elastic half-space models or any other model that permits radiation of the waves. In addition to the radiation damping, material damping will, in general, exist.

The radiation damping is highly frequency dependent. Because of that, the stiffness components are usually written as

$$K_{ij} = K_{0ij}(\ k_{ij} + i\ a_0\ c_{ij}\ ) \tag{24}$$

where K_{0ii} is the static value of the "ij" stiffness component, k_{ij} and c_{ij} are frequency-dependent coefficients, $a_0 = \omega B/c_s$, B is a characteristic length of the foundation, and c_s is the shear wave velocity of the soil.

When material damping exists an attempt to isolate the effect of that damping is done by writing the dynamic stiffness in the following form:

$$K_{ij} = K_{0ij}(\ k_{ij} + i\ a_0\ c_{ij}\ )(1 + 2i\beta\ ) \tag{25}$$

where β is the damping ratio. The coefficients k_{ij} and c_{ij} still depend on the material damping; however, for deep soil deposits and typical values of β, this dependence is small.

In a similar way to that used to define the stiffness matrix, one may define its inverse by

$$\boldsymbol{u} = \boldsymbol{F}\ \boldsymbol{R} \tag{26}$$

The matrix $\boldsymbol{F}$, frequently used instead of the stiffness matrix, is known as the dynamic compliance matrix or the dynamic flexibility matrix. In the following, both the dynamic stiffness matrix and the dynamic compliance matrix will be used. Following complex notation, the terms of the dynamic compliance matrix may be written as

$$F_{ij}(\ \omega\ ) = Re(\ F_{ij}\ ) + i\ Im(\ F_{ij}\ ) \tag{27}$$

2.1. THREE-DIMENSIONAL FOUNDATIONS

As was said above, Boundary Elements are well suited for 3-D dynamic analysis of foundations since they can represent in a simple manner the half-space under the footing. Dynamic stiffnesses of surface and embedded, square and rectangular, foundations computed using the frequency domain BEM formulation, are presented in this section.

Due to the use of a fundamental solution corresponding to the complete space, not only the soil-foundation interface, but also the soil free-surface, should be discretized. However, in practice only a small region around the foundation has to be included in the model since there is a small effect of the free-surface far away from the foundation on the computed values of the stiffness coefficients. Some authors have used the half-space fundamental solution for the BE analysis of problems involving a half-space (Luco and Apsel, 1983). In such a case, the soil free-surface is automatically taken into account and no discretization of this part of the boundary is required. However, there is a price paid for the simplification. Since there is no closed form expression for the half-space fundamental solution, which depends on unbounded integrals, rather involved approximate procedures are required for its evaluation. The author's experience shows that the use of a full-space fundamental solution is very simple and produces accurate results with discretizations of the free surface restricted to a rather small region around the foundation.

For all the problems analyzed in this section, constant rectangular boundary elements are used. Constant elements produce enough accuracy for problems which do not include flexure. Figure 10 shows the very simple BE discretizations used for stiffness computation. Results obtained for surface and embedded foundations using this kind of discretization are accurate for low frequencies and small embedment. A more refined mesh is required for high frequencies and large embedment ratio as shown in Figure 11.

One column of the foundation stiffness matrix is obtained prescribing a unit rigid body motion of the foundation following a certain co-ordinate. Zero tractions are prescribed on the soil free-surface.

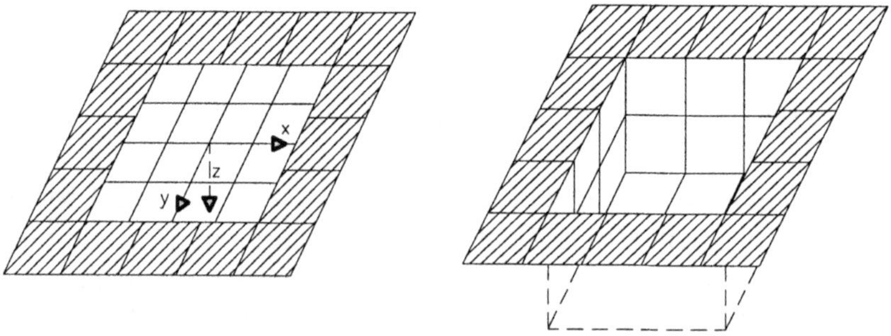

Figure 10. Simple BE meshes for surface and embedded foundation.

Even though the stress distribution under the footing has sharp peaks, the BE mesh for the soil-foundation interface does not have to be very dense (Domínguez, 1978a) since the stress resultants over the foundation, and not the stress distributions, are needed. A study of the number of elements required under the footing and on the soil free-surface can be found in Domínguez (1993). A discretization like the one shown in Figure 11 with A > 8E yields accurate results for embedded foundations.

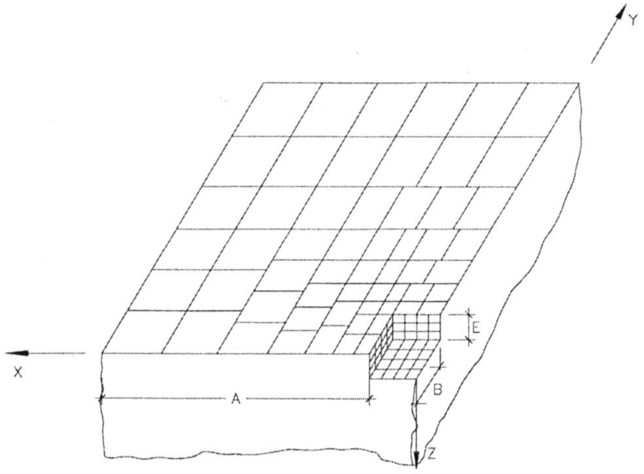

Figure 11. Discretization of one quarter of a square embedded foundation.

The following approximate formulae for the static stiffnesses of square embedded foundations obtained using constant BE were proposed by Domínguez and Abascal (1987):

Quadratic approximation:

$$K_{0xx} = 9.41\frac{\mu B}{2-\upsilon}\{1+1.13\frac{E}{B}-0.16(\frac{E}{B})^2\}$$

$$K_{0zz} = 4.75\frac{\mu B}{1-\upsilon}\{1+0.47\frac{E}{B}-0.05(\frac{E}{B})^2\}$$

$$K_{0\phi_1\phi_1} = 4.38\frac{\mu B^3}{1-\upsilon}\{1+0.98\frac{E}{B}+1.13(\frac{E}{B})^2\}$$

$$K_{0\phi_3\phi_3} = 8.71\,\mu B^3 \{1 + 2.8\frac{E}{B} - 0.19(\frac{E}{B})^2\} \tag{28}$$

The values of the dynamic stiffness components may be computed without special difficulties for different values of the dimensionless frequency a_o. Dynamic stiffness coefficients for square and rectangular, surface and embedded foundations may be found in the works of Domínguez (1978a and 1993). In the following, only a small number of results taken from those references are shown.

Figure 12 shows the variation with frequency of the dynamic horizontal, vertical and rocking stiffness coefficients for several values of *E/B*. Except for the horizontal component, the real parts normalized with respect to the corresponding static values present a variation with a_0 that is almost linear and independent of *E/B*. The imaginary horizontal and vertical coefficients are highly dependent on *E/B* but remain almost constant with frequency a_0; that is, the imaginary part of the dynamic stiffness varies almost linearly with a_0.

2.2. STRIP FOUNDATIONS

The frequency domain formulation of the BEM for 2-D regions may be used to compute dynamic stiffnesses of strip footings in the same way as was done for 3-D foundations. The number of unknowns is smaller and the discretization and treatment of the data, simpler. In the following, results for a homogeneous viscoelastic soil model are analyzed. Particular attention is also paid to foundations resting on non-homogeneous soils.

Earthquake damage observation shows that local soil mechanical properties, underground and surface topography, and foundation geometry have an important effect on the dynamic behaviour of structures. The complexity of the system to be modelled has made numerical methods the most suitable way to deal with the problem. After the development of energy absorbing boundaries by Waas (1972) and Kausel (1974), finite elements became a widely used technique for this kind of problem. However, finite element models present two unavoidable requirements that may be difficult to satisfy in certain cases. First, the model must be bounded at the bottom by a rigid bedrock and second, the soil away from the foundation must be represented by parallel layers unbounded in the horizontal direction. These two conditions are not always close to reality.

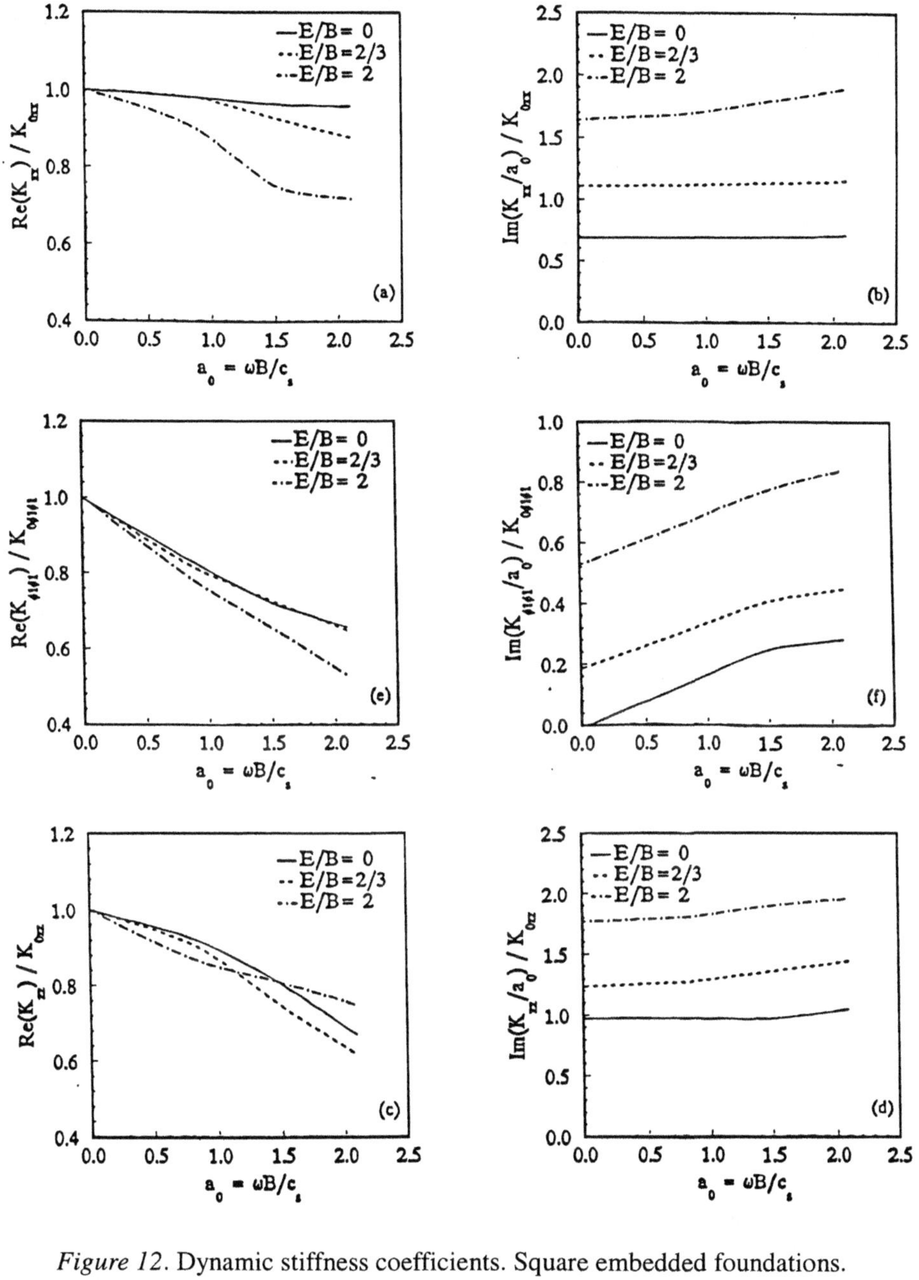

Figure 12. Dynamic stiffness coefficients. Square embedded foundations.

There are cases where the base under the soil deposit is not very rigid or the soil geometry is far from being horizontally layered. The BEM is a good alternative for those problems. It permits an easy representation of soils with irregular shape (Figure 13) and the modelling of soils bounded at the bottom by a compliant half-space.

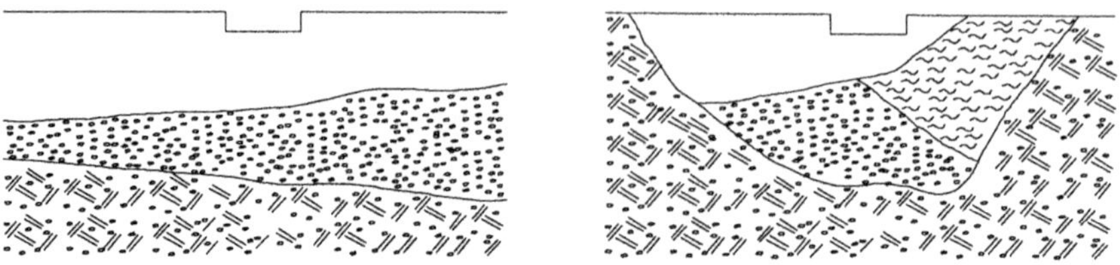

Figure 13. Zoned soils.

In the following, compliances of surface massless foundations are presented. The compliances of surface strip footings resting on a viscoelastic half-plane are studied first. Secondly, compliances of foundations resting on the surface of a soil deposit which is on the top of a viscoelastic half-plane model are studied. The analysis is done for a soil layer on top of the viscoelastic half-plane and also for a semi-elliptical soil deposit included in the half-plane (Figure 14). In order to check the importance of a deformable lower bedrock, the rigidity of the half-plane

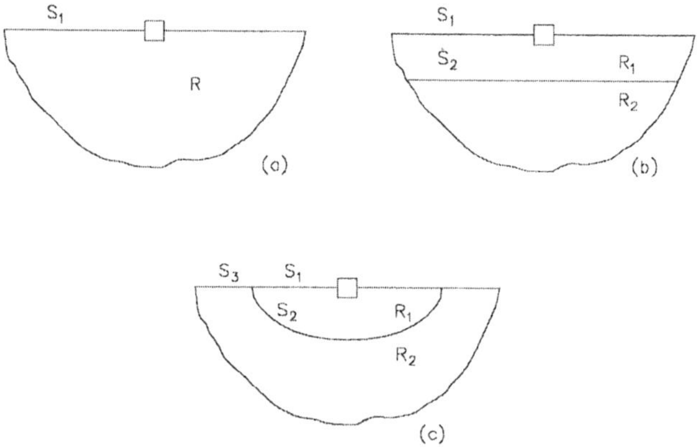

Figure 14. Homogeneous and zoned viscoelastic soil models.

takes values going from that of the soil deposit (homogeneous half-plane) to infinity (rigid bedrock). The axis ratio (D/H) of the ellipse takes several values from unity to infinity (horizontal layer) to show the influence of the soil deposit not being a horizontal layer. A more complete study of the use of the BEM for dynamic stiffnesses of foundations on zoned viscoelastic soils was presented by Abascal and Domínguez (1986).

To obtain compliances of a surface strip footing resting on a viscoelastic half-plane, half of the soil surface is discretized into 43 elements, ten of which are under the foundation (Figure 15). An amount of free-field equal to ten times the foundation width is discretized. This amount is not necessary for this particular case but it has been used to maintain the same surface discretization of the layered model.

The soil model of Figures 14b and 15b has been used to show the influence of non-infinitely rigid bedrock on the foundation compliance. A parametric study showing this influence can be found in Domínguez (1993).

As has been said above, the hypothesis of horizontal soil layers boundless in the horizontal direction may not be in correspondence with reality. However, this hypothesis has to be made in Finite Elements. If the BEM is used, the above limitation does not exist and arbitrarily shaped soil profiles may be modelled. An analysis of the influence of the shape of the soil deposit can be done using the model of Figures 14c and 15c.

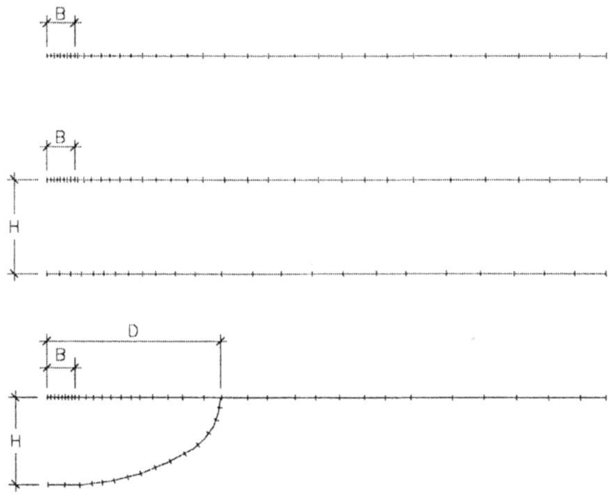

Figure 15. BE discretization for models of Figure 14.

It has been shown in Domínguez (1993) that the hypothesis of layered soil and rigid bedrock may lead to erroneous values of the compliances if the base is not very rigid or the soil deposit is not wide enough.

The problems presented are examples of the capabilities of the BEM for the computation of dynamic stiffnesses of foundations, for arbitrary underground geometry and properties

2.3. AXISYMMETRIC FOUNDATIONS

The first formulations of the BEM for axisymmetric elastostatic problems were done by Mayr (1975) and Kermanidis (1975). The integral representation in cylindrical coordinates has the same expression of 2-D and 3-D problems when one uses a fundamental solution corresponding to ring loads following the radial, tangential and axial directions. Those fundamental solutions for elastostatics are written in terms of Legendre functions or elliptic integrals (Kermanidis, 1975, Cruse et al., 1977), which makes their integration along the boundary elements rather involved. The harmonic ring load fundamental solution may be obtained in terms of an infinite line integral of Hankel functions (Domínguez and Abascal, 1984) and its integration along the elements is again complicated. On the contrary, the 3-D static or time-harmonic point load solution may be easily integrated over axisymmetric surface elements. Because of that they were used by Domínguez (1993) for the BEM treatment of axisymmetric problems. The geometry and field variables are axisymmetric. The boundary of the generating surface of the body is discretized into line elements and the point load collocated at each node. The 3-D fundamental solution, in terms of cylindrical coordinates, is integrated along the boundary elements of the generating surface and along the azimuthal co-ordinate. The results shown in the present section were obtained following this approach.

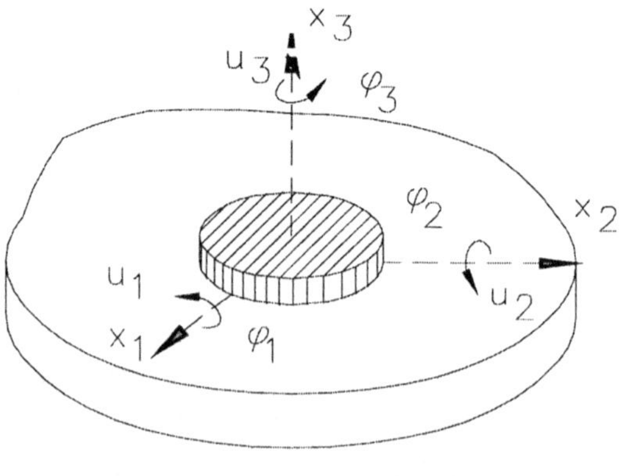

Figure 16. Circular foundation on half-space.

The generalized frequency-dependent force-displacement relationship (stiffness matrix) for a massless rigid axisymmetric foundation can be written as (see Figure 16)

$$\begin{bmatrix} F_1 \\ M_1 \\ F_2 \\ M_2 \\ F_3 \\ M_3 \end{bmatrix} = \begin{bmatrix} K_{hh} & -K_{hr} & 0 & 0 & 0 & 0 \\ -K_{rh} & K_{rr} & 0 & 0 & 0 & 0 \\ 0 & 0 & K_{hh} & K_{hr} & 0 & 0 \\ 0 & 0 & K_{rh} & K_{rr} & 0 & 0 \\ 0 & 0 & 0 & 0 & K_{vv} & 0 \\ 0 & 0 & 0 & 0 & 0 & K_{tt} \end{bmatrix} \begin{bmatrix} u_1 \\ \phi_1 \\ u_2 \\ \phi_2 \\ u_3 \\ \phi_3 \end{bmatrix} \tag{29}$$

The stiffness functions are written as

$$K_{ij} = K_{0\,ij}(k_{ij} + i\,a_0\,c_{ij}),\ (i, j = h, r, v, t) \tag{30}$$

where $K_{0\,ij}$ is the static value, k_{ij} and c_{ij} are the dynamic stiffness and damping coefficients, respectively, and $a_0 = \omega R/c_s$ is the dimensionless frequency. R is the characteristic foundation radius and c_s the shear-wave velocity in the material under the foundation. Each material is defined by a complex modulus $\mu^* = \mu(1 + 2i\xi)$ in which ξ is the material damping; the density ρ and the Poisson's ratio υ. Each column of the stiffness matrix is obtained prescribing a unit displacement or rotation following one of the co-ordinates and computing the resultant force and moment at the foundation centre point.

A first test of the B.E. approach is obtained by the comparison of the calculated dynamic stiffness and damping coefficients, of a circular foundation on a half-space, with analytical solutions obtained assuming relaxed contact conditions. (Veletsos and Wei; 1971; Luco and Westmann, 1971). The material is assumed to be perfectly elastic with a Poisson's ratio $\upsilon = 1/3$. Welded contact conditions between the soil and the foundation are used in the present study. The boundary element discretization under the foundation consists of eight constant elements of variable length (Figure 17). Even though a complete space fundamental solution is used, the soil free surface is not discretized because the effect of those elements on the stiffness of surface foundation is very small. In fact, if "smooth" contact between the foundation and the soil were assumed, the free-surface element would not have any effect at all on the equations of the interface elements (Domínguez, 1978a).

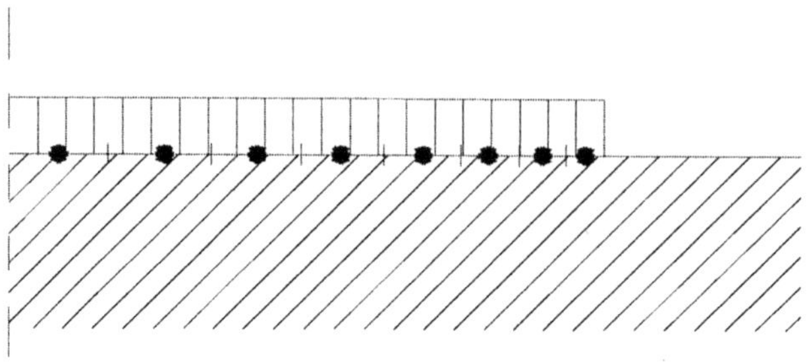

Figure 17. Boundary discretization for circular foundation on half-space.

Figure 18 shows a comparison between the stiffness coefficients computed by the proposed approach and the analytical solutions published by Veletsos and Wei (1971) and by Luco and Westmann (1971). The agreement between the results can be considered as good, in particular when the simplicity of the mesh (only eight constant elements) and the kind of contact conditions used in each study are taken into account. A zero damping factor has been considered for half-space. Results for values of a_0 as low as 0.01 have been obtained without any numerical difficulty. Results for $a_0 = 0$ have also been computed.

Stiffnesses of circular foundations on multi-layered soils were computed by Alarcón et al. (1989) using the above procedure. Additional results for this problem can be found in Domínguez (1993). This type of problems shows the ability of the proposed approach to compute dynamic stiffness coefficients in cases where there are resonance peaks due to the existence of a soil layer on a stiffer bedrock. The results obtained were in good agreement with those obtained by Chapel (1981) using a boundary element approach and with the results presented by Luco (1974).

A first estimation of the capabilities of the BEM for embedded foundations is obtained by computation of the dynamic stiffness coefficients of cylindrical foundations bounded by a uniform viscoelastic half-space. A discretization used for one of these foundations is shown in Figure 20.

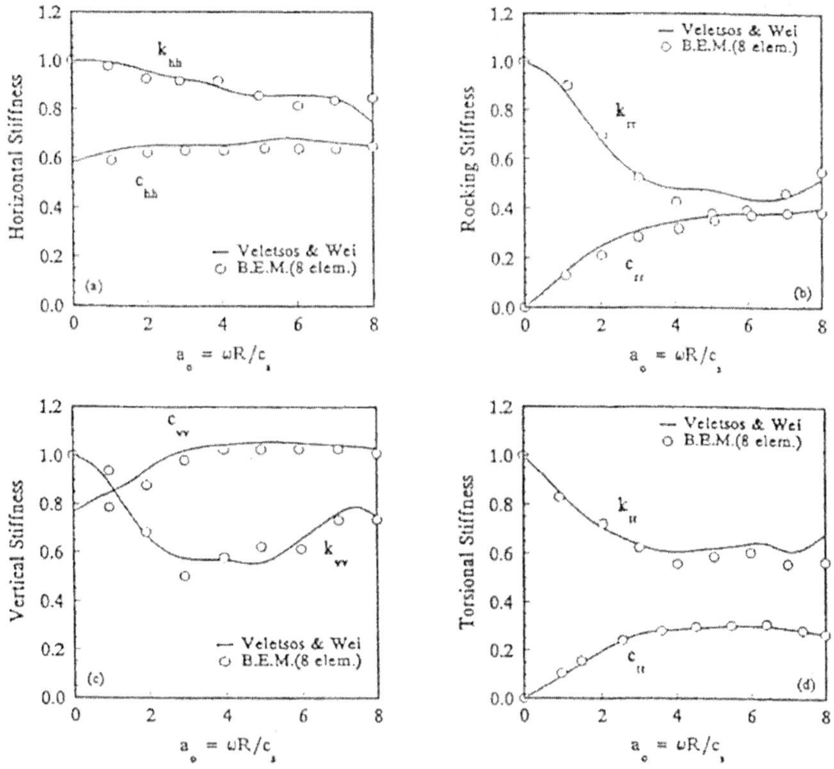

Figure 18. Stiffness coefficients for circular foundation on an elastic half-space (υ=1/3).

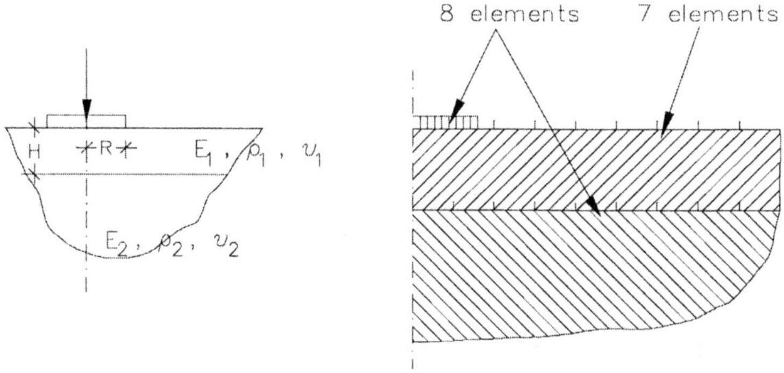

Figure 19. Circular foundation on a soil layer on a half-space. Model and boundary element discretization.

Results for the dynamic stiffness components of cylindrical foundations or axisymmetric foundations of any other shape can be obtained by using BE models as the one shown in Figure 21. Results obtained with this kind of discretizations are proved to be accurate (Domínguez, 1993).

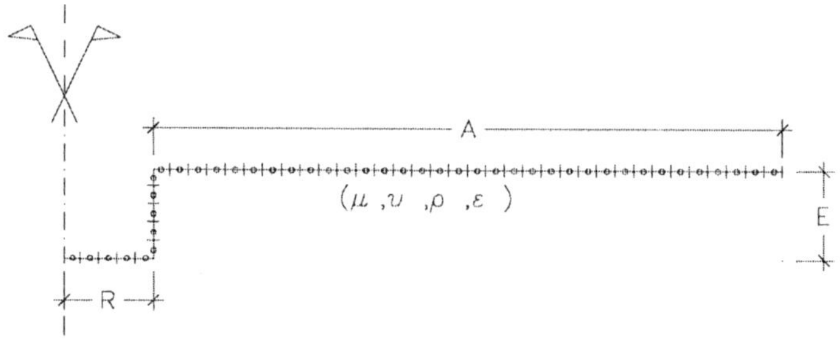

Figure 20. Boundary discretization for cylindrical foundation embedded in a uniform half-space.

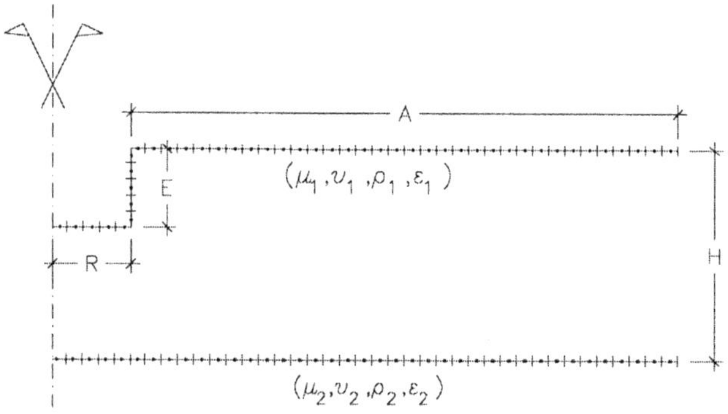

Figure 21. Boundary discretization for cylindrical foundation embedded in a soil layer on compliant bedrock.

2.4. FOUNDATIONS ON SATURATED POROELASTIC SOILS

In the present section the use of the BEM to obtain dynamic stiffness coefficients of strip foundations on two-phase poroelastic soils is shown. The technique is based on the Boundary Element formulation for poroelastic media obtained by Domínguez (1991, 1992) and Cheng et al. (1991) after developing an integral equation formulation from Biot's differential equations. The soil under the foundation may consist of poroelastic and viscoelastic zones, the interaction between them being rigorously represented. The contact condition between the soil and foundation may be of any type (pervious or impervious, smooth or welded). Although the BE models presented are only for foundations on a half-space or on a stratum based on rigid or compliant bedrock, the technique is very versatile and the analysis of embedded foundations and more complicated underground geometry only require a different Boundary Element mesh.

The behaviour of fluid-filled poroelastic regions can be represented using a boundary element model. The integral equation formulation obtained by Domínguez (1991, 1992) from Biot's equations is discretized into quadratic elements to obtain the traditional boundary element (BE) equation

$$\boldsymbol{H}\,\boldsymbol{u} = \boldsymbol{G}\,\boldsymbol{p} \tag{31}$$

where $\boldsymbol{u}$ is a $3N$ vector including all nodal values of displacement components of solid phase and the fluid stress; $\boldsymbol{p}$ is a $3N$ vector containing boundary tractions on solid phase and the pore fluid displacement normal to boundary at N nodes of mesh; and $\boldsymbol{H}$ and $\boldsymbol{G}$ are $3N \times 3N$ matrices whose terms are obtained by integration of fundamental solution components times shape functions along boundary elements. The previous system of equations together with the boundary conditions permits one to solve the boundary value problems to obtain all the unknown solid phase boundary displacements or tractions, and all the unknown fluid stresses or normal displacements. Once the boundary value problem is solved, values of $\boldsymbol{u}$ and $\boldsymbol{p}$ at internal points may be easily obtained. Simple numerical examples of the BE solution of dynamic poroelastic problems using constant elements may be found in Domínguez (1992, 1993).

Should the domain of interest consist of coupled poroelastic saturated zones and viscoelastic zones, the boundaries of each zone can be discretized into elements. Equation (31) can be written for the poroelastic zones and a well-known relation of the same type, relating nodal values of the boundary displacements $\boldsymbol{u}$ and tractions $\boldsymbol{p}$, written for the viscoelastic zones. Establishing the compatibility and equilibrium conditions along the interfaces and the external boundary conditions, the coupled dynamic problem can be solved.

The equilibrium and compatibility conditions along the interface between an pervious solid zone (denoted by superscript s) and a poroelastic zone (denoted by superscript p) are:

(1) Equilibrium of the normal traction on the solid p_n^s and the total normal traction on the poroelastic medium T_n^p

$$p_n^s + T_n^p = 0 \tag{32}$$

with $T_n^p = t_n^p + \tau^p = t_n^p - \phi^p p^p$, t_n^p being the normal traction on the skeleton, p^p the pore pressure and ϕ^p the porosity of the poroelastic medium.

(2) Equilibrium between the tangential traction on the solid p_t^s and on the porous medium T_t^p

$$p_t^s + T_t^p = 0 \tag{33}$$

where $T_t^p = t_t^p$ is the tangential traction on the skeleton.

(3) Compatibility of displacements along the interface

$$u_n^s = u_n^p = U_n^p \tag{34}$$

$$u_t^s = u_t^p \tag{35}$$

where u_n^s and u_t^s denote the displacement components of the solid zone, u_n^p and u_t^p the displacement components of the skeleton of the porous zone and U_n^p the normal displacement of the pore fluid.

If the solid material is pervious, the above compatibility and equilibrium conditions along the interface become as follows: equation (32) remains the same with $T_n^p = t_n^p$, ($\tau^p = p^p = 0$), equations (33) and (35) do not change and equation (34) becomes $u_n^s = u_n^p$. Note that a condition on the normal displacement of the pore fluid has been substituted by a condition on the pore pressure.

Quadratic boundary elements are used to discretize the soil surface and the interface of a massless rigid foundation on a homogeneous saturated poroelastic soil subjected to a harmonic load (Figure 22). The discretization, shown in Figure 22, is symmetric. The number of elements of the same size of each part of the boundary is shown in brackets.

Using the BE discretizations of Figures 22 and 24, Japon et al. (1997) studied the effects of the contact conditions, the seepage force, and the added density of the poroelastic material on the foundation dynamic stiffness. Results for the problems shown in Figures 22 and 24 can be found in their paper. The BE model may include easily the poroelastic material properties and the underground topography as required to obtain reliable results.

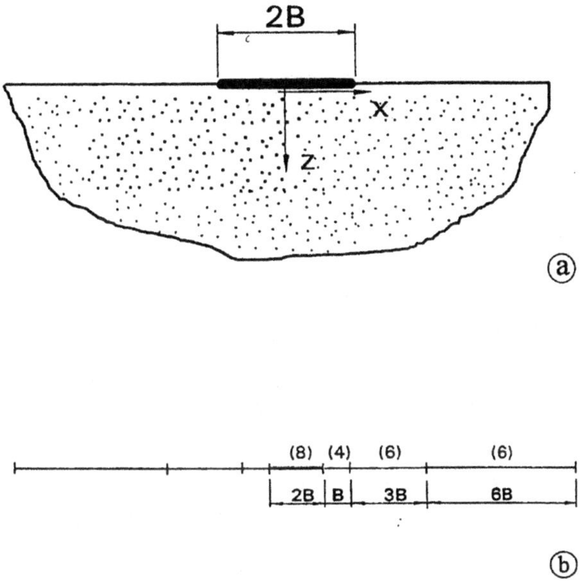

Figure 22. Strip footing on saturated poroelastic half-space: (a) geometry; (b) Boundary Element discretization.

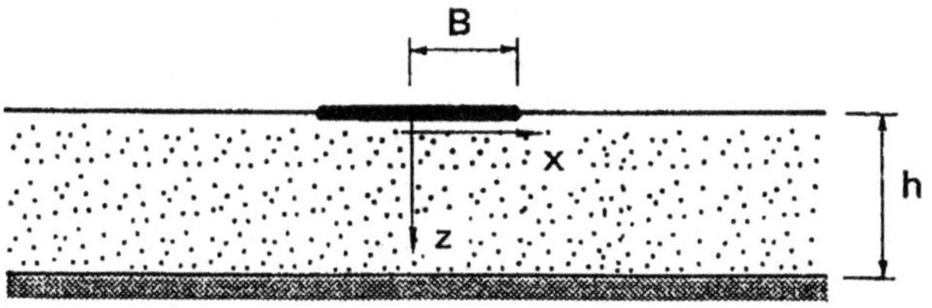

Figure 23. Strip footing on saturated poroelastic stratum.

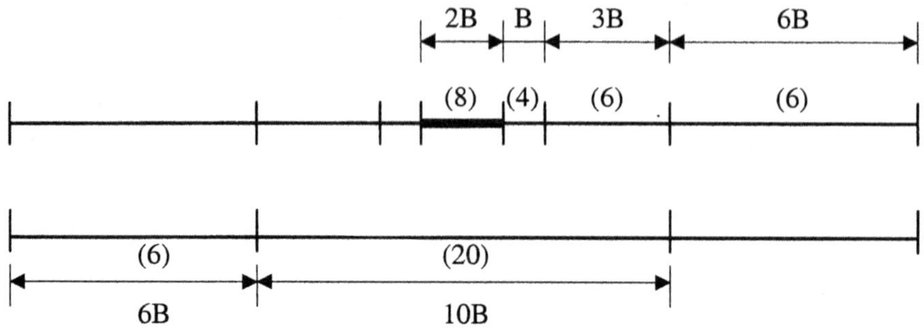

Figure 24. Boundary Element discretization for strip footing on saturated poroelastic stratum.

3. Seismic Response of Foundations

Diffraction problems dealing with infinite or semi-infinite regions are usually formulated by decomposing the total displacement and stress fields into two parts. One is the undisturbed (free) field ($\boldsymbol{u}^{(f)}$, $\sigma^{(f)}$) and the other the scattered field ($\boldsymbol{u}^{(s)}$, $\sigma^{(s)}$). This decomposition permits the use of the displacement integral representation of the scattered field for an external region, which consists of integrals that only extend over the internal boundaries since the radiation and regularity conditions are satisfied.

The integral representation of the scattered field for points on the soil free surface or on the soil-foundation interface when the soil is a homogeneous viscoelastic half-space may be written as

$$c_{lk}\,[u_k(\xi) - u_k^{(f)}(\xi)] = \int_{S_I} u_{lk}^{*}\,[p_k - p_k^{(f)}]\,dS - \int_{S_I} p_{lk}^{*}\,[u_k - u_k^{(f)}]\,dS \qquad (36)$$

where S_I represents (Figure 14) the soil-foundation interface plus the soil free surface, $u_k^{(f)}$ and $p_k^{(f)}$ are known and all the other symbols have the usual meaning. Once S_I has been discretized and the boundary conditions applied, the resulting system of equations gives the unknown values u_k and p_k.

In the case of a boundless horizontally layered soil (Figure 14b), the undisturbed free field variables correspond to the free field of the layered soil without foundation. For a non-homogeneous half-space including finite zones (Figure 14c), the $u_k^{(f)}$, $p_k^{(f)}$ exist only for the outermost zone and correspond to a uniform half-space with the properties of that zone. For instance, in Figure 14c the total fields are:

$$u(x) = u^{(f)}(x) + u^{(s)}(x) \qquad x \in R_2$$

$$\sigma(x) = \sigma^{(f)}(x) + \sigma^{(s)}(x)$$

$$u(x) = u^{(s)}(x) \qquad x \in R_1$$

$$\sigma(x) = \sigma^{(s)}(x) \tag{37}$$

The system of equations for the soil model of Figure 14b, once all the boundaries are discretized, is composed of: the integral representation of the scattered displacements for nodes on S_2 as boundaries of zone R_2, plus the integral representation of the scattered displacements for nodes on $S_1 + S_2$ as boundaries of zone R_1. In the case of Figure 14c, the system of equations is obtained from the integral representation of the scattered displacements for nodes on S_2 as boundaries of zone R_2 plus the integral representation of the total displacements for nodes on $S_1 + S_2$ as boundaries of zone R_1. The scattered fields are written as the difference between the total fields and the known free-fields. Compatibility and equilibrium conditions of the total fields are established along the internal boundary (S_2) and boundary conditions are prescribed for the total displacements or tractions along the external boundaries (S_1 *or* $S_1 + S_3$). Thus, the total displacements and tractions over the boundaries can be computed by means of the system of equations.

The motion of a rigid massless foundation induced by incident waves is computed following two steps. For the first step, zero tractions at the free-surface and zero displacements under the footing are prescribed. The solution of the system gives the tractions under the footing and its resultant $\boldsymbol{R}$ may be easily computed. The second step is the determination of the rigid body motion of the footing by solving the system $\boldsymbol{Ku^r} = -\boldsymbol{R}$, where $\boldsymbol{K}$ is the foundation stiffness matrix that is required for the soil-structure interaction analysis and, in any case, may be computed with little extra effort using the same integrals along the elements of the first step.

The free field motions are known in terms of exponential functions. The tractions are obtained by differentiation of those displacements.

The analysis of the response of foundations to incoming waves taking into account the interaction between the soil and the footing is a problem of diffraction of elastic waves. This kind of problem for inclusions and cavities has been treated by numerous authors since the early 1970´s. Most of the existing exact solutions correspond to 2-D antiplane models (Wong and Trifunac, 1974; Sanchez-Sesma and Rosenblueth 1979). Wong and Luco (1976) studied the response of 3-D surface foundations to travelling waves using the same above-mentioned boundary method they had used to compute foundation stiffnesses. Kobori, Minai and Shinozaki (1973) and Luco (1976) studied the torsional response of axisymmetric structures resting on the surface, to obliquely incident SH waves, using similar analytical procedures.

The BEM was applied to the diffraction of seismic waves by foundations by Domínguez (1978b). He studied the response of 3-D surface and embedded foundations to incident SH, SV and P waves, assuming a homogeneous viscoelastic soil. Karabalis and Spyrakos (1984) studied the response of foundations to travelling waves using the time domain BEM formulations and assuming a homogeneous elastic soil. Seismic response of foundations on non-homogeneous viscoelastic soils have been studied in the frequency domain by Domínguez and Abascal (1989).

The same BE discretizations used for dynamic stiffness computation can be used to obtain the seismic response of foundations. Different geometries for two and three-dimensional foundations can be found in Domínguez (1993). The BEM allows not only for the analysis of the seismic response of foundations but also for the study of the site amplification due to particular soil profiles. An example of this type of problem can be seen in Figure 25. Effects of the actual geometry of the site and material properties of the sub-regions can be studied with models like the one used for the surface stiffness computation shown in Figure 15.

The combined influence of the site amplification and the kinematic interaction due to a rigid foundation can be studied by including in the same model the foundation and the local underground topography as done in models like the one shown in Figure 20 for a cylindrical foundation, or the one shown in Figure 26 for a strip footing in an alluvial deposit.

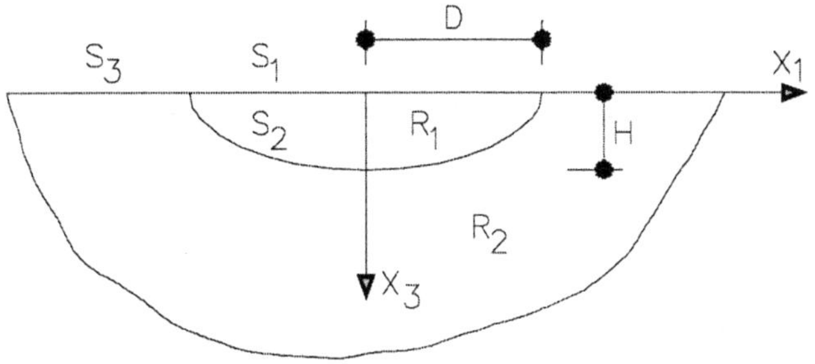

Figure 25. Semi-elliptical soil deposit included in viscoelastic half-space.

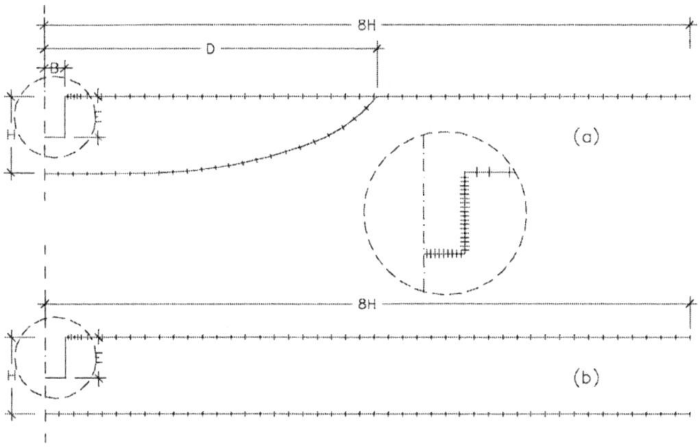

Figure 26. Boundary element discretization for foundations embedded in alluvial deposits on compliant bedrock: (a) Semi-elliptical deposit; (b) Layered deposit.

4. Dynamic Soil-Water-Structure Interaction. Seismic Response of Dams.

The analysis of the seismic response of dams is a very important problem within dynamic soil-structure interaction. In this case there is not only the soil, but also a second, very large region -the water- with a great influence on the earthquake response of the system. The effect of dam-water, dam-foundation and water-foundation interaction makes necessary the use of models including the three media and the interactions between them (notice that the word "foundation" is not used here in the sense of footing, as it was sometimes used in the previous sections, but in the sense of an underground region, normally rock, on which the structure is based). When an earthquake takes place, the different media (Figure 27) interact forming a coupled system. No single domain should be excluded from the model. The effect of the foundation and the water domain, both being very large or infinite regions, makes the dynamic analysis complicated. Variations of the soil properties or of the reservoir geometry far from the dam may have an important effect on its seismic response. It is obvious that in such a context the boundary techniques are very advantageous and, as shown below, permit the consideration of important factors which cannot be modelled at present by the domain techniques.

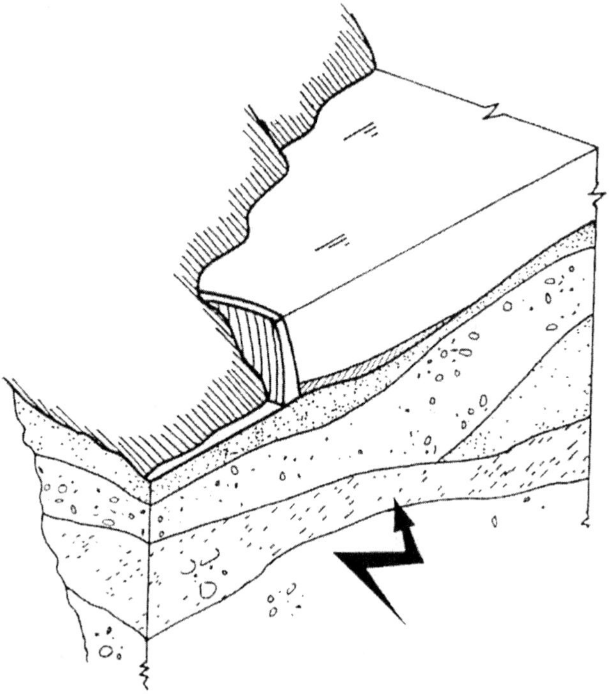

Figure 27. Dam-foundation-reservoir system.

In the case of gravity dams, the dam-water-foundation system behaves to a large extent as a plane one and therefore, a two-dimensional model can be used in most cases. Particularly notable in this context is the research conducted by Chopra and his co-workers using the FEM (Chopra and Chakrabarti, 1981; Hall and Chopra, 1982 a and b; Fenves and Chopra, 1985). Lotfi et al. (1987) presented an FE approach in which the interaction effects were taken into account more rigorously than in the previous models. The latter authors treated the foundation as a deep layered stratum on a rigid base, while, in the research by Chopra and his co-workers, the foundation was a homogeneous half-space.

Boundary Elements have been used by some authors to study fluid-solid interaction in two-dimensions. Kakuda and Tosaka (1983) did studies of coupled fluid-solid systems in the frequency domain while Antes and Von Estorff (1987) analyzed the effects of soil-fluid and dam-fluid interaction in the time domain. More recently, Medina and Domínguez (1989) and Domínguez and Medina (1989), presented a coupled two-dimensional BE model for the frequency domain analysis of gravity dams. In this model the proper equilibrium and compatibility conditions at fluid solid interfaces are taken into account. Irregular topographies of the bottom, underground inhomogeneities, bottom sediments or simple uniform viscoelastic soils can be represented. Domínguez et al. (1997) analyzed the effects of porous sediments on the seismic response of gravity dams.

Arch dams cannot be studied as two-dimensional structures and require a three-dimensional model. There have been different numerical studies of the earthquake behaviour of dams. Most of them deal with the evaluation of the hydrodynamic pressure on the dam using Finite Elements (Hall and Chopra, 1983), Boundary Elements (Aubry and Crepel, 1986; Tsai and Lee, 1987; Jablonsky and Humar, 1990) or Finite Differences (Wang and Hung, 1990). There are three-dimensional Finite Element models which include the three domains (dam, water and foundation rock) and take into account the water compressibility, the foundation rock flexibility and the dynamic interaction effects, Fok and Chopra (1986, 1987). However, most FE models contain important simplifications which may give rise to unrealistic results. Some of these simplifications are: a massless foundation, a bottom absorption coefficient and a uniform cross section of the reservoir from the vicinity of the dam to infinity.

Domínguez and Maeso (1993) and Maeso and Domínguez (1993) presented a coupled BE model which includes the three domains (dam-water-foundation) and the interaction between them. Due to the BE characteristics, this model is able to rigorously represent: the actual topography of the reservoir and the foundation free surface up to a significant distance from the dam, the foundation as a viscoelastic solid and the interaction effects. A study of the factors having an important influence on the earthquake response of arch dams may be seen in Maeso and Domínguez (1993) and Domínguez and Maeso (1993). A study of the bottom sediment effects on the seismic response of arch dams using the

BEM was presented by Maeso et al. (1999). The same author studied (Maeso et al. 2000) the effects of the space distribution of the excitation on the seismic response of arch dams.

4.1. FLUID-SOLID INTERFACES

The BE model for the dam-water-foundation system requires of the modelling of the water which is assumed to be a zero viscosity fluid under a time-harmonic small amplitude irrotational motion governed by the scalar wave propagation equation. The fluid boundary is discretized into BE. Solid-fluid interfaces require the shear tractions on the boundary of the solid in contact with water to be zero. The equilibrium condition indicates that the fluid pressure should be equal and opposite to the normal traction on the solid. The compatibility condition leads to equal normal displacements of both media at the points on the interface. These conditions can be written as:

$$0 = p^s \tau \tag{38}$$

$$- p^w = \mathbf{p}^s \mathbf{n} \tag{39}$$

$$u_n^w = \mathbf{u}^s \mathbf{n} \tag{40}$$

where $\boldsymbol{n}$ is the outward unit normal from the solid, τ is the unit vector tangent to the interface for two dimensions or any unit tangent vector for three dimensions, p^w is the water pressure, $u_n{}^w$ the fluid displacement along $\boldsymbol{n}$ and $\boldsymbol{p}^s$ and $\boldsymbol{u}^s$ are the tractions and displacement vectors respectively, at the boundary points of the solid. In the case of dynamic problems considered here the pressure, tractions and displacements are dynamic magnitudes in excess of the static ones which were already in equilibrium. The fluid normal velocity is related to the pressure derivatives as:

$$\rho \frac{\partial \dot{\boldsymbol{u}}}{\partial t} = -\Delta \boldsymbol{p}^w \tag{41}$$

where ρ is the water density.

Multiplying by a direction vector $\boldsymbol{n}$ and taking into account the value of the time derivatives for time-harmonic problems, one gets the following expression for the water displacement along $\boldsymbol{n}$

$$u_n^w = \frac{1}{\rho\,\omega^2}\frac{\partial p^w}{\partial n} \tag{42}$$

Substituting equation (42) into (40) one obtains

$$\frac{\partial p^w}{\partial n} = \rho\,\omega^2\,\boldsymbol{u}^s\,\boldsymbol{n} \tag{43}$$

There are six unknowns at the interface points of a two-dimensional problem, namely: two solid displacement components, two solid traction components, the water pressure and its normal derivative. There are also six equations: two BE equations for the solid, one for the fluid, and the interface equations (38), (39) and (43). In the three-dimensional case the number of unknowns is eight since displacements and tractions have three components. The eight equations are: three BE equations for the solid, one for the fluid, equation (38) for two-perpendicular tangential vectors, equation (39) and equation (43).

5. Gravity Dams

The study of gravity dams is normally done assuming that the dam behaves as a two-dimensional system. This assumption is based on the construction procedure of this kind of dams and on the observation of the Koyna Dam during the earthquake of 1967. A discussion of this simplification may be found in Rea et al. (1975) and Chopra and Chakrabarti (1981).

The analysis of the dam should be done for different foundation assumptions and different levels of water. Figure 28 shows the system analyzed. The height of the dam, its base width, the depth of the water, and the depth of the foundation are denoted by H, B, H_w, and H_f, respectively. The dam cross section is triangular with $B/H = 0.8$. Several cases are examined with respect to the reservoir: empty ($H_w/H = 0$), full of water ($H_w/H = 1$). Further, the analysis is carried out with a rigid foundation as well as a deep-stratum ($H_f/H = 4$) and half-space idealizations of a flexible foundation. The material properties are: (dam) modulus of elasticity = 27.5 GPa, Poisson's ratio = 0.2, unit weight = 2,480 Kg/m^3; (foundation) modulus of elasticity = 27.5 GPa, Poisson's ratio = 0.333333, unit weight = 2,640 Kg/m^3. For the dam and the foundation, material damping is assumed to be of the hysteretic type; the damping ratio is taken equal to 0.05, that is., the imaginary part of the modulus of elasticity is taken as 1/10 of the real part.

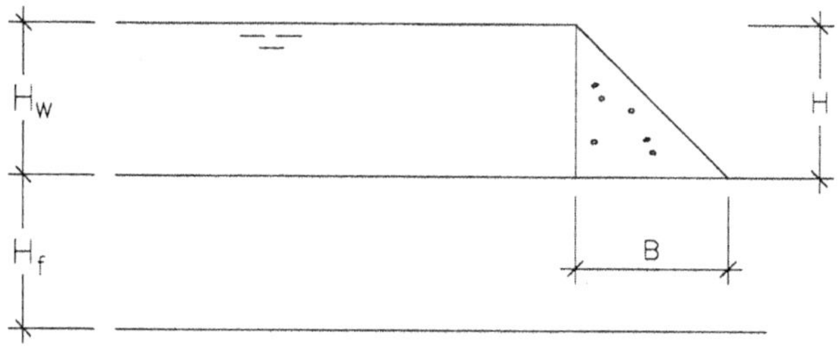

Figure 28. Dam-water-foundation system idealization.

The existence of sediments at the bottom of the reservoir may have an important effect on the seismic response of the dam. These effects are also analysed in the present section.

5.1. DAM ON RIGID FOUNDATION. EMPTY RESERVOIR

The dam is first assumed to be perfectly bonded to a rigid foundation and the reservoir assumed to be empty. The dynamic response of the dam is evaluated by the amplitude of the complex-valued frequency response function that represents the relative acceleration at the dam crest due to a unit free-field ground acceleration. Figure 29 shows the B.E. model for the dam. Quadratic elements are used. It should be noticed that some preliminary studies were done using constant elements and the results were very poor even for a large number of elements; however only a small number of quadratic elements were required to obtain accurate results.

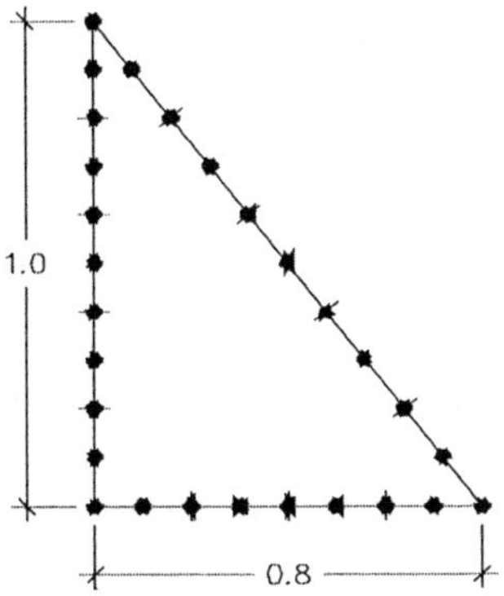

Figure 29. Gravity dam cross section and boundary discretization.

5.2. DAM ON RIGID FOUNDATION. RESERVOIR FULL OF WATER

The water region is assumed to be a constant depth channel extending to infinity. This assumption is made in order to compare with existing F.E. results. The B.E. discretization used for this case is shown in Figures 30 and 31.

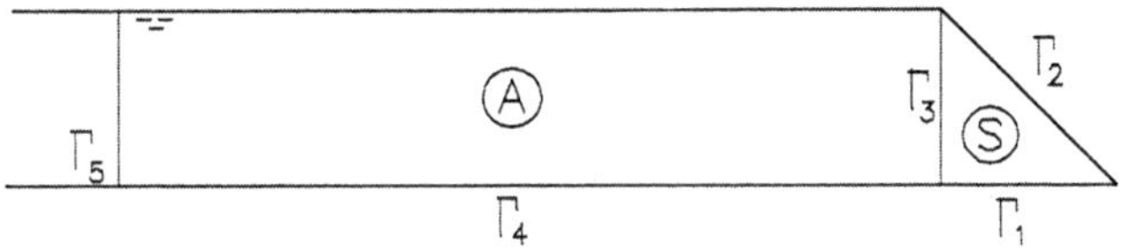

Figure 30. Model for reservoir on rigid foundation.

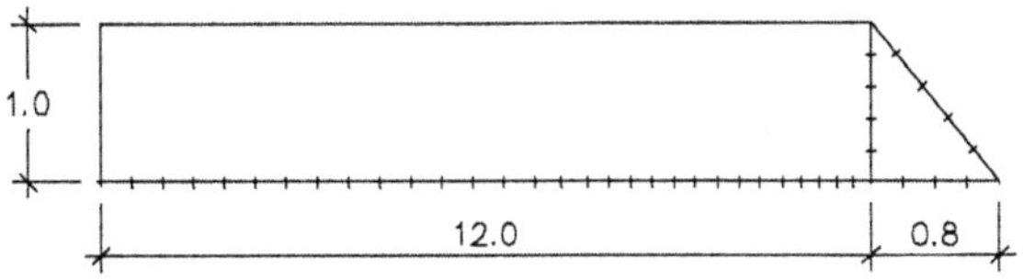

Figure 31. BE model (rigid foundation).

The discretization of the water channel cannot be truncated as was done in the previous sections for the foundation stiffness problems. For frequencies of excitation higher than a certain value, there will be standing waves along the channel, which do not damp out with the distance to the dam.

A BE discretization of both the fluid and the solid regions using quadratic elements is carried out. The BE equation for time-harmonic motion of viscoelastic media is written for nodes on the boundaries of the dam (Γ_1 , Γ_2 and Γ_3). The BE equation for Helmholtz equation problems is written for nodes on the boundaries of the fluid (Γ_3 , Γ_4 and Γ_5) using a double source fundamental solution. Compatibility and equilibrium conditions for nodes along the solid-fluid interface (Γ_3), prescribed boundary conditions for nodes along Γ_1 , Γ_2 and Γ_4, and the known pressure-displacement relation along Γ_5 obtained form the standing modes of a constant depth water channel complete the system of equations (see, Domínguez 1993).

5.3. DAM ON FLEXIBLE FOUNDATION. EMPTY RESERVOIR.

The dam-foundation system is assumed to be under the effects of vertical SV or P-waves that produce a unit acceleration on the far field free surface. The prescribed free field stress and displacement at any depth, if the dam and reservoir did not exist, can be easily computed. These stress and displacement fields are the same both in the upstream and the downstream directions at long distances from the dam. In the near field the stress and displacements are modified by the scattering effects of the dam-reservoir system. Displacements and stresses in the soil can be written as the sum of the unknown scattered field and the known far field solution.

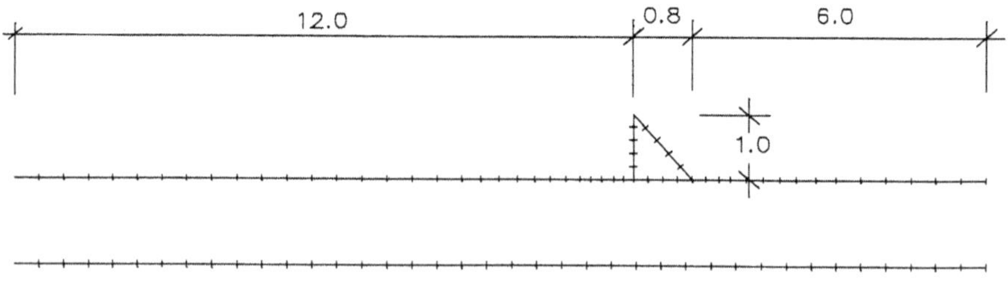

Figure 32. BE model for dam on flexible foundation.

Figure 32 shows the quadratic BE model for the problem being considered The lower boundary is only used when the foundation is assumed to be a deep stratum. This boundary does not exist for the half-space. BE equations are established for the scattered field in the soil regions

5.4. DAM ON FLEXIBLE FOUNDATION. RESERVOIR FULL OF WATER.

The same gravity dam on a deep stratum or a half-space is considered. The reservoir is full of water and it is assumed to extend up to infinity. When the system is under the effects of vertical SV-waves producing upstream motion, the free field motion is the same as in the empty reservoir situation at both sides of the dam. The BE equations for the foundation are written for the scattered field as above. The water region is closed by a vertical boundary at a distance from the dam equal to *12 H*. The use of this relation implies that the behaviour of the water channel outside the discretized zone is approximated by the behaviour of a rigid bottom channel of the same depth. This simplification has little effect on the dam motion.

When the dam-soil-reservoir is under the effects of vertical P-waves (producing vertical free-field motion), the soil far field motion is not the same in the upstream and the downstream direction. In the downstream direction, the far field motion corresponds to a stratum on a compliant bedrock, whereas in the upstream direction, it corresponds to the same foundation profile with a water layer on the top.

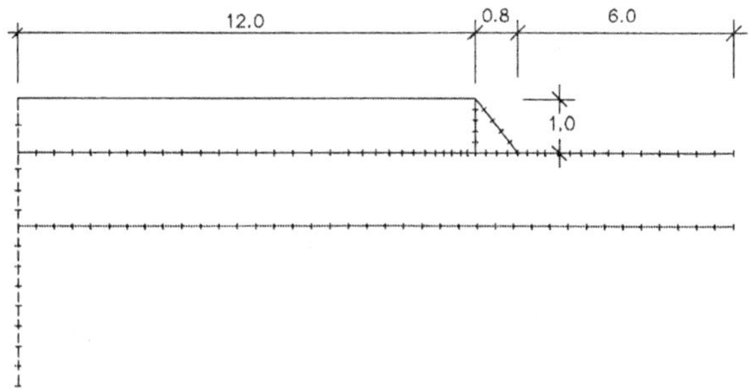

Figure 33. BE model for dam on flexible foundation and reservoir full of water. (the vertical boundary of the foundation is only needed for incident fields with vertical component).

This fact is taken into account introducing a vertical boundary going to infinity at the same distance from the dam of the water artificial boundary

(Figure 33). No unknowns exist on this boundary where both displacements and tractions are assumed to be those of the far field in the upstream direction. Since the field in the downstream direction has been taken as a reference, the prescribed conditions on the vertical boundary of the soil are the difference between the upstream free-field motion and the downstream free-field motion taken as a reference. It is worth noticing that this closing boundary is not needed for 3-D models (Domínguez, 1993) like those shown in Section 6.

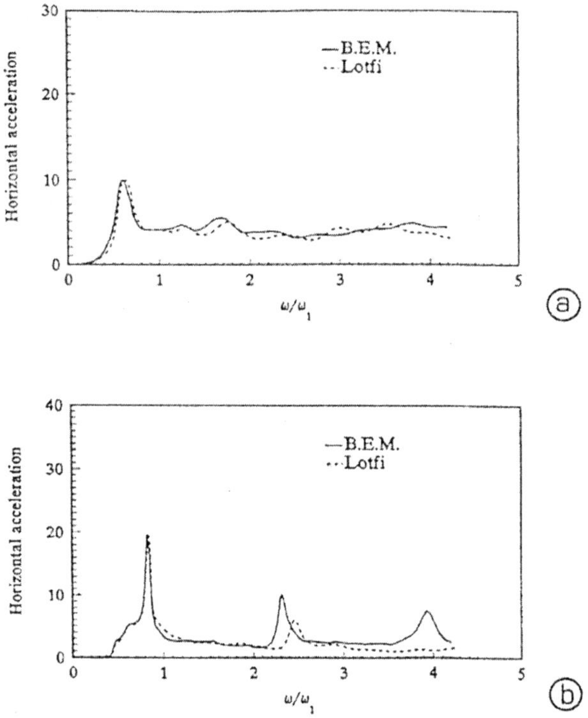

Figure 34. Response at dam crest to ground motion. Flexible deep-stratum foundation. Reservoir full of water:
(a) horizontal ground motion; (b) vertical ground motion.

Figures 34 (a and b) and 35 (a and b) show the results for a dam on a deep stratum and a half-space, respectively, with the reservoir full of water. The results shown in Figures 34 (a and b) can be compared directly with those obtained by Lotfi (for a deep stratum using a FE model); the agreement is excellent. Fenves and Chopra (1984) simulate water-foundation interaction by means of an approximate absorbing boundary condition on the reservoir bottom. Thus, some care should be exercised in comparing the results in Figures 35 (a and b) with those of Fenves and Chopra (half-space). The condition involves the so-called "wave reflection coefficient" α, which can be calculated from the properties of the materials that constitute the bottom of the reservoir.

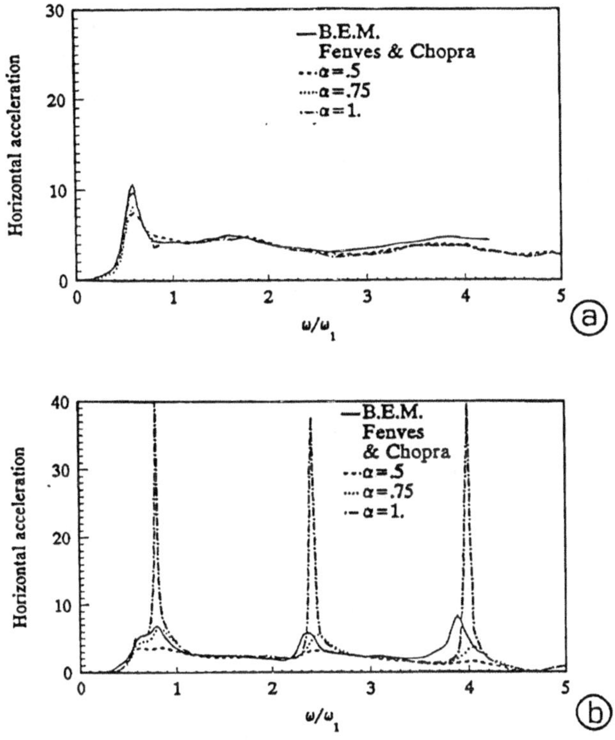

Figure 35. Response at dam crest to ground motion. Flexible half-space foundation. Reservoir full of water: (a) horizontal ground motion; (b) vertical ground motion.

For a reservoir without sediments and the foundation considered in the present study, α is about 0.71. The response to vertical ground motion for $\alpha = 0.75$, at low frequencies, is in good agreement with the present Boundary Element results. The differences at high frequencies are apparently due to the coarse meshes of Finite Elements employed for the dam in both of the earlier studies.

5.5. BOTTOM SEDIMENT EFFECTS

The model used for the dynamic analysis of concrete gravity dams taking into account effects of dam-water-sediment-foundation interaction should be able to represent the dynamic behaviour of compressible water regions, viscoelastic solid regions, poroelastic regions and also, the interaction between any two of these domains at the interfaces. The model should be able to accommodate infinite and very large regions using a reasonable number of unknowns and represent the radiation damping properly.

There are three kinds of interfaces in the problem being considered: poroelastic-viscoelastic, water-poroelastic and water-viscoelastic. The compatibility and equilibrium conditions along these interfaces are can be found in Domínguez et al. (1997).

Figure 36 shows the boundary-element discretization used for the case of a uniform viscoelastic half-space foundation. The foundation horizontal boundary in the upstream direction is an interface with the sediment or with the water region, and it is discretized using the same elements as for the water and bottom sediment. A portion of free surface extending up to six times the dam height is discretized into 15 elements in the downstream direction. The boundary-element equations corresponding to the viscoelastic foundation are written for the scattered field, that is, the difference between the total and the free field. Thus, the

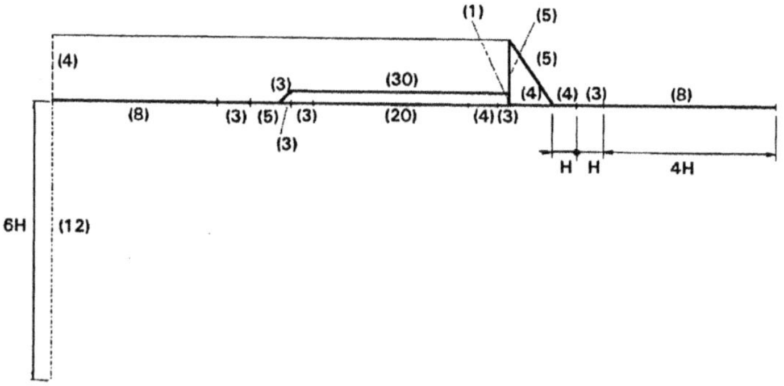

Figure 36. Boundary-Element model for dam-reservoir-sediment system on flexible foundation.

radiation conditions are satisfied and the boundary of the foundation half-space can be left open. These conditions are satisfied at both ends when the incident field is a vertical shear wave. However, when the incident field is a vertical pressure wave, the free field conditions are not the same in the upstream and the downstream direction. One corresponds to a half-space. One of the free fields (the downstream free field) must be taken as a reference.

The effects of a sediment layer with thickness $h = 0.1H$ are shown in Figures 37 and 38 for horizontal and vertical excitation,

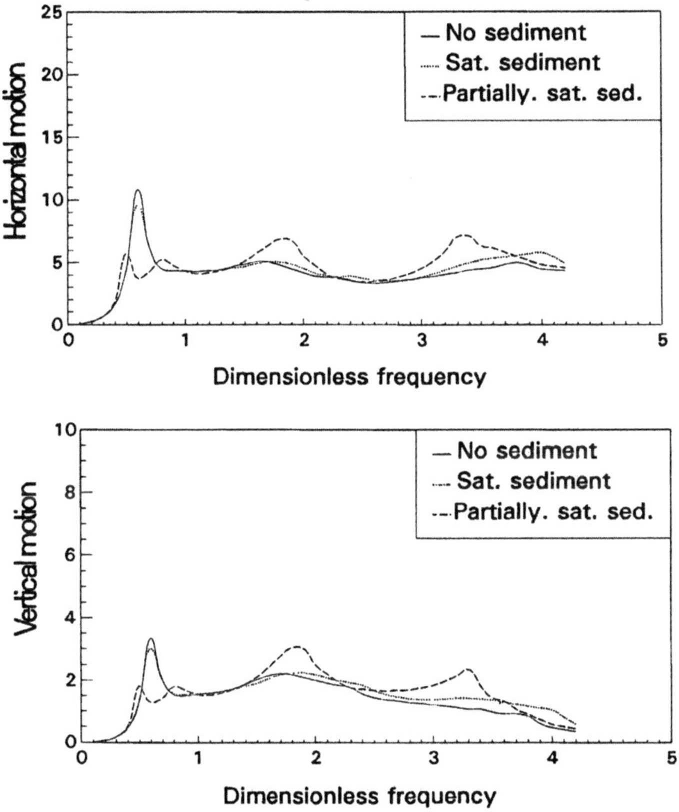

Figure 37. Response at dam crest to horizontal ground motion: Flexible foundation, full reservoir, and sediment layer with thickness $h = 0.1H$.

respectively. The fully saturated sediment has little influence on the dam response, particularly when the excitation is a horizontal base motion. Figure 37 shows that the partially saturated sediment lowers the first natural frequency of the system and reduces substantially the response at this frequency. The response at the second characteristic frequency becomes clear but is not as large as in the rigid bedrock case. Other peaks corresponding to higher natural frequencies became more prominent than in the case of fully The effects of partially saturated sediment shown in Figure 37 are as those reported by Bougacha and Tassoulas (1991a,b) for a deep stratum foundation. Partially saturated sediment also produces changes in the dam response to vertical ground motion (Figure 38) for the lower part of the frequency range analyzed.

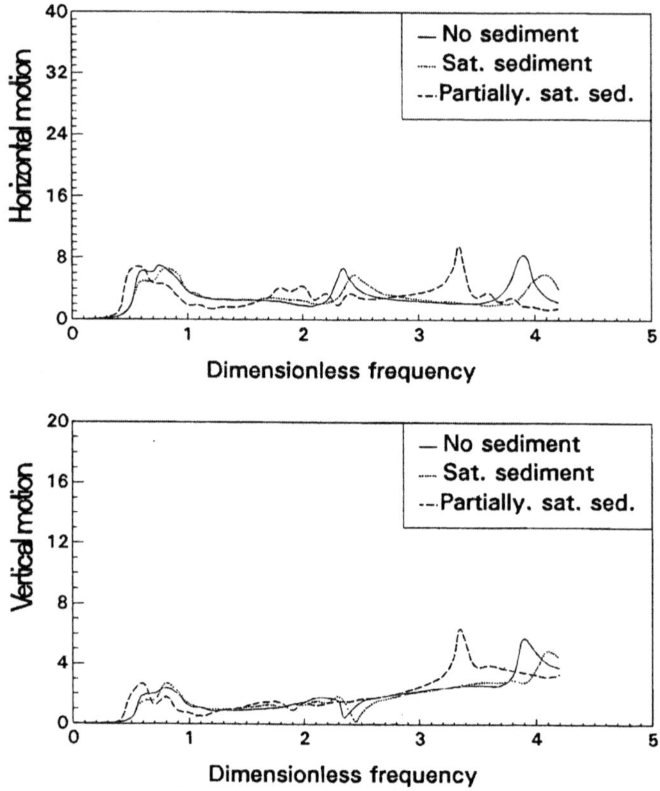

Figure 38. Response at dam crest to vertical ground motion: Flexible foundation, full reservoir, and sediment layer with thickness $h = 0.1H$.

6. Arch Dams

As in the case of gravity dams, the earthquake response of arch dams is conditioned by a series of characteristics of the foundation-dam-reservoir system in addition to the dam geometry and its material properties. The importance of some of these factors, which show their influence through the interaction effects, has already been pointed out for the case of gravity dams.

Three kinds of factors can be mentioned: first, those related to the location (geological and geotechnical characteristics of the site, and local topography) second, factors having a direct influence on the hydrodynamic pressure (water compressibility, reservoir geometry and bottom sediments) and third, the space distribution of the excitation. An analysis of the influence of these factors was done by Maeso and Domínguez (1993), Domínguez and Maeso (1993) and Maeso et al. (1999 and 2000). In any case, it should be said that their influence on the seismic response is normally more important for arch dams than for gravity dams.

The Morrow Point arch dam has been selected for this presentation following the work of Hall and Chopra (1983) and Fok and Chopra (1986 and 1987). The amplitude of the upstream complex-valued frequency response functions are analyzed. They are accelerations of the dam crest centre point on the upstream face of the dam due to harmonic waves. A complete BE analysis of the seismic response of this dam can be seen in Maeso and Dominguez (1993), Dominguez and Maeso (1993) and Maeso et al. (1999 and 2000).

6.1. DAM ON RIGID FOUNDATION. EMPTY RESERVOIR

The Boundary Element idealization of the dam is presented in Figure 39. It consists of generalized quadrilateral and triangular elements with a quadratic variation in the two directions both for the geometry and the boundary variables. Boundary Element results and the Finite Element ones, for this case, are in very good agreement. In this simple case no travelling wave or interaction effects exist.

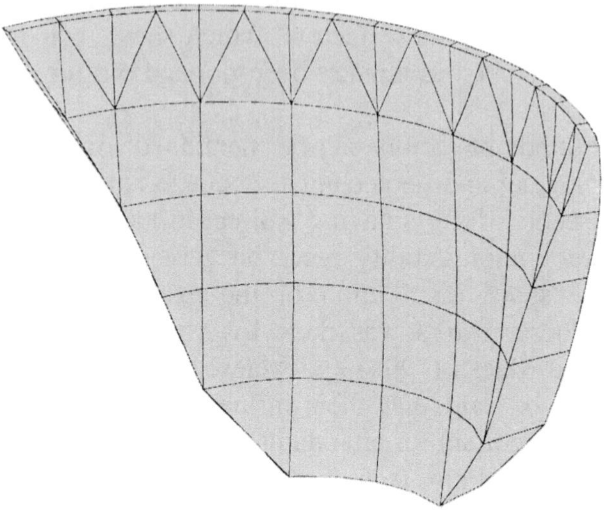

Figure 39. Boundary Element discretization of the Morrow Point dam on rigid foundation.

6.2. DAM ON FLEXIBLE FOUNDATION. EMPTY RESERVOIR.

The foundation rock is assumed to be a linear elastic solid. Considering the foundation rock as a proper solid allows for the inertial, travelling wave and damping effects in the soil to be represented. The ground horizontal surface and the canyon are discretized using the same kind of Boundary Elements as the dam. The dam and foundation rock discretization is shown in Figure 40. The shape of the canyon is the same as that assumed by Chopra and his co-workers.

The use of a full space fundamental solution requires discretization of the foundation rock free-surface. The boundary discretization extends to a certain distance from the dam as shown by Figure 40. The truncation of the surface discretization, usual in Boundary Element representations for soil-structure interaction problems, produces good results for a uniform boundless foundation rock or for non-uniform foundation domains as long as the external region where the discretization is truncated is an infinite region and the underground zones close to the dam are included in the model (Dominguez and Meise, 1991). Preliminary analyses carried out with several free surface discretizations show that a BE mesh extending to a distance equal to 2.5 times the dam height ensures a good representation of the dam-foundation rock interaction. The exact topography of the canyon at the distance from the dam where the boundary discretization is truncated does not affect the earthquake response of the dam when the reservoir is

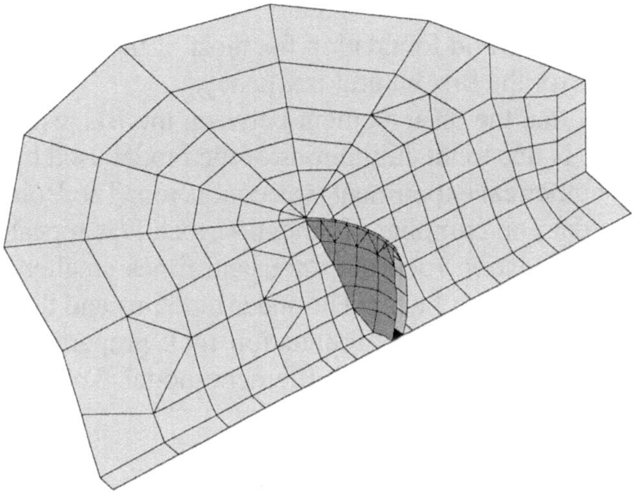

Figure 40. Discretization of the Morrow Point dam on flexible foundation.

assumed to be empty. The elements within a distance from the dam equal to its height must be equal or smaller than one half of the wavelength. The size of the elements can be gradually increased for larger distances.

The travelling wave and the soil structure interaction effects produce an important variation of the earthquake response of the dam. The results show a decrease of the fundamental frequency of the system. However, the most important change as compared to the rigid foundation rock situation is a decrease of the response over most of the frequency range. The decrease of the response for intermediate and high frequencies is, to a large extent, due to the effect of the space distribution of the excitation. The importance of this effect can be explained by the fact that the frequency range considered includes wavelengths in the foundation rock going from infinity to 113 m for the S-wave. Taking into account that the dam has a maximum height of 141 m, the space distribution of the excitation effect should be noticeable from the first resonant frequency for which the height of the dam is of the same order of magnitude as 1/4 of the wavelengths in the foundation rock.

The decrease of the amplification functions is consistent with the gravity dam results but it is in clear disagreement with the results obtained by Fok and Chopra (1986b) for the same dam. The results presented by these authors did not show a decrease in the resonance peaks due to the compliant foundation rock effect. On the contrary, the results obtained by Fok and Chopra presented a greater amplification for a compliant foundation than for a rigid foundation for most of the peaks of the response and in particular for the first natural frequency.

To show that the disagreement between the BE results and those of Fok and Chopra is due to the limitations of the model used by these authors, the following numerical experiment was conducted: The Boundary Element analysis of the dam on compliant foundation rock was repeated assuming a density of the foundation rock one thousand times smaller than that used above and a zero damping factor. The shear modulus and the Poisson's ratio remained the same. These new foundation rock properties are almost the same as those assumed in the Finite Element model. The only difference is now the extension of the foundation rock.

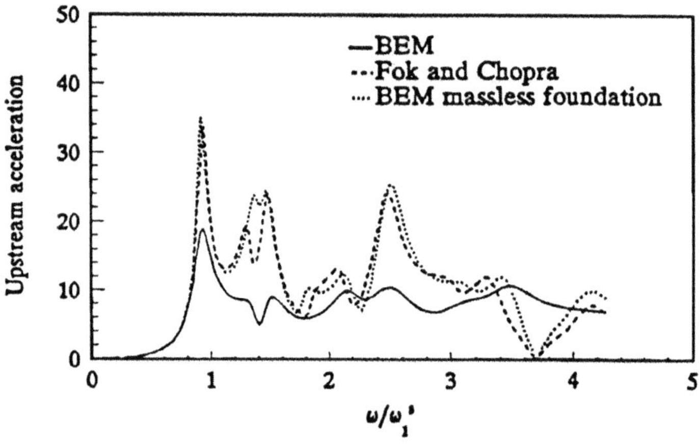

Figure 41. Comparison with FE results for the response at dam crest to ground upstream excitation. Flexible foundation. Empty reservoir.

Figure 41 shows the frequency response functions obtained using the three different models; namely, the BE model proposed in this paper, the FE model used by Fok and Chopra and the BE model simulating a massless undamped foundation rock and hence, a spatially-uniform excitation at the dam foundation rock interface. The results presented in the figure show that when the rock properties are altered in the BE model the results basically agree with those obtained using the FE model. The resonance peaks obtained with the fictitious foundation rock properties are very close to those obtained with FE: much higher than they should be.

6.3. DAM ON FLEXIBLE FOUNDATION. RESERVOIR FULL OF WATER

The same dam and the same reservoir geometry of the previous analysis are assumed in this case. The foundation is considered to be a linear viscoelastic solid with the same properties as in the empty reservoir study. The boundary discretization of the dam and the foundation are similar to those used for the empty reservoir study (Figure 40). It also includes now a discretization of the water-foundation interface and a closing boundary. Thus, the BE discretization including dam, water and foundation is as shown in Figure 42a. The discretization includes a closing boundary where infinite channel boundary conditions are applied as carried out in the 2-D case of Section 5. The B.E. discretization of the coupled system used for the analysis of a closed reservoir is shown in Figure 42b. This discretization is only different from that of the open reservoir case in the zone that closes the reservoir. The rest of the dicretization is the same.

The boundary integral equations are now written twice for the elements on the reservoir boundaries. Once, as part of the rock or the dam, and another time, as part of the fluid domain. The water free surface boundary conditions are satisfied by the half-space fundamental solution used for the fluid. The maximum size of the elements in the solid-water interface is determined by the wavelength of the water waves. The BE equations for the foundation rock are written in terms of the scattered field in order to satisfy the radiation conditions.

The BE results for a full reservoir and a compliant foundation are compared with the FE results obtained by Fok and Chopra (1986b) for the same dam. Figure 43 shows this comparison. The FE results have been taken directly from the figures of the paper by Fok and Chopra (1986b) and may have small discrepancies with the exact numerical values. They correspond to an absorption coefficient $\alpha = 0.75$ very close to that obtained for the assumed foundation rock properties.

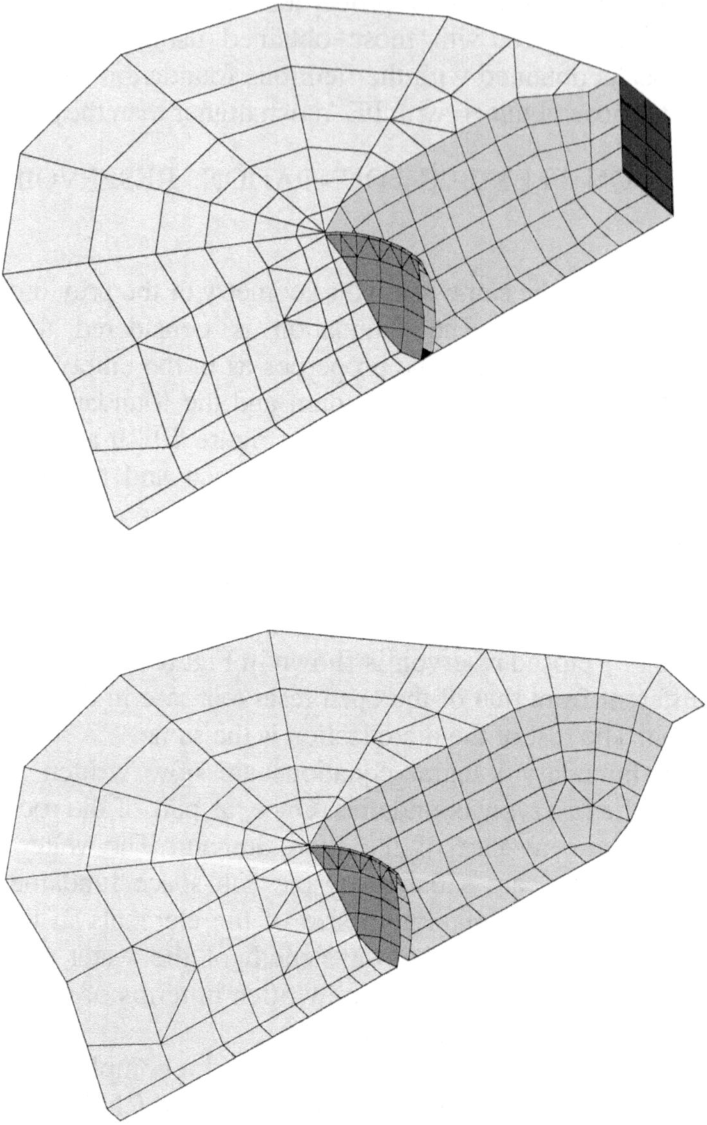

Figure 42. Discretization for dam-reservoir-flexible foundation system: (a) open fluid domain; (b) closed fluid domain.

Important differences between the BE and the FE results are apparent in Figure 43. These differences are due to the travelling wave effects which are not taken into account in the FE model and by the water-foundation-rock interaction which is approximated by the one-dimensional theory in that model. To verify the validity of the above statement, a numerical experiment was conducted. The BE analysis was repeated assuming a density of the foundation rock one thousand times smaller than that used before and a zero damping factor. The shear modulus and the Poisson's ratio remained the same. The approximate boundary conditions for the reservoir bottom and banks assumed in the FE model were introduced in the BE model for the fluid (Hall and Chopra, 1983). This distorted BE analysis does not take into account travelling wave effects and uses the one-dimensional theory for the water-foundation-rock interaction as the existing FE models do. The results obtained are also presented in Figure 43. They show a remarkable agreement between the BE results obtained using massless foundation rock and absorption coefficient and the FE results.

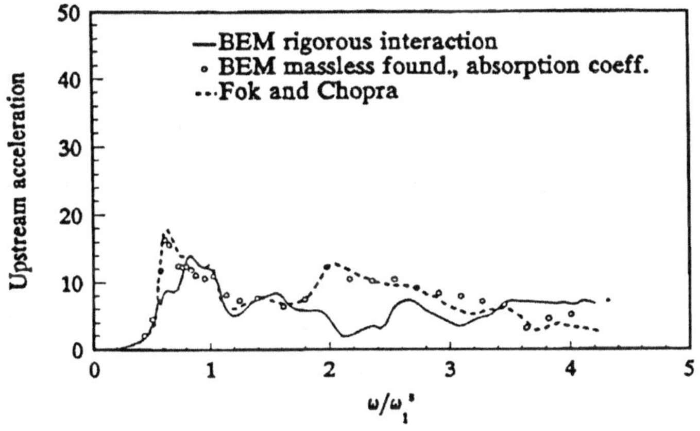

Figure 43. Response at dam crest to upstream excitation. Flexible foundation. Reservoir full of water.

6.4. TRAVELLING WAVE EFFECTS

In many cases the size of a dam may be close to the length of the seismic waves that would arrive to the dam site in the event of an earthquake. As a consequence, when seismic waves impinge on a large dam, the excitation of the dam-foundation rock interface is not uniform. Different points along the interface are under the effects of different foundation acceleration values at the same time. In other words, the seismic waves travel along the dam-foundation rock interface. The importance of this effect depends on the dam size, the length of the seismic waves and its direction of propagation but it seems clear that assuming a uniform excitation along the dam-foundation rock interface may lead to erroneous consequences.

Figure 44 shows the kind of problem studied to show the importance of the travelling wave effect. The figure shows an arch dam closing a canyon with a geometry which is arbitrary in a large region close to the dam. The reservoir may be filled to any given level. The excitation is a harmonic wave (SH, SV, P, Rayleigh) which impinges on the dam site from any direction. Dam, water and foundation rock are coupled in a three dimensional boundary element model which takes rigorously into account these three media.

The Morrow Point dam has been chosen for the study as in the previous sections of this chapter. The boundary element discretization is shown in Figure 45. The elements are nine node quadrilaterals and six node triangles with a cubic representation of the geometry and the boundary

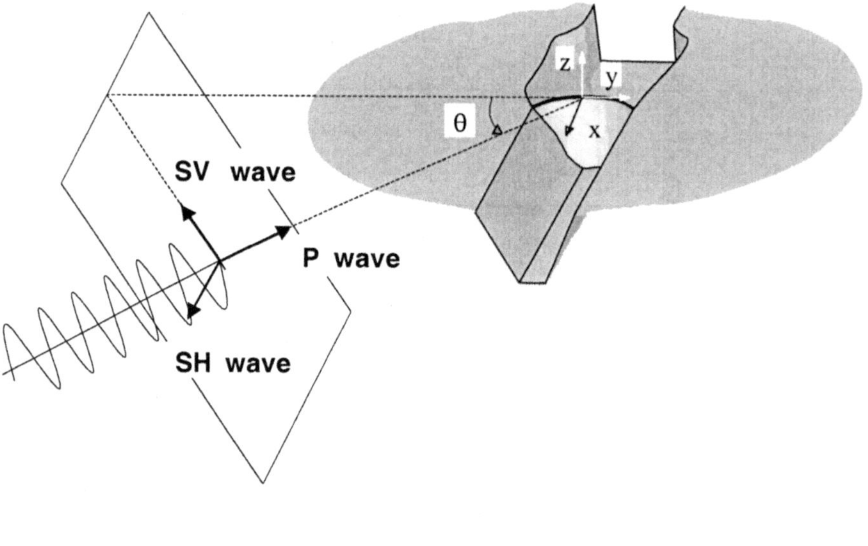

Figure 44. Description of the problem.

variables. The dam is modelled as a viscoelastic medium with the discretization of the soil free surface extending up to a distance from the dam equal to 2.5 times the dam height. The impounded water is considered as a compressible inviscid fluid. Its boundary element representation is done using the same elements as for the canyon upstream of the dam. The geometry of the reservoir is assumed to vary smoothly within the mode and to be uniform from the limits of the discretized zone to infinity. Reservoirs which are not very long in the direction perpendicular to the dam, may be represented using a realistic geometry fully modelled by boundary elements.

The seismic excitation has been assumed to be a time harmonic plane wave coming from infinity. To satisfy the radiation conditions, the problem has been solved in terms of scattered wave fields. The total displacements and stresses are the superposition of the incident field corresponding to a uniform half-space and the field scattered by the dam-reservoir-canyon system. The incident field, for which the analytic solution is known, becomes part of the right hand side vector in the system of equations.

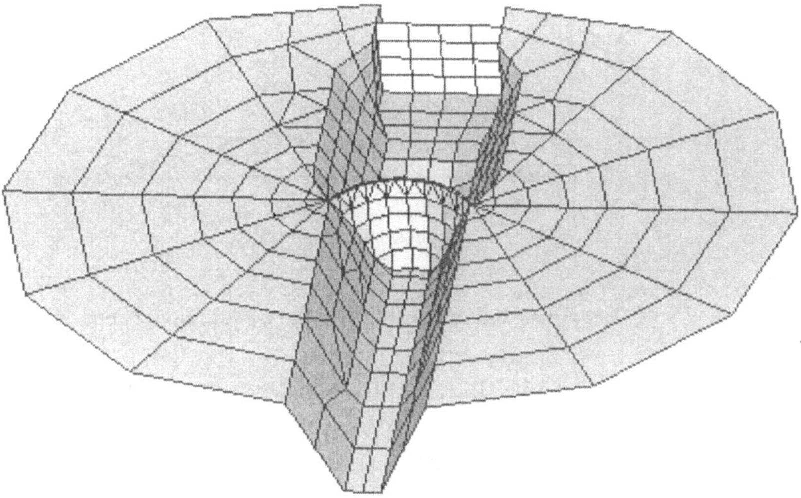

Figure 45. Boundary Element discretization of Morrow Point dam on compliant foundation rock.

Several numerical examples are presented to show the influence of the angle of incidence of the waves on the seismic response of the dams. SH, P and SV waves propagating in the plane y-z perpendicular to the cannon axis of symmetry x are considered (Figure 44).

The first case analyzed corresponds to the reservoir empty of water and SH waves arriving to the dam site with several different angles θ. Figure 46 shows the amplitude of the displacement at the dam crest mid-point normalized by the amplitude of the motion at the same point in the case of a uniform half-space (that is., without canyon, dam or reservoir). The response is represented versus frequency for four different angles of incidence. The response for rigid foundation conditions is also included in the figure. Frequencies are normalized by the first natural frequency of the dam on rigid foundation. It is clear from Figure 46 that foundation rock flexibility reduces the amplitude of the resonance peaks and the first natural

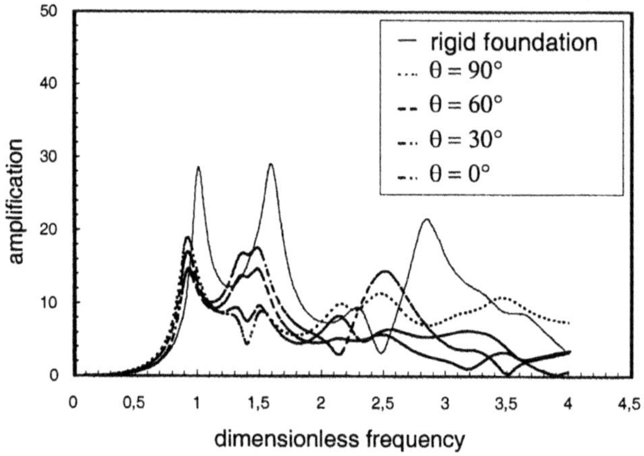

Figure 46. Response at dam crest to SH wave excitation. Empty reservoir.

frequency. In addition to that, the travelling wave effect and the angle of incidence of the wave modify the response in particular for dimensionless frequencies higher than one. Similar conclusions can be drawn from the response in the case of reservoir full of water (Figure 47) and also for cases of incident P or SV waves. Additional results may be found in Maeso et al. (2000).

The results show the important effects that the foundation rock flexibility and the space distribution of the excitation have on the seismic response of arch dams. These effects have been shown to be relevant in the case of full reservoir and also when the reservoir is empty. The influence of the angle of incidence of the waves is important for all different kinds of waves considered.

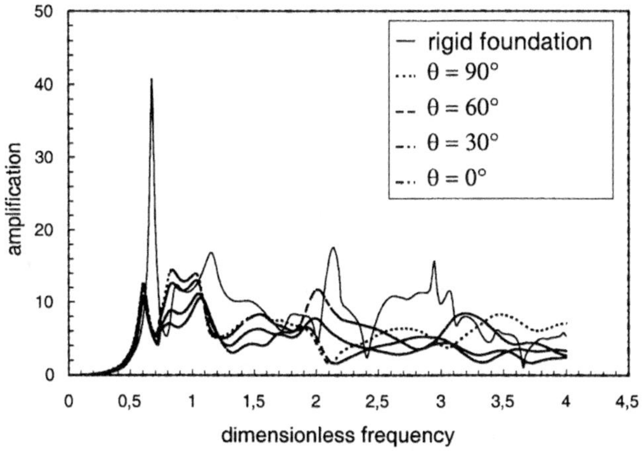

Figure 47. Response at dam crest to SH wave excitation. Full reservoir.

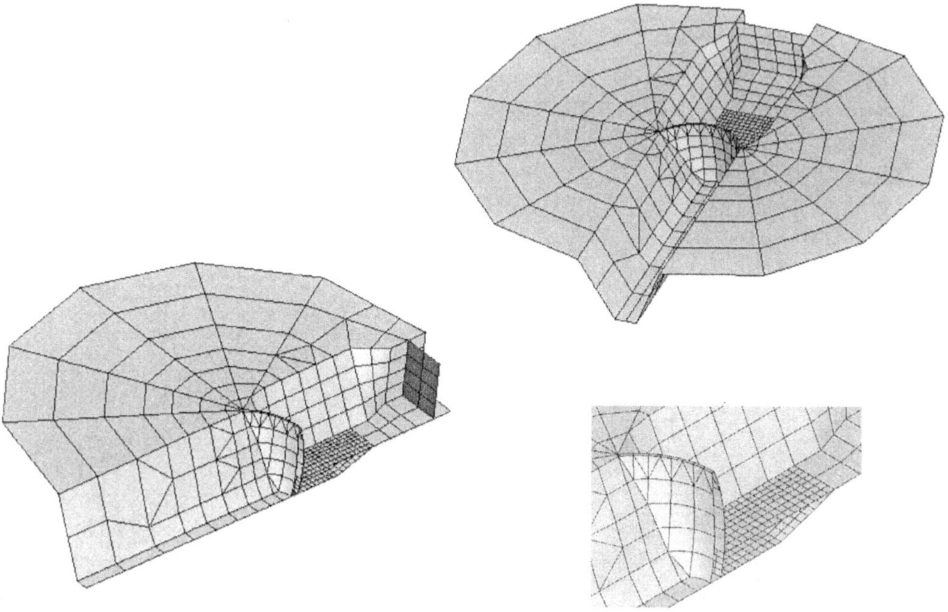

Figure 48. Morrow point dam-reservoir full of water, compliant foundation rock and porous sediment.

6.5. POROELASTIC SEDIMENT EFFECTS

In spite of the great progress in the knowledge of earthquake behaviour of arch dams achieved in recent years, some important effects which may influence this behaviour are not well evaluated yet. One of the important matters which require some additional research effort is the effect of porous bottom sediments on the seismic response.

The Boundary Element representation for the Morrow Point Dam, which has been studied in previous sections, is shown in Figure 48. The model includes now a sub-region where the BE equations for poroelastic media are written. The elements for this region are smaller than for any other since there are very short waves in this type of materials. Equilibrium and compatibility conditions among the different regions are written as in the two-dimensional case studied in Section 5.5.

Assuming the same type of porous sediment material as in the 2-D case, the results shown in Figure 49 are obtained for a vertical incident S-wave producing upstream free-field motion. The important influence of the porous sediments can be seen from this figure. Single phase models yield accurate results for certain cases (see, Maeso et al. 1999).

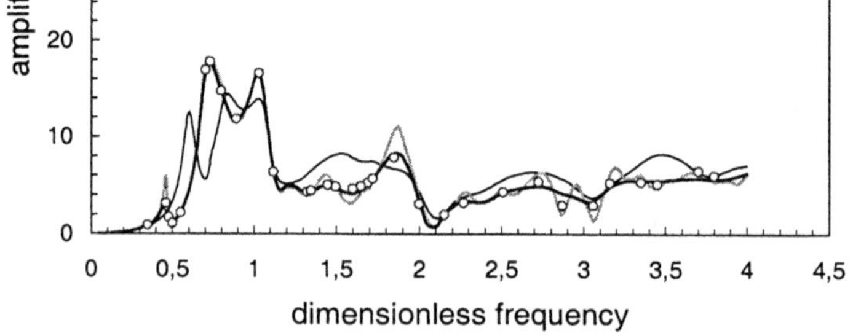

Figure 49. Response at dam crest to SV wave excitation. Full reservoir.

7. References

1. Abascal, R. and Domínguez, J., 1986, "Vibrations of Footings on Zoned Viscoelastic Soil", J. Engrg. Mech., ASCE, Vol. 112, pp. 433-447.
2. Abascal, R. and Domínguez, J., 1990, "Dynamic Response of Two-dimensional Foundations Allowed to Uplift", Computer and Geotechnics, Vol. 9, pp. 113-129.
3. Alarcón, E., Cano, J.J. and Domínguez, J., 1989, "Boundary Element Approach to the Dynamic Stiffness Functions of Circular Foundations", Int. J. Numer. Anal. Methods Geomechs., Vol. 13, pp. 645-664.
4. Antes, H. and Von Estorff, O., 1987, "Analysis of Absorption Effects on the Dynamic Response of Dam Reservoir Systems by Element Methods", Earthquake Engineering and Structural Dynamics, Vol. 15, pp. 1023-1036.
5. Apsel, R.J. and Luco, J.E., 1983, "On Green's Functions for a Layered Half-Space: Part II", Bull. Seism. Soc. Am., Vol. 73, pp. 931-951.
6. Arnold, R.N., Bycroft, G.N. and Warburton, G.B., 1965, "Forced Vibration of a Body on a Infinite Elastic Solid", J. Appl. Mech., ASME, Vol. 22, pp. 391-400.
7. Aubry, D., and Crepel, J.M., 1986, "Interaction Sismique Fluide-Structure. Application aux Barrages Voute", Congres Francais Genú Parasismique, A.F.P.S., Paris, France, pp.1-12, (in French).
8. Bougacha, S., and Tassoulas, J. L. (1991a), "Seismic Analysis of Gravity Dams. I: Modeling of Sediments", J. Engrg. Mech., ASCE, 117(8), 1826-1837.
9. Bougacha, S., and Tassoulas, J. L. (1991b), "Seismic Response of Gravity dams. II: Effects of Sediments", J. Engrg. Mech., ASCE, 117(8), 1839-1850.
10. Bycroft, O.N., 1956, "Forced Vibration of a Rigid Circular Plate on a Semi -Infinite Elastic Space or a Elastic Stratum", Phil. Trans. Royal Soc., Vol. 248, pp. 327-368.
11. Chapel, F., 1981, "Application de la Méthode des Équations Intégrales á la Dynamique de Sols. Structures sur Pieux", Ph. D. Thesis, Ecole Central des Arts et Manufactures, Paris.
12. Cheng, A. J.-D., Badmus, T., and Beskos, D.E. (1991), "Integral Equation for Dynamic Poroelasticity in Frequency Domain with Boundary Element Solution", J. Engrg. Mech., ASCE, 117(5), 1136-1157.
13. Chopra, A.K., and Chakrabarti, P., 1981, "Earthquake Analysis of Concrete Gravity Dams Including Dam-Water-Foundations Rock Interaction", Earthquake Eng. and Struct. Dyn.,Vol. 9, pp. 363-383.
14. Cruse, T.A., Snow, D.A. and Wilson, R.B., 1977, "Numerical Solutions in Axisymmetric Elasticity", Computer and Structures, Vol. 7, pp. 445-451.
15. Domínguez, J., 1978a, "Dynamic Stiffness of Rectangular Foundations", Research Report R78-20, Dept. Civ. Engrg., Massachusetts Inst. of Tech., Cambridge, Mass.
16. Domínguez, J., 1978b, "Response of Embedded Foundations to Travelling Waves", Research Report R78-24, Dept. Civ. Engrg., Massachusetts Inst. of Tech., Cambridge, Mass.
17. Domínguez, J. (1991), "An Integral Formulation for Dynamic Poroelasticity", J. Appl. Mech., 58(June), 588-591.
18. Domínguez, J. (1992), "Boundary Element Approach for Dynamic Poroelastic Problems", Int. J. Numer. Methods in Engrg., 35(2), 307-324.

19. Domínguez, J. (1993), "Boundary Elements in Dynamics", Computational Mechanics Inc., Elsevier Applied Science, New York, N.Y.
20. Domínguez, J. and Abascal, R., 1984, "On Fundamental Solutions for the BIEM in Static and Dynamic Elasticity", Eng. Analysis, Vol. 1, pp. 128-134.
21. Domínguez, J. and Abascal, R., 1987, "Dynamics of Foundations", in Topics in Boundary Element Research, C.A. Brebbia (Ed.), Vol. 4, Springer-Verlag, Berlin.
22. Domínguez, J. and Abascal, R., 1989, " Seismic Response of Strip Footings on Zoned Viscoelastic Soils", J. Eng. Mech., ASCE, Vol. 115, pp. 913-934.
23. Domínguez, J., and Meise, T., 1991, "On the Use of the BEM for Wave Propagation in Infinite Domains", Eng.Anal. with B.E., Vol.8, pp. 132-138.
24. Domínguez, J. and Maeso, O., 1993, "Earthquake Analysis of Arch Dams. II: Dam-Water-Foundation Interaction", J. of Eng. Mech., ASCE, Vol.119, pp.513-530.
25. Domínguez, J., and Medina, F., 1989, "Boundary Elements for the Analysis of the Seismic Response of Dams Including Dam-Water-Foundation Interaction Effects.II", Engrg.Anal, Vol.6, pp.158-163.
26. Domínguez, J., Gallego, R. And Japón, B.R., 1997, "Effects of Porous Sediments on the Seismic Response of Concrete Gravity Dams", J. of Eng. Mech., ASCE, Vol. 123, pp. 302-311.
27. Emperador, J. M. and Domínguez, J., 1989, "Dynamic Response of Axisymmetric Embedded Foundations", Earthquake Eng. Struct. Dyn., Vol. 18, pp. 1105-1117.
28. Fenves, G, and Chopra, A.K., 1984, "Earthquake Analysis of Concrete Gravity Dams Including Reservoir Bottom Absorption and Dam-Water-Foundation Rock Interaction", Earthquake Eng. and Struct. Dyn.,Vol. 12, pp. 663-680.
29. Fenves, G, and Chopra, A.K., 1985, "Effects of Reservoir Bottom Absorption and Dam-Water-Foundation Rock Interaction on Frequency Response Functions for Concrete Gravity Dams", Earthquake Eng. and Struct.Dyn., Vol.13, pp.13-31.
30. Fok, K., and Chopra, A.K., 1986, "Earthquake Analysis of Arch Dams Including Dam-Water-Interaction Reservoir Boundary Absorption and Foundation Flexibility", Earthquake Eng. Struc.Dyn., Vol.14, pp.155-184.
31. Fok, K., and Chopra, A.K., 1987, "Water Compressibility in Earthquake Response of Arch Dams", J. of Struct.Eng., Vol.113, pp. 958-975.
32. Hall, J.F., Chopra, A.K., 1982a, "Two-Dimensional Dynamic Analysis of Concrete Gravity and Embankment Dams Including Hydrodynamic Effects", Earthquake Eng. and Struct.Dyn., 10, pp. 305-332.
33. Hall, J.F., and Chopra, A.K., 1982b, "Hydrodynamic Effects in the Response of Concrete Gravity Dams", Earthquake Eng. and Struct. Dyn.,Vol. 10, pp. 333-345.
34. Hall, J.F., and Chopra, A.K., 1983, "Dynamic Analysis of Arch Dams Including Hydrodynamic Effects", J.Eng.Mech.Div., ASCE,Vol.109, pp.149-163.
35. Hillmer, P. and Schmid, G, 1988, "Calculation of Foundation Uplift Effects Using a Numerical Laplace Transform", Earthquake Eng. Struct. Dyn., Vol. 16, pp. 789-801.
36. Jablonsky, A.M., and Humar, J.L., 1990, "Three-Dimensional Boundary Element Reservoir Model for Seismic Analysis of Arch and Gravity Dams", Earthquake Eng. and Struct.Dyn., Vol.19, pp.359-376.
37. Japón, B.R., Gallego, R. and Domínguez, J., 1997, "Dynamic Stiffness of Foundations on Saturated Poroelastic Soils", J. of Eng. Mech., ASCE, Vol. 123, pp. 1121-1129.

38. Kakuda, K. and Tosaka, N., 1983, "Numerical Analysis of Coupled Fluid-Elasticity Systems using the BEM", in Boundary Elements (Brebbia, C.A. et al., Ed.), Springer-Verlag, Berlin.
39. Karabalis, D.L. and Beskos, D.E., 1984, "Dynamic Response of 3-D Rigid Surface Foundations by Time Domain Boundary Element Method", Earthquake Eng. Struct. Dyn., Vol. 12, No. 1, pp. 73-94.
40. Karabalis, D.L. and Spyrakos, C.C., 1984, "Dynamic Response of Surface Foundations by Time Domain Boundary Element Method". Int. Symposium on Dynamic Soil-Struct. Int., Minneapolis, Balkema, Rotterdam.
41. Kausel, E., 1974, "Forced Vibrations of Circular Foundations on Layered Media". Research Report R74-11, Dept. Civ. Engrg., Massachusetts Inst. of Tech., Cambridge, Mass.
42. Kermanidis, T.A., 1975, "Numerical Solution for Axially Symmetrical Elasticity Problems", Journal of Solids and Structures, Vol. 11, pp. 493-500.
43. Kobori, R., Minai, R. and Shinozaki, Y., 1973, "Vibrations of a Rigid Circular Disc on a Elastic Half-Space Subjected to Plane Waves", Theoretical and Applied Mechanics, Vol. 21, Univ. of Tokyo Press.
44. Lamb, H., 1904, "On the propagation of Tremors over the Surface of an Elastic Solid", Philos.Trans.Royal Soc., London, Ser. A203,pp.1-42.
45. Lotfi, V., Roesset, J.M., and Tassoulas, J., 1987, "A Technique for the Analysis of the Response of Dams to Earthquakes", Earthquake Eng. and Struct. Dyn.,Vol. 15, pp.463-490.
46. Luco, J.E., 1974, "Impedance Functions for a Rigid Foundation on a Layered Medium", Nuclear Eng. Des., Vol. 31, pp. 204-217.
47. Luco, J.E., 1976, " Torsional Response of Structures for SH-Waves: The Case of Hemispherical Foundations", Bull. Seism. Soc. Am., Vol. 66, pp. 109-124.
48. Luco, J.E. and Westmann, R.A., 1971, "Dynamic Response of Circular Foundations", J. Eng. Mech. Div., ASCE, Vol. 97, pp. 1381-1395.
49. Luco, J.E. and Apsel, R.J., 1983, " On Green's Functions to a Layered Half- Space: Part I", Bull. Seism. Soc. Am., Vol. 73, pp. 209-229.
50. Maeso, O., and Dominguez, J., 1993, "Earthquake Analysis of Arch Dams. I: Dam-Foundation Interaction", J.of Eng.Mech., ASCE,Vol.119, pp.496-512.
51. Maeso, O., Aznares, J.J. and Domínguez, J., 1999, "A 3-D Model for the Seismic Analysis of Concrete Dams Including Poroelastic Sediment Effects", 13th Eng. Mech. Conference, ASCE, Johns. Hopkins Univ., N. Jones and G Ghanem (Eds.). CD-ROM.
52. Maeso, O., Aznares, J.J. and Domínguez, J., 2000, "Travelling Wave Effects on the Seismic Response of Arch Dams", 14th Eng. Mech. Conference, ASCE, Univ. Of Texas, Austin, J.L. Tassoulas (Ed.), DC-ROM.
53. Mayr, M., 1975, "Ein Integralgleichungsverdahren zur Lösung Rotationssymmetrischer Elastizitätsprobleme", Dissertation, T.U., München.
54. Medina, F., and Dominguez, J. 1989, "Boundary Elements for the Analysis of Dams Including Dam-Water-Foundation Interaction Effects.I", Eng.Analysis with B.E.,Vol. 6, pp.151-157.
55. Ottenstreuer, M. and Schmid, G, 1981, "Boundary Elements Applied to Soil-Foundation Interaction", Proc. 3rd Int. Sem. on Recent Advances in Boundary Element Methods, Irvine, Calif., Springer-Verlag, pp. 293-309.

56. Ottenstreuer, M., 1982, "Frequency Dependent Dynamic Response of Footings", in Proc. Soil Dynamics and Earth. Engng. Conf., Southampton, England, Springer-Verlag, pp. 799-809.
57. Paul, H.S., 1967, "Vibration of a Rigid Circular Disk on an Infinite Elastic Plate", J. Acoust. Soc. Am., Vol. 42, pp. 412-416.
58. Rea, D., Liaw, C-Y and Chopra, A.K., 1975, "Mathematical Models for the Dynamic Analysis of Concrete Gravity Dams", Earthquake Eng. Struct. Dyn., Vol.9 pp. 249-258.
59. Reissner, E., 1936, "Stationäre, Axialsymmetrische, Durch Eine Schüttelnde Masse Erregte Schwingung Eines Homonenen Elastischen Halbraunes", Ingenieur Archiv., Vol. 7, pp. 381-396.
60. Richart, F.E., Woods, R.D. and Hall, J.R., 1970, "Vibrations of Soils and Foundations", Prentice-Hall.
61. Roesset, J.M., 1980, "Stiffness and Damping Coefficients of Foundations", Dyn.Resp.Pile Found.,ASCE 1-30,O'Neal and Dobry, Eds.
62. Roesset, J. M. and Tassoulas, J.L., 1982, "Non Linear Soil-Structure Interaction: An Overview", Earthquake Ground Motion and its Effects on Structures, S. K. Data Ed., A.S.M.E., AMD, Vol. 53.
63. Sanchez-Sesma, F.J. and Rosenblueth, E., 1979, "Ground Motion at Canyons of Arbitrary Shape Under Indicent SH Waves", Earthquake Eng. and Struct. Dyn., Vol. 7, pp. 441-450.
64. Tsai, Chogn-Shien, and Lee, G.C., 1987, "Arch Dam-Fluid Interactions by FEM-BEM and Substructure Concept", Int.J.Num.Meth.Eng.,Vol. 24, pp.2367-2388.
65. Veletsos, A.S. and Wei, Y.T., 1971, "Lateral and Rocking Vibration of Footings", J. Soil Mech. Found. Engrg. Div., ASCE, Vol. 97, pp. 1227-1248.
66. Veletsos, A.S. and Verbic, Y.T., 1973, "Vibration of Viscoelastic Foundation", J. Geo. Eng. Div., ASCE, Vol. 100, pp. 225-246.
67. Waas, G, 1972, "Linear Two-Dimensional Analysis of Soil Dynamics Problems in Semi-Infinite Layered Media", Ph. D. Thesis, University of California, Berkeley.
68. Wang, M.H., and Hung, T.K., 1990, "Three-Dimensional Analysis of Pressures on Dams", J. of Eng.Mech., ASCE., Vol.116, pp. 1290-1304.
69. Whitman, R.V. and Richart, F.E., 1967, "Design Procedures for Dynamically Loaded Foundations", J. Soil Mech. Found. Engrg. Div., ASCE, Vol. 93, SM6, pp. 169-193.
70. Wong, H.L. and Trifunac, M.D., 1974, "Scattering of Plane SH Waves by a Semi-Elliptical Canyon", Earthquake Eng. Struct. Dyn., Vol. 3, pp. 157-169.
71. Wong, H.L. and Luco, J.E., 1976, "Dynamic Response of Rigid Foundations of Arbitrary Shape", Earthquake Eng. Struct. Dyn., Vol. 4, pp. 579-586.

CHAPTER 2.
COMPUTATIONAL SOIL-STRUCTURE INTERACTION

DIDIER CLOUTEAU
Laboratoire de Mécanique de Sols-Structures-Matériaux, CNRS UMR 8370
École Centrale de Paris, Châtenay Malabry, France
clouteau@mss.ecp.fr
DENIS AUBRY
Laboratoire de Mécanique de Sols-Structures-Matériaux, CNRS UMR 8370
École Centrale de Paris, Châtenay Malabry, France
aubry@mss.ecp.fr

1. Introduction

Dynamic Soil-Structure Interactions [54, 192] play a key role when assessing the seismic safety of either large structures such as nuclear power plant [73], dams or smaller ones when built on soft soils [146, 102]. At a larger scale, this interaction can even modify the seismic free field giving rise to the so called city-site effect [26, 191, 60]. As far as site effects are concerned, the interactions between the bedrock and the upper geological structure: mountains, valleys and lakes give rise to significant modifications of the seismic signal either in terms of amplitude, frequency spectrum and duration [86, 155, 24, 25, 160, 159, 81, 92, 161]. In the particular case of large arch dams, site effects and Soil-Fluid-Structure Interactions cannot be distinguished and have to be accounted for at the same time. As reported in the literature [76, 52, 53, 109, 98, 190], Soil-Structure Interactions also play a key role when analyzing vibrations induced by vehicle traffic either in the environment or inside the vehicle itself. During the last decade, SSI has also been studied in the petroleum industry to understand the wavefield propagating in the vicinity of bore-holes [27, 164, 80, 10, 19] during seismic experiments.

These aspects have often been considered separately. Our opinion is that SSI has now reached a high level of maturity so that they can be presented in the same framework.

1.1. PHYSICAL MODELS

Modelling of such phenomena is an involved task since the soil, the fluid and the structure have very different geometrical features and mechanical behaviour : unboundedness of the soil and/or the fluid, slenderness of the structure, irrotational flows, several orders of magnitude on the stiffnesses. Compared to classical fluid-structure interaction accounted for in structural dynamics, the heterogeneity

W.S. Hall and G. Oliveto (eds.), Boundary Element Methods For Soil-Structure Interaction, 61–125.

of the soil formation along the vertical and the horizontal directions is a major source of concern.

Non-linear behaviour of the soil, the structure and the interfaces is another important issue especially when strong motions are dealt with. However as far as amplifications due to resonance are concerned, linear or equivalent linear models have proved their efficiency in quantifying these effects. Moreover, most design methods are based on linear assumptions, giving credit to linear models. The last argument in favour of linear models is the uncertanty on the data. Indeed in practical situations the standard deviation on the elastic parameters and the damping ratio can often reach 50% or 100%. Dealing with more complex constitutive models is often out of reach since expensive laboratory experiments to calibrate these laws cannot be performed in current practice.

Uncertanties play a key role in earthquake engineering as earthquakes are random events not only regarding their occurrence and location -not addressed here- but also during the event itself. Indeed recordings of strong motion arrays [2, 103, 104, 132, 197] show a strong variability of the seismic field both in the time and space variables even over short distances[77, 180].

The variability of the incident field is partly due to the variability of the soil itself [183, 171, 97]. As a consequence the structural response is affected not only by the uncertanties of the incident field [114] but also by the uncertainties of the soil characteristics [153, 89]. This effect is usually neglected since it is assumed that these fluctuations induce an additional damping due to diffusion phenomena. However local heterogeneities can also focus the energy in the vicinity of the structure giving rise to localization phenomena [50, 51, 148, 9, 49, 124, 125, 130]. Accounting for these uncertainties is then of primary importance to give a safety margin on existing or planned structures. Very few attempts have been made in the past [137, 140, 138] to quantify these effects. As a matter of fact analytical models used in other fields [111] cannot account for the complexity of real applications and especially complex boundary conditions. Thanks to the exponential growth of the performance of presentday computational facilities, these physical phenomena can now be quantified using efficient and validated numerical models.

1.2. NUMERICAL MODELS

Classical numerical methods in structural dynamics [143, 110] and particularly the Finite Element Method [198, 107] are currently able to account for the complexity of structures. Many extensions of these methods to account for unbounded domains [118, 119, 170], including absorbing boundaries [85, 29], have been proposed in the literature. Specific developments dedicated to non-linear soil dynamics have been carried out by the geotechnical community [108, 154]. However refined three-dimensional non-linear studies are still expensive [17, 149]. To overcome these computational limitations, substructuring techniques have been proposed [142] originally for linear analyses. More recently domain decomposition techniques [177, 88, 74] have been introduced to take advantage of parallel computers. Spectral Finite Elements classically used in fluid dynamics [45] have also shown their efficiency in modeling unbounded elastic domains [87] even though their coupling with non-linear standard Finite Elements is still an involved task [30]. Uncertanties can be accounted for in the Finite element framework using the so-called Stochastic Finite Element Method [93, 95, 141, 94] coupled with either Neumann expansions [131, 94, 195] or Monte Carlo simulations [147, 158, 168].

Boundary Element Methods [23, 42] have also been widely used in the fields of earthquake engineering and seismology (see [13, 38, 59, 55, 58, 79, 120, 135, 162, 184, 194, 6, 7, 173, 174, 91, 115, 185, 193, 163, 182] and reviews [71, 31, 32]) since they can easily account for seismic wave propagation in unbounded media. In order to deal with the heterogeneity of the geological formation (velocity contrasts often reach a ratio of 10 in practical applications) many improvements of the classical Boundary Element Method have been proposed in the literature and subsection 1.3. gives a brief review of them.

1.3. HETEROGENEITIES IN THE BEM

Accounting for material heterogeneities is a major issue in BEM since Boundary Integral Equations are based on fundamental solutions having analytical expressions only for very simple and often homogeneous cases. Depending on the type of heterogeneities the following techniques have been proposed :

- For piecewise constant and bounded heterogeneities, substructuring is usually employed. The domain under consideration is split into several homogeneous subdomains on which the classical BEM is applied using either collocation techniques [3, 79] or variational principles [13].

- For non-constant bounded heterogeneities the aforementioned technique is still applicable when using standard FEM for complex structures and BEM for homogeneous unbounded domains. This technique has been used in many fields and especially for dynamic soil-structure [99, 134] and soil-fluid-structure interaction [13, 55, 64, 58]. Dealing with such techniques, the variational BEM formulation proposed in [145] for Laplace problems is very attractive since it emphasises the link between volume and boundary variational formulations as proposed in [113] and [28] (see [35] for a review on recent developments on the Symmetric Galerkin Boundary Element Method).

  In the same situation many authors have proposed to keep on using BEM adding integration cells on the heterogeneous region. This technique has been extensively used for elasto-plastic analyses [4, 126]. However it usually requires the computation of the derivatives of the Green's function leading to highly singular integrals.

- For unbounded heterogeneities none of the above techniques is suitable for dynamic analyses at least from the mathematical point of view. The only rigorous solution in that case is to look for analytical or numerical Green's functions that account for the heterogeneities outside of a bounded region.

The Green's functions of a layered half space have first been proposed by Luco and Apsel [135] to perform SSI analyses. Computing these kind of Green's functions accurately enough to be used in BEM [121, 184] is a major issue since Cauchy principal values and singular integrals have to be computed. Special regularisation techniques for the direct boundary integral equation have to be employed [12] for BEM meshes crossing the layers or reaching the free surface. To the author's knowledge, such regularisation techniques are not available for hypersingular integral operators.

- When the BEM mesh lies inside an homogeneous region, the procedure becomes simpler since the Green's function can be written as the sum of the singular homogeneous Green's function and a regular term that can be computed using any convenient numerical technique. The procedure is not restricted to horizontally layered domains and much more complex situations may be accounted for. In particular for acoustic or elastodynamic analyses ray methods or substructuring techniques can be used to compute this regular part [65, 165] as long as the heterogeneities are not too close to the BEM mesh.

This review shows that two basic ingredients have to be mixed : substructuring techniques for local strong fluctuations of the mechanical properties - including perturbations due to structures - and numerical Green's functions for unbounded heterogeneities. These two items are dealt with in detail in Sections 3, 4 and 6.

1.4. TIME DOMAIN BEM/FREQUENCY DOMAIN BEM

The time domain BEM [71, 21, 5, 115] has become very popular in recent years [152, 167] even for anisotropic materials [186]. Compared to more classical frequency or Laplace domain approaches [139], TD-BEM can account for non linear problems either in the BEM framework [4, 126] or coupled with non linear FEM [136].

Another advantage of TD-BEM compared to FD-BEM is the reduced computational effort when assembling and solving the final linear system. Indeed the TD-Green's function has a bounded support while the FD-Green's function never vanishes. Due to the bounded support of the TD Green's functions, TD-BEM requires only the inversion of one sparse matrix while a full complex one has to be inverted at each frequency in the FD-BEM. However for a seismic signal of long duration and wide frequency spectrum, the computation of the convolution product in the TD-BEM becomes very expensive and sometimes unstable [112].

As far as Green's functions of a layered half-space are concerned, the frequency domain approach remains very attractive as these Green's functions are usually computed in the frequency domain and are much less singular than in the time domain. Moreover, due to dispersive waves the support of the time domain fundamental solutions is not bounded anymore. Finally, when parametric or stochastic analyses on the incident field are performed, FD approaches are very competitive since the computational cost is almost insensitive to the number of loading cases once the system is inverted whereas it increases linearly with the number of loading cases for time domain approaches.

1.5. STOCHASTIC APPROACH

As mentioned previously, giving error bounds for the numerical simulations of Soil-Structure Interaction is a major issue in earthquake engineering since large uncertainties are associated with the seismic incident fields and the mechanical and geometrical parameters of the model. A chapter of this book is dedicated to this subject and only the consequences on deterministic numerical modeling will be briefly discussed here.

As far as linear model and stochastic loads are concerned the linear filtering theory applies [127]. Performing stochastic analyses simply consists in the computation of an appropriate set of deterministic problems on a given numerical model and thus boils down to a particular parametric study. Having an efficient multiple right hand side deterministic solver is then of primary importance for these applications.

When dealing with large fluctuations of the elastic properties of the soil using Monte-Carlo simulations [147, 158, 168] , the main concern is to account for the heterogeneity of the soil and the efficiency. This can be achieved using reduction techniques either in physical space [20] using Ritz-Galerkin projection and in the parameter space using Karhunen-Loeve expansions [178]. Some practical results proposed in [166, 69, 166, 165, 66, 123] will be summarized in Section 7.3..

1.6. UNBOUNDED STRUCTURES

A basic assumption in the above mentioned deterministic and stochastic techniques is that the structure and the fluctuating heterogeneous soil region remain bounded. Thus for very long structures such as tunnels, boreholes, railway track, roads or sheet piles some additional developments have to be made.

In current practice these very long structures are assumed to be translation invariant and thus only two-dimensional models are considered. However the incident fields are not translation invariant. The importance of inclined incident fields with respect to the invariant axis has been reported by geophysicists [150, 151]. Moreover, these inclined incident fields have a strong impact on the seismic behaviour of networks because these waves generate differential displacements in the longitudinal direction [196] and thus additional stresses. The finite correlation length of recorded seismic motion indicates that these inclined incident waves always exist in practice. For traffic induced vibrations or borehole geophysics the load can be modelled as a still or moving source and the problem is then far from being translationally invariant. The extension of the preceding numerical techniques to translationally invariant geometries with non translationally invariant loads has been initiated in [176] and applied in [14, 19, 133]

Translation invariance of the geometrical and mechanical properties of the model is often a coarse approximation and for many applications a periodic assumption is better suited. The proposed numerical techniques have been extended to periodic models in [144, 83, 82, 62] with stochastic loadings [61] or stochastic constitutive parameters [60]. A summary of these developments will be given in Section 5.

1.7. GUIDELINES

Section 3 deals with a generic dynamic Soil-Fluid-Structure Interaction problem and points out the basic physical assumptions regarding the geometry, the constitutive laws and the applied loads together with the uncertainties attached to them.

Assuming a linear behaviour and deterministic loads and properties, Section 3 presents a general dynamic domain decomposition technique [13] based on either BEM or FEM for each sub-domain. Reduction techniques for large models are also discussed.

Sections 4 and 6 are devoted to the BEM for a layered half-space with particular attention being given to the numerical implementation of the Green's functions in terms of efficiency and regularisation.

Extensions of these techniques for unbounded structures are proposed in Section 5 with a particular attention to periodic structures.

Finally Section 7 presents some applications of these techniques :

- Soil-Fluid-Structure Interaction is considered in Sections 7.1. for dams and 7.4. for quay-walls,
- classical and advanced SSI studies of large reactor buildings are described in Sections 7.2. and 7.3.,
- site and city-site effects are studied in Sections 7.5. and 7.6.,
- borehole geophysics is finally addressed in Section 7.7.

2. Physical and Mathematical Models

This aim of this section is to present a generic Soil-Fluid-Structure Interaction problem and the associated parameters, unknowns and equations.

2.1. GEOMETRY

The physical domain is denoted by Ω and is decomposed into three subdomains: the unbounded soil denoted by Ω_s, the bounded fluid denoted by Ω_f and the bounded structure denoted by Ω_b as shown in Figure 1. The interfaces between these domains are denoted respectively by Σ_{bs}, Σ_{bf} and Σ_{sf}. On the other parts of their boundaries denoted by Γ_{ba}, Γ_{sa} et Γ_{fa} free surface boundary conditions are assumed. The boundary of $\Omega = \Omega_s \cup \Omega_f \cup \Omega_b$ is denoted by $\Gamma_a = \Gamma_{ba} \cup \Gamma_{sa} \cup \Gamma_{fa}$. The interfaces between Ω_s (resp. Ω_f) and the other subdomains is denoted by Σ_s (resp. Σ_f). Finally Σ will denote the generalised interfaces and will be equal to $\Omega_b \cup \Sigma_s \cup \Sigma_f$.

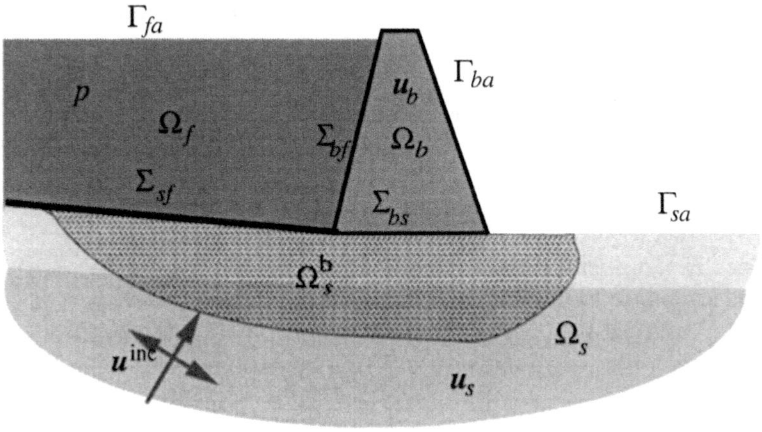

Figure 1. Model layout and notation

2.2. THE UNKNOWN FIELDS

The permanent displacement fields on Ω_s and Ω_b due to static loads (the weight or the hydrostatic pressure $p_o(\boldsymbol{x})$) are denoted by $\boldsymbol{u}_{so}(\boldsymbol{x})$ and $\boldsymbol{u}_{bo}(\boldsymbol{x})$. These fields are assumed to be known in the following and will play the role of parameters.

The dynamic perturbations of these fields due to dynamic loadings are denoted by $\boldsymbol{u}_s(\boldsymbol{x},t)$, $\boldsymbol{u}_b(\boldsymbol{x},t)$ and $p(\boldsymbol{x},t)$. They are assumed to be small enough to allow for a linear approximation of the constitutive and equilibrium equations in the vicinity of the static state $(\boldsymbol{u}_{so}, \boldsymbol{u}_{bo}, p_o)$. Thus, the dynamic perturbations of the stress tensors denoted by $\boldsymbol{\sigma}_s(\boldsymbol{u}_s)$ and $\boldsymbol{\sigma}_b(\boldsymbol{u}_b)$ can be expressed as linear functions of the dynamic fluctuation of the strain tensors denoted by $\boldsymbol{\epsilon}(\boldsymbol{u}_s)$ and $\boldsymbol{\epsilon}(\boldsymbol{u}_b)$ using the classical Hooke's Law :

$$\boldsymbol{\sigma}_\alpha(\boldsymbol{u}_\alpha) = \lambda_\alpha(\mathrm{div}\boldsymbol{u})\mathbf{I}_d + 2\mu_\alpha\boldsymbol{\epsilon}(\boldsymbol{u}) \tag{1}$$

$$\epsilon_{ij} = \frac{\partial_i u_j + \partial_j u_i}{2} \qquad \mathrm{div}\boldsymbol{u} = \sum_{i=1}^{3} \epsilon_{ii} \tag{2}$$

λ_α and μ_α being the classical Lamé parameters and $\mathbf{I}_d$ being the 3×3 identity matrix. The traction vectors applying on a given interface oriented by the outer normal vector $\boldsymbol{n}$ are denoted by $\boldsymbol{t}_s(\boldsymbol{u}_s)$ and $\boldsymbol{t}_b(\boldsymbol{u}_b)$:

$$\boldsymbol{t}_\alpha(\boldsymbol{u}_\alpha) = \boldsymbol{\sigma}_\alpha(\boldsymbol{u}_\alpha)\boldsymbol{n} \tag{3}$$

and $u_n = \boldsymbol{u} \cdot \boldsymbol{n}$ is the normal component of any vector $\boldsymbol{u}$. Following the same lines for the fluid and assuming that the flow is irrotational, the dynamic pressure increment p can be related to the divergence of the flow :

$$\dot{p} = \lambda_f \mathrm{div}\boldsymbol{v}_f \tag{4}$$

where λ_f is the bulk modulus of the fluid and $\boldsymbol{v}_f$ is the flow velocity.

2.3. LOADS

Loads for Soil-Structure Interaction problems are either incident fields, applied forces and tractions inside or on the boundaries of the domain and non homogeneous initial conditions.

2.3.1. *Incident Fields*

Classically the seismic loading is accounted for by defining inside Ω_s a given incident field denoted by $\boldsymbol{u}^{\text{inc}}$. This incident field can be seen as a parameter of the dynamic interaction problem that has to satisfy some constraints. Firstly, it is assumed that $\boldsymbol{u}^{\text{inc}}(\boldsymbol{x},t)$ vanishes before a given finite time t_o in the vicinity of the structure and the fluid :

$$\boldsymbol{u}^{\text{inc}}(\boldsymbol{x},t) = \boldsymbol{0} \qquad \forall t < t_o \qquad \forall \boldsymbol{x} \in \Omega_s^{\text{b}} \tag{5}$$

where Ω_s^{b} is a bounded subset of Ω_s in the vicinity of the structure. Without any loss of generality t_o can be set to 0 and this will be assumed in the following.

It must be noticed that Ω_s^{b} has been introduced to allow for a simple definition of the incident field $\boldsymbol{u}^{\text{inc}}$ especially when the soil around the structure is heterogeneous. Indeed it has to be assumed for the following developments that $\boldsymbol{u}^{\text{inc}}$ satisfies the Navier equation and free surface boundary conditions outside of Ω_s^{b} :

$$\begin{array}{rll} -\mathbf{Div}\boldsymbol{\sigma}_s(\boldsymbol{u}^{\text{inc}}) + \rho_s\partial_{tt}\boldsymbol{u}^{\text{inc}} = \boldsymbol{0} & \text{in} & \Omega_s \setminus \Omega_s^{\text{b}} \\ \boldsymbol{t}_s(\boldsymbol{u}^{\text{inc}}) = \boldsymbol{0} & \text{on} & \Gamma_{sa} \setminus \partial\Omega_s^{\text{b}} \end{array} \tag{6}$$

When Ω_s^{b} is not empty the following seismic force $\boldsymbol{f}^{\text{inc}}$ in the soil and seismic traction $\boldsymbol{t}^{\text{inc}}$ can be defined as the lack of balance due to $\boldsymbol{u}^{\text{inc}}$ inside Ω_s^{b} and on its free surface :

$$\begin{array}{rll} -\mathbf{Div}\boldsymbol{\sigma}_s(\boldsymbol{u}^{\text{inc}}) + \rho_s\partial_{tt}\boldsymbol{u}^{\text{inc}} = -\boldsymbol{f}^{\text{inc}} & \text{in} & \Omega_s^{\text{b}} \\ \boldsymbol{t}_s(\boldsymbol{u}^{\text{inc}}) = -\boldsymbol{t}^{\text{inc}} & \text{in} & \partial\Omega_s^{\text{b}} \cap \partial\Omega_s \end{array} \tag{7}$$

Since Ω_s^{b} is bounded, $\boldsymbol{f}^{\text{inc}}$ and $\boldsymbol{t}^{\text{inc}}$ have bounded supports and vanish for negative times. Additional dynamic forces and applied tractions sharing the same properties can also be defined inside Ω_s^{b}.

Finally, $\boldsymbol{u}^{\text{inc}}_{|\Sigma_{bs}}$ and $\boldsymbol{u}^{\text{inc}}_{|\Sigma_{sf}}$ the incident field on the soil-structure interface and on the soil-fluid interface will also play the role of applied loads in the following. It should be noticed that thanks to the hypothesis on the incident field, these fields vanish for negative times.

2.3.2. *Initial Conditions*

Initial conditions $\boldsymbol{u}(\boldsymbol{x},t=0) = \boldsymbol{u}_o(\boldsymbol{x})$ and $\partial_t\boldsymbol{u}(\boldsymbol{x},t=0) = \boldsymbol{v}_o(\boldsymbol{x})$ defined for the problem can be transformed to usual applied body forces $\boldsymbol{f}_{\text{o}}(\boldsymbol{x},t)$ as follows using the classical distribution framework :

$$\boldsymbol{f}_{\text{o}}(\boldsymbol{x},t) = -\rho(\boldsymbol{v}_o\delta(t) - \boldsymbol{u}_o\partial_t\delta(t)) \tag{8}$$

and it is assumed in the following that $\boldsymbol{u}_o$ and $\boldsymbol{v}_o$ have bounded support.

2.3.3. *Applied Forces and Tractions*

Applied forces and tractions are denoted by $\boldsymbol{f}_v(\boldsymbol{x},t)$ and $\boldsymbol{t}_o(\boldsymbol{x},t)$ respectively and are assumed to have bounded support inside Ω and on $\partial\Omega$ respectively.

2.4. LINEAR EQUATIONS

The fields defined in Section 2.2. have to satisfy field equations and boundary conditions inside and on the boundary of each subdomain and coupling equations along the interfaces.

2.4.1. *Field Equations*

Using the definitions of the incident field, a new auxiliary field, namely the *diffracted* or *scattered field* , denoted by $\boldsymbol{u}_{\mathrm{d}}$ is defined in the soil as a function of the total field $\boldsymbol{u}_s$:

$$\boldsymbol{u}_{\mathrm{d}} = \boldsymbol{u}_s - \boldsymbol{u}^{\mathrm{inc}} \qquad \text{in } \Omega_s \tag{9}$$

Accounting for equations (6), $\boldsymbol{u}_{\mathrm{d}}$ has to satisfy the Navier equation in Ω_s for any positive time t with $\boldsymbol{f}_s = \boldsymbol{f}^{\mathrm{inc}} + \boldsymbol{f}_{\mathrm{o}} + \boldsymbol{f}_v$ as a source term and $\boldsymbol{t}_{so} = \boldsymbol{t}^{\mathrm{inc}} + \boldsymbol{t}_o$ as boundary conditions on Γ_{sa} :

$$\begin{array}{rll} -\mathbf{Div}\boldsymbol{\sigma}_s(\boldsymbol{u}_{\mathrm{d}}) + \rho_s \partial_{tt}\boldsymbol{u}_{\mathrm{d}} = \boldsymbol{f}_s & \text{in} & \Omega_s \\ \boldsymbol{t}_s(\boldsymbol{u}_{\mathrm{d}}) = \boldsymbol{t}_{so} & \text{on} & \Gamma_{sa} \end{array} \tag{10}$$

$\boldsymbol{u}_b$, the displacement in the structure has to satisfy :

$$\begin{array}{rll} -\mathbf{Div}\boldsymbol{\sigma}_b(\boldsymbol{u}_b) + \rho_b \partial_{tt}\boldsymbol{u}_b = \boldsymbol{f}_b & \text{in} & \Omega_b \\ \boldsymbol{t}_b(\boldsymbol{u}_b) = \boldsymbol{t}_{bo} & \text{on} & \Gamma_{ba} \end{array} \tag{11}$$

with $\boldsymbol{f}_b = \boldsymbol{f}_{\mathrm{o}} + \boldsymbol{f}_v$ and $\boldsymbol{t}_{bo} = \boldsymbol{t}_o$ on Ω_b and Γ_{ba} respectively.

The pressure field p satisfies the wave equation inside Ω_f and the free surface conditions :

$$\begin{array}{rll} c^2\Delta p - \partial_{tt}p = 0 & \text{in} & \Omega_f \\ p = 0 & \text{on} & \Gamma_{fa} \end{array} \tag{12}$$

Moreover these fields have to satisfy homogeneous initial conditions :

$$\begin{array}{rl} \boldsymbol{u}_{\mathrm{d}}(\boldsymbol{x},0) = \partial_t \boldsymbol{u}_{\mathrm{d}}(\boldsymbol{x},0) = \mathbf{0} & \forall \boldsymbol{x} \in \Omega_s \\ \boldsymbol{u}_b(\boldsymbol{x},0) = \partial_t \boldsymbol{u}_b(\boldsymbol{x},0) = \mathbf{0} & \forall \boldsymbol{x} \in \Omega_b \\ p(\boldsymbol{x},0) = \partial_t p(\boldsymbol{x},0) = \mathbf{0} & \forall \boldsymbol{x} \in \Omega_f \end{array} \tag{13}$$

2.4.2. *Coupling Equations*

The fields $\boldsymbol{u}_{\mathrm{d}}$, $\boldsymbol{u}_b$, p have to satisfy the following local equilibrium conditions along the interfaces Σ_{sf}, Σ_{bf} and Σ_{bs} :

$$\boldsymbol{t}_b(\boldsymbol{u}_b) - p\,\boldsymbol{n} = \mathbf{0} \qquad \text{on} \quad \Sigma_{bf} \tag{14}$$

$$\boldsymbol{t}_s(\boldsymbol{u}_{\mathrm{d}}) - p\,\boldsymbol{n} = \boldsymbol{t}^{\mathrm{inc}} \qquad \text{on} \quad \Sigma_{sf} \tag{15}$$

$$\boldsymbol{t}_b(\boldsymbol{u}_b) + \boldsymbol{t}_s(\boldsymbol{u}_{\mathrm{d}}) = \boldsymbol{t}^{\mathrm{inc}} \qquad \text{on} \quad \Sigma_{bs} \tag{16}$$

the flux conditions along Σ_{sf} and Σ_{bf} :

$$\rho_f \partial_{tt} \boldsymbol{u}_{\rm d} \cdot \boldsymbol{n} + \partial_n p = -\rho_f \partial_{tt} u_n^{\rm inc} \quad \text{on} \quad \Sigma_{sf} \tag{17}$$

$$\rho_f \partial_{tt} \boldsymbol{u}_b \cdot \boldsymbol{n} + \partial_n p = 0 \quad \text{on} \quad \Sigma_{bf} \tag{18}$$

and the kinematical condition on Σ_{bs} :

$$\boldsymbol{u}_b - \boldsymbol{u}_{\rm d} = \boldsymbol{u}^{\rm inc} \quad \text{on} \quad \Sigma_{bs} \tag{19}$$

where $\boldsymbol{n}$ is the outer normal vector of the fluid domain Ω_f.

Equations (11–19) form a well posed boundary value problem with homogeneous intial conditions, $\boldsymbol{u}_{\rm d}(\boldsymbol{x}, t)$ and $\boldsymbol{u}_b$ depending linearly on $(\boldsymbol{u}^{\rm inc}, \boldsymbol{f}_v, \boldsymbol{t}_o, \boldsymbol{u}_o, \boldsymbol{v}_o)$. Sections 3 to 4 will be dedicated to the numerical approximation of the associated convolution operators.

For the sake of simplicity it will be assumed in the following that $\boldsymbol{u}^{\rm inc}$ is the only non vanishing driving term since the work is very similar for the other terms.

2.5. VARIABILITY ON THE PARAMETERS

The unknown fields $\boldsymbol{u}_{\rm d}$ and $\boldsymbol{u}^{\rm inc}$ in equations (11–19) depend :

- linearly on the applied loads $\boldsymbol{\Phi}^{\rm inc} = (\boldsymbol{f}^{\rm inc}, \boldsymbol{t}^{\rm inc}, \boldsymbol{u}^{\rm inc}{}_{|\Sigma_{bs}}, u_n^{\rm inc}{}_{|\Sigma_{sf}})$, e.g. space time fields with bounded support in space.
- non linearly on the mechanical properties $\boldsymbol{m}_\alpha = (\lambda_\alpha, \mu_\alpha, \rho_\alpha), \alpha \in \{s, f, b\}$ e.g. time independent fields with either bounded or unbounded support.

In practical situations large uncertainties are attached to the effective values of these parameters and especially to the applied loads $\boldsymbol{\Phi}^{\rm inc}$ and the elastic parameters in the soil $\boldsymbol{m}$. These uncertainties can be accounted for by modeling these fields as second order stochastic fields.

2.5.1. *Stochastic Model of the Soil Parameters*

In order to allow for a numerical approximation of the stochastic fields $\boldsymbol{m}$ associated with the soil parameters, it is assumed that these parameters fluctuate only in a bounded region near the structure, denoted by $\Omega_s^{\rm b}$. Although questionable this assumption is justified since even if large, the uncertainties far from the structure may modify the incident field but have a limited impact on the structural response itself. This influence can then be accounted for in an average sense using *effective* elastic properties far from structure instead of *mean* properties. Anyhow based on this assumption $\boldsymbol{m}$ can be decomposed into a known *mean* or *effective* field $\underline{\boldsymbol{m}} = (\underline{\lambda}, \underline{\mu}, \underline{\rho})$ defined on Ω_s and a centred second order stochastic field $\boldsymbol{m}$ defined on $\Omega_s^{\rm b}$ with known covariance tensor $\boldsymbol{C}_m(\boldsymbol{x}, \boldsymbol{x}')$:

$$\boldsymbol{C}_m(\boldsymbol{x}, \boldsymbol{x}') = E < \boldsymbol{m}(\boldsymbol{x}') \boldsymbol{m}(\boldsymbol{x})^T > \tag{20}$$

$E <>$ standing for the mathematical expectation.

The numerical approximation of $\boldsymbol{m}$ is obtained using the following Karhunen Loeve expansion on $\boldsymbol{m}$ [172] :

$$\boldsymbol{m}(\boldsymbol{x}) = \underline{\boldsymbol{m}}(\boldsymbol{x}) + \sum_{k=1}^{N_k} \theta_k \boldsymbol{m}_k(\boldsymbol{x}) \tag{21}$$

where θ_k are uncorrelated random variables with standard deviation σ_k defined as follows:

$$\theta_k = \int_{\Omega_s^b} \boldsymbol{m}_k(\boldsymbol{x}) \cdot \boldsymbol{m}(\boldsymbol{x}) dV(\boldsymbol{x}) \tag{22}$$

$\boldsymbol{m}_k(\boldsymbol{x})$ being the N_k first eigenvectors of the covariant operator, e.g. satisfying:

$$\int_{\Omega_s^b} \boldsymbol{C}_m(\boldsymbol{x}, \boldsymbol{x}') \boldsymbol{m}_k(\boldsymbol{x}') dV(\boldsymbol{x}') = \sigma_k^2 \boldsymbol{m}_k(\boldsymbol{x}) \tag{23}$$

Using Monte Carlo simulation for the random variables θ_k, the Karhunen Loeve expansion of the soil parameters (21) provides a simple way to build a set of mechanical models for the soil in accordance with the available knowledge on the physical parameter distribution e.g. $\underline{\boldsymbol{m}}$ and $\boldsymbol{C}_m$. Finally, this set of models can be used to perform statistical analyses on the structural response.

2.5.2. *Stochastic Model for the Applied Loads*

Accounting for the uncertainties in the applied loads $\boldsymbol{\Phi}^{\text{inc}}$ is in some sense easier than accounting for the soil variability since the structural response depends linearly on these loads. The remaining difficulty consists in modelling these uncertainties and in particular:

1 the dependency on time of $\boldsymbol{\Phi}^{\text{inc}}$,

2 the cross correlation between elements of $\boldsymbol{\Phi}^{\text{inc}}$ since $\boldsymbol{f}^{\text{inc}}$, $\boldsymbol{t}^{\text{inc}}$, $\boldsymbol{u}^{\text{inc}}{}_{|\Sigma_{bs}}$ and $u_n^{\text{inc}}{}_{|\Sigma_{sf}}$ may have different spatial supports,

3 the approximation of the associated stochastic fields.

As long as a single point and a single component are considered, the time dependency is classically accounted for by assuming that the field is stationary with respect to time or is a deterministic modulation of a stationary process e.g. at a given point $\boldsymbol{x}$, the jth component of $\boldsymbol{u}^{\text{inc}}(\boldsymbol{x}, t)$ reads :

$$u_j^{\text{inc}}(\boldsymbol{x}, t) = f_j(\boldsymbol{x}, t) u_o^{\text{inc}}(\boldsymbol{x}, t) \tag{24}$$

$u_o^{\text{inc}}(\boldsymbol{x}, t)$ being stationary with respect to time and $f_j(\boldsymbol{x}, t)$ being a given deterministic window. Using this technique a set a synthetic seismograms fitting a given power spectrum can be built. Stochastic results can also be obtained defining a time dependent power spectral density [63]. When several points or surfaces have to be considered this approach is much more difficult since due to propagation effects the deterministic modulation $f_j(\boldsymbol{x}, t)$ cannot be the same at each point $\boldsymbol{x}$. The cross-correlation has to be analysed.

Accounting for cross correlation seems rather easy to do since $\boldsymbol{f}^{\mathrm{inc}}$, $\boldsymbol{t}^{\mathrm{inc}}$, $\boldsymbol{u}^{\mathrm{inc}}{}_{|\Sigma_{bs}}$ and $u_n^{\mathrm{inc}}{}_{|\Sigma_{sf}}$ depend linearly on $\boldsymbol{u}^{\mathrm{inc}}$ defined on Ω_s (see equations (7). However very little statistical information is available on $\boldsymbol{u}^{\mathrm{inc}}$ inside Ω_s. Moreover this field cannot be modelled as a purely random field since it has to satisfy some constraints given by equations (6). In fact the only available information on $\boldsymbol{u}^{\mathrm{inc}}$ is its covariance on the free surface [77]. Finding the statistical properties of $\boldsymbol{u}^{\mathrm{inc}}$ inside Ω_s based on its statistical properties on the free surface and the elastodynamic constraints is called the stochastic deconvolution problem and has been proposed in [116].

In particular, assuming an horizontal free surface denoted by S and stationary random field with respect to time, $\boldsymbol{C}_S^{\mathrm{inc}}$, the cross spectral density of $\boldsymbol{u}_{|S}^{\mathrm{inc}}$ is given as a function of the frequency as follows :

$$\boldsymbol{C}_S^{\mathrm{inc}}(\boldsymbol{x}',\boldsymbol{x},\omega) = \boldsymbol{S}^{\mathrm{inc}}(\omega)\gamma(\frac{\omega r}{c})e^{i\frac{\omega}{c_a}\boldsymbol{d}_a\cdot(\boldsymbol{x}-\boldsymbol{x}')} \tag{25}$$

where $\boldsymbol{S}^{\mathrm{inc}}$ is a cross-spectral density at a given reference point, γ is a normalized coherency function depending on the dimensionless parameter $\frac{\omega r}{c}$, r is the distance between $\boldsymbol{x}$ and $\boldsymbol{x}'$, c is a characteristic velocity in the soil and $\boldsymbol{k}_a = \frac{\omega}{c_a}\boldsymbol{d}_a$ is an apparent wavenumber along the free surface corresponding to waves propagating in direction $\boldsymbol{d}_a$ with a velocity equal to c_a.

Using a Fourier transform with respect to the two horizontal variables, $\boldsymbol{C}_S^{\mathrm{inc}}$ reads :

$$\begin{aligned}
\boldsymbol{C}_S^{\mathrm{inc}}(\boldsymbol{x}',\boldsymbol{x},\omega) &= \sum_{l=1}^{3}\int_{\mathbb{R}^2} \widehat{\boldsymbol{u}_{Sl}^{\mathrm{inc}}}(\boldsymbol{k},\omega)\otimes\widehat{\boldsymbol{u}_{Sl}^{\mathrm{inc}}}^*(\boldsymbol{k},\omega)e^{i\boldsymbol{k}\cdot(\boldsymbol{x}-\boldsymbol{x}')}d\boldsymbol{k} &(26)\\
\widehat{\boldsymbol{u}_{Sl}^{\mathrm{inc}}}(\boldsymbol{k},\omega) &= \boldsymbol{d}_l(\omega)\sqrt{\frac{\omega}{c}s_l(\omega)\widehat{\gamma}(\boldsymbol{k}\frac{c}{\omega}-\frac{\boldsymbol{d}_a c}{c_a})} &(27)
\end{aligned}$$

where $\boldsymbol{d}_l(\omega)$ and $s_l(\omega)$ are respectively the deterministic eigenvectors and the eigenvalues of $\boldsymbol{S}^{\mathrm{inc}}$, * denoting the conjugate and $\widehat{\cdot}$ the Fourier transform. $\widehat{\boldsymbol{u}_{Sl}^{\mathrm{inc}}}(\boldsymbol{k},\omega)e^{i\boldsymbol{k}\cdot\boldsymbol{x}}$ is a deterministic field on S with a given wave number $\boldsymbol{k}$. A standard plane-wave deconvolution in a layered media allows for the extension of this field in the half-space. The corresponding plane wave field is denoted by $\widehat{\boldsymbol{u}_{Sl}^{\mathrm{inc}}}(\boldsymbol{x},\boldsymbol{k},\omega)$.

The polarisation of these plane waves is characterized by $\boldsymbol{d}_l(\omega)$ and their frequency contents by $s_l^{1/2}(\omega)$. The dispersion around the mean propagation characterized by $\boldsymbol{d}_a$ and c_a is controlled by $\widehat{\gamma}^{1/2}$. Thus, $\boldsymbol{C}_{\Omega_s}^{\mathrm{inc}}(\boldsymbol{x}',\boldsymbol{x},\omega)$ the covariance of the incident field inside Ω_s is given as the superposition of these plane waves :

$$\boldsymbol{C}_{\Omega_s}^{\mathrm{inc}}(\boldsymbol{x}',\boldsymbol{x},\omega) = \sum_{l=1}^{3}\int_{\mathbb{R}^2} \widehat{\boldsymbol{u}_l^{\mathrm{inc}}}(\boldsymbol{x}',\boldsymbol{k},\omega)\otimes\widehat{\boldsymbol{u}_l^{\mathrm{inc}}}^*(\boldsymbol{x},\boldsymbol{k},\omega)d\boldsymbol{k} \tag{28}$$

The applied loads denoted by $\widehat{\boldsymbol{\Phi}_l^{\mathrm{inc}}}(\boldsymbol{k}) = (\widehat{\boldsymbol{f}_k^{\mathrm{inc}}},\widehat{\boldsymbol{t}_k^{\mathrm{inc}}},\widehat{\boldsymbol{u}_k^{\mathrm{inc}}}_{|\Sigma_{bs}},\widehat{u_{nk}^{\mathrm{inc}}}_{|\Sigma_{sf}})$ associated to these incident plane waves can be computed using equations (7) and thus the covariance of $\boldsymbol{\Phi}_l^{\mathrm{inc}}(\boldsymbol{k})$ in the frequency domain formally reads :

$$C_{\Phi}^{\text{inc}}(\omega) = \sum_{l=1}^{3} \int_{\mathbb{R}^2} \widehat{\boldsymbol{\Phi}_l^{\text{inc}}}(\boldsymbol{k},\omega) \otimes \widehat{\boldsymbol{\Phi}_l^{\text{inc}}}^{*}(\boldsymbol{k},\omega) d\boldsymbol{k} \tag{29}$$

This expression is a little misleading since it is an expansion of the covariance on a non countable set of incident plane waves. From the numerical point of view this means that when performing a stochastic analysis on the incident field, an infinite number of incident plane waves have to be accounted for. Fortunately this is not true since these loads have a bounded support and thus the Karhunen-Loeve expansion technique applies either at each frequency or in a given frequency range.

Finally let us remark that incident fields $\widehat{\boldsymbol{u}_l^{\text{inc}}}(\boldsymbol{x}', \boldsymbol{k}, \omega)$ can be used as deterministic loads in parametric studies.

2.6. SUMMARY OF THE MODELLING SECTION

It has been shown in this section that the generic Soil-Structure Interaction problem presented in Section 2.1. can be analysed by solving the set of linear equations (11–19) for fields $\boldsymbol{u}_{\text{d}}$, $\boldsymbol{u}_b$ and p.

Section 2.5. has also shown that accounting for the uncertainties in the data leads to the computation of a set deterministic generic problems with a set of applied loads and parameters.

Due to the linearity of equations (11–19) a Fourier or Laplace transform can be applied to these equations to solve the problem in the frequency domain. In the following the same notation will be used for time domain functions and frequency domain ones and in particular equations (11–19) now read :

$$\textbf{Div}\boldsymbol{\sigma}_s(\boldsymbol{u}_{\text{d}}) + \boldsymbol{f}_s = -\rho_s \omega^2 \boldsymbol{u}_{\text{d}} \quad \text{in} \quad \Omega_s \tag{30}$$

$$\boldsymbol{t}_s(\boldsymbol{u}_{\text{d}}) = \boldsymbol{t}^{\text{inc}} \quad \text{on} \quad \Gamma_{sa} \tag{31}$$

$$\textbf{Div}\boldsymbol{\sigma}_b(\boldsymbol{u}_b) + \boldsymbol{f}_b = -\rho_b \omega^2 \boldsymbol{u}_b \quad \text{in} \quad \Omega_b \tag{32}$$

$$\boldsymbol{t}_b(\boldsymbol{u}_b) = \boldsymbol{0} \quad \text{on} \quad \Gamma_{ba} \tag{33}$$

$$c^2 \Delta p + \omega^2 p = 0 \quad \text{in} \quad \Omega_f \tag{34}$$

$$p = 0 \quad \text{on} \quad \Gamma_{fa} \tag{35}$$

$$\boldsymbol{t}_b(\boldsymbol{u}_b) - p\,\boldsymbol{n} = \boldsymbol{0} \quad \text{on} \quad \Sigma_{bf} \tag{36}$$

$$\boldsymbol{t}_s(\boldsymbol{u}_{\text{d}}) - p\,\boldsymbol{n} = \boldsymbol{t}^{\text{inc}} \quad \text{on} \quad \Sigma_{sf} \tag{37}$$

$$\boldsymbol{t}_b(\boldsymbol{u}_b) + \boldsymbol{t}_s(\boldsymbol{u}_{\text{d}}) = \boldsymbol{t}^{\text{inc}} \quad \text{on} \quad \Sigma_{bs} \tag{38}$$

$$-\rho_f \omega^2 \boldsymbol{u}_{\text{d}} \cdot \boldsymbol{n} + \partial_n p = \rho_f \omega^2 u_n^{\text{inc}} \quad \text{on} \quad \Sigma_{sf} \tag{39}$$

$$-\rho_f \omega^2 \boldsymbol{u}_b \cdot \boldsymbol{n} + \partial_n p = 0 \quad \text{on} \quad \Sigma_{bf} \tag{40}$$

$$\boldsymbol{u}_b - \boldsymbol{u}_{\text{d}} = \boldsymbol{u}^{\text{inc}} \quad \text{on} \quad \Sigma_{bs} \tag{41}$$

2.6.1. *Wellposedness and Approximation*

As long as damping is present in each domain this set of equations has a unique weak solution in proper functional spaces. To keep this property when no damping is present, the Laplace transform instead of the Fourier transform has to be used, giving thus a small imaginary part to the frequency ω.

Based on classical results on separable Hilbert spaces this set of equations can be approximated on a finite dimension vector space for the unknown fields with *a priori* error estimates.

This approximation is achieved in two steps : the approximation of the interaction between the structure and the other domains using domain decomposition techniques proposed in Section 3 and approximation of local boundary value problems in the soil and in the fluid using Boundary Elements as given in Section 4.

3. Domain Decomposition

The basic ideas of domain decomposition techniques are :

- to define new unknown fields on the interfaces, either displacements or tractions, so that one of the two coupling equations on each interface holds *a priori*,
- to solve Boundary Value Problems in each subdomain using these new unknown fields as boundary conditions,
- to enforce the other coupling equation in a weak sense, e.g. for any trial admissible fields on the interfaces.

The numerical approximation simply consists in taking these new unknowns in given finite dimension spaces. Primal Domain Decomposition techniques consist in using the displacements as coupling variables while tractions are used in dual approaches. In hybrid Domain Decomposition techniques both displacements and tractions are used as coupling variables and the two coupling equations are written in a weak form.

3.1. COUPLING FIELDS

In the present approach, FEM being used for the structure, a primal approach is best suited. Moreover, to prevent any technical difficulty in matching the displacement field inside the structure and on the soil-fluid interface, the coupling variable denoted by $\boldsymbol{u}_\Sigma$ is defined as a displacement field on Ω_b and on Σ_{sf}.

3.2. LOCAL BOUNDARY VALUE PROBLEMS

When a given field $\boldsymbol{u}_\Sigma$ is enforced on interfaces Σ_{bs} and Σ_{sf}, a displacement field denoted by $\boldsymbol{u}_\mathrm{d}(\boldsymbol{u}_\Sigma)$ is radiated in the soil. This field satisfies the following Boundary Value Problems :

$$\begin{aligned} \mathbf{Div}\boldsymbol{\sigma}_s(\boldsymbol{u}_\mathrm{d}) &= -\rho_s\omega^2\boldsymbol{u}_\mathrm{d} && \text{in } \Omega_s \\ \boldsymbol{u}_\mathrm{d} &= \boldsymbol{u}_\Sigma && \text{on } \Sigma_s \\ \boldsymbol{t}_s(\boldsymbol{u}_\mathrm{d}) &= \mathbf{0} && \text{on } \Gamma_{sa} \end{aligned} \qquad (42)$$

When interfaces $\Sigma_{sf} \cup \Sigma_{bs}$ are kept fixed, the incident field $\boldsymbol{u}^{\text{inc}}$ generates a *local scattered field* denoted by $\boldsymbol{u}_{\text{do}}$ and satisfying :

$$\begin{aligned} \mathbf{Div}\boldsymbol{\sigma}_s(\boldsymbol{u}_{\text{do}})+\boldsymbol{f}_s &= -\rho_s\omega^2\boldsymbol{u}_{\text{do}} && \text{in } \Omega_s \\ \boldsymbol{u}_{\text{do}} &= -\boldsymbol{u}^{\text{inc}} && \text{on } \Sigma_s \\ \boldsymbol{t}_s(\boldsymbol{u}_{\text{do}}) &= \boldsymbol{t}_{so} && \text{on } \Gamma_{sa} \end{aligned} \tag{43}$$

When the normal flux $-\rho_f \boldsymbol{u}_\Sigma\cdot\boldsymbol{n}$ is applied on the interfaces a pressure field denoted by $p_{\text{d}}(\boldsymbol{u}_\Sigma)$ is radiated into the fluid. It satisfies the following acoustic problem :

$$\begin{aligned} c^2\Delta p_{\text{d}} + \omega^2 p_{\text{d}} &= 0 && \text{in } \Omega_f \\ \partial_n p_{\text{d}} &= \rho_f \boldsymbol{u}_\Sigma \cdot \boldsymbol{n} && \text{on } \Sigma_f \\ p_{\text{d}} &= 0 && \text{on } \Gamma_{fa} \end{aligned} \tag{44}$$

By construction, fields $\boldsymbol{u}_{\text{d}}(\boldsymbol{u}_\Sigma) + \boldsymbol{u}_{\text{do}}$ or $\omega^2 p_{\text{d}}(\boldsymbol{u}_\Sigma)$ satisfy field equations (30) or (34) and kinematical conditions (39-41) on the interfaces.

3.3. VARIATIONAL FORMULATIONS

In order to become a solution of the interaction problem, $\boldsymbol{u}_\Sigma$ and the radiated fields have to satisfy the equilibrium equation in the structure (32) and along the soil-fluid interface (38) together with reciprocity conditions (36-37) on the interfaces between the structure on one hand and the soil and the fluid on the other hand. These equations are written in a weak sense.

Multiplying the equilibrium equation (32) by any virtual field $\boldsymbol{v}_\Sigma$, integrating over Ω_s, then integrating by parts and accounting for (36) and (37) gives :

$$\begin{aligned} \int_{\Omega_b} \{\boldsymbol{\sigma}_b(\boldsymbol{u}_\Sigma) : \boldsymbol{\epsilon}(\boldsymbol{v}_\Sigma) - \omega^2 \rho_b \boldsymbol{u}_\Sigma \cdot \boldsymbol{v}_\Sigma\} dV &- \omega^2 \int_{\Sigma_{bf}} p_{\text{d}} \boldsymbol{v}_\Sigma \cdot \boldsymbol{n} dS \\ + \int_{\Sigma_{bs}} \boldsymbol{t}_s(\boldsymbol{u}_{\text{d}} + \boldsymbol{u}_{\text{do}}) \cdot \boldsymbol{v}_\Sigma dS &= \int_{\Omega_b} \boldsymbol{f}_b \cdot \boldsymbol{v}_\Sigma dV \end{aligned}$$

Then multiplying equation (38) by the same virtual field $\boldsymbol{v}_\Sigma$, integrating on Σ_{bs} and adding to the previous equation gives the following variational formulation of the interaction problem for all $\boldsymbol{v}_\Sigma$ in $\boldsymbol{V}_\Sigma$:

$$\left\{\mathrm{K}_b + \mathrm{K}_s(\omega) - \omega^2 \left((\mathrm{M}_b + \mathrm{M}_f(\omega)\right)\right\} (\boldsymbol{u}_\Sigma, \boldsymbol{v}_\Sigma) = \mathrm{f}_s(\omega; \boldsymbol{v}_\Sigma) + \mathrm{f}_b(\boldsymbol{v}_\Sigma) \tag{45}$$

where $\boldsymbol{V}_\Sigma$ is the functional space containing the restrictions on Ω_b and Σ_{sf} of fields belonging to $\boldsymbol{V}$, $\boldsymbol{V}$ being the functional space containing fields having a finite energy on Ω and satisfying kinematical conditions (39-41). $\mathrm{K}_b(\boldsymbol{u}_\Sigma, \boldsymbol{v}_\Sigma)$, $\mathrm{M}_b(\boldsymbol{u}_\Sigma, \boldsymbol{v}_\Sigma)$ are the classical stiffness and mass bilinear forms arising in Finite

Elements and $\mathrm{f}_b(\boldsymbol{v}_\Sigma)$ is the linear form induced by the applied forces $\boldsymbol{f}_b$:

$$
\begin{aligned}
\mathrm{K}_b(\boldsymbol{u}_\Sigma,\boldsymbol{v}_\Sigma) &= \int_{\Omega_b} \boldsymbol{\sigma}_b(\boldsymbol{u}_\Sigma) : \boldsymbol{\epsilon}(\boldsymbol{v}_\Sigma) dV \\
\mathrm{M}_b(\boldsymbol{u}_\Sigma,\boldsymbol{v}_\Sigma) &= \int_{\Omega_b} \rho_b \boldsymbol{u}_\Sigma \cdot \boldsymbol{v}_\Sigma dV \\
\mathrm{f}_b(\boldsymbol{v}_\Sigma) &= \int_{\Omega_b} \boldsymbol{f}_b \cdot \boldsymbol{v}_\Sigma dV
\end{aligned} \quad (46)
$$

The bilinear form $\mathrm{M}_f(\boldsymbol{u}_\Sigma,\boldsymbol{v}_\Sigma)$ represents the added mass induced by the fluid while $\mathrm{K}_s(\boldsymbol{u}_\Sigma,\boldsymbol{v}_\Sigma)$ stands for the dynamic stiffness of the soil . Finally, the linear form $\mathrm{f}_s(\boldsymbol{v}_\Sigma)$ accounts for the applied seismic force induced by the incident field :

$$
\begin{aligned}
\mathrm{M}_f(\omega;\boldsymbol{u}_\Sigma,\boldsymbol{v}_\Sigma) &= \int_{\Sigma_f} p_{\mathrm{d}} \boldsymbol{v}_\Sigma \cdot \boldsymbol{n} dS \\
\mathrm{K}_s(\omega;\boldsymbol{u}_\Sigma,\boldsymbol{v}_\Sigma) &= \int_{\Sigma_s} \boldsymbol{t}_s(\boldsymbol{u}_{\mathrm{d}}) \cdot \boldsymbol{v}_\Sigma dS \\
\mathrm{f}_s(\omega;\boldsymbol{v}_\Sigma) &= -\int_{\Sigma_s} \boldsymbol{t}_s(\boldsymbol{u}^{\mathrm{inc}} + \boldsymbol{u}_{\mathrm{do}}) \cdot \boldsymbol{v}_\Sigma dS
\end{aligned} \quad (47)
$$

As a consequence of the Betti-Maxwell reciprocity theorem, these bilinear forms are symmetric but not hermitian since damping or radiation conditions are accounted for. Moreover the seismic force takes the following equivalent expression being independant of $\boldsymbol{u}_{\mathrm{do}}$:

$$
\mathrm{f}_s(\boldsymbol{v}_\Sigma) = \int_{\partial\Omega_s} \{\boldsymbol{u}^{\mathrm{inc}} \cdot \boldsymbol{t}_s(\boldsymbol{v}_{\mathrm{d}}) + \boldsymbol{t}^{\mathrm{inc}} \cdot \boldsymbol{v}_{\mathrm{d}}\} dS + \int_{\Omega_s^b} \boldsymbol{f}_s \cdot \boldsymbol{v}_{\mathrm{d}} dS \quad (48)
$$

where $\boldsymbol{v}_{\mathrm{d}} = \boldsymbol{u}_{\mathrm{d}}(\boldsymbol{v}_\Sigma)$ is the radiated field in the soil when $\boldsymbol{v}_\Sigma$ is applied on the interfaces.

3.4. THE SFSI EQUATION

The numerical approximation of the variational interaction problem (45) is obtained by looking for $\boldsymbol{u}_\Sigma$ in $\boldsymbol{V}_\Sigma^H$ a finite dimension subspace of $\boldsymbol{V}_\Sigma$ with given basis $\{\boldsymbol{w}_m\}_{m=1,M}$ as follows :

$$
\boldsymbol{u}_\Sigma(\boldsymbol{x}) = \sum_{m=1}^{M} \boldsymbol{w}_m(\boldsymbol{x}) q_m \quad (49)
$$

$\mathbf{q}$ being the vector of generalized degrees of freedom associated with $\boldsymbol{u}_\Sigma$. Taking for $\boldsymbol{v}_\Sigma$ the basis function $\boldsymbol{w}_n$ with $n = 1, M$, in the weak formulation (45) leads to the following linear system :

$$
\{\mathbf{K}_b + \mathbf{K}_s(\omega) - \omega^2(\mathbf{M}_b + \mathbf{M}_f(\omega))\}\mathbf{q} = \mathbf{f}_b + \mathbf{f}_s(\omega) \quad (50)
$$

where the generic terms of the $M \times M$ matrices $\mathbf{K}_b$ and $\mathbf{M}_b$ and vector $\mathbf{f}_b$ are computed using expressions (46) of the bilinear forms K_b and M_b and the linear form f_b, taking $\boldsymbol{u}_\Sigma = \boldsymbol{w}_m$ and $\boldsymbol{v}_\Sigma = \boldsymbol{w}_n$. The generic term of the $M \times M$ matrices $\mathbf{K}_s$ and $\mathbf{M}_f$ and vector $\mathbf{f}_s$ are computed using expressions (47) of the bilinear forms K_s and M_f and the linear form f_s, taking $\boldsymbol{u}_\mathrm{d} = \boldsymbol{u}_{\mathrm{d}m}$, $p_\mathrm{d} = p_{\mathrm{d}m}$ and $\boldsymbol{v}_\Sigma = \boldsymbol{w}_n$, $\boldsymbol{u}_{\mathrm{d}m}$ and $p_{\mathrm{d}m}$ satisfying the following local Boundary Value Problems :

$$\begin{aligned} \mathbf{Div}\boldsymbol{\sigma}_s(\boldsymbol{u}_{\mathrm{d}m}) &= -\rho_s\omega^2\boldsymbol{u}_{\mathrm{d}m} && \text{in } \Omega_s \\ \boldsymbol{u}_\mathrm{d} &= \boldsymbol{w}_m && \text{on } \Sigma_s \\ \boldsymbol{t}_s(\boldsymbol{u}_{\mathrm{d}m}) &= \mathbf{0} && \text{on } \Gamma_{sa} \end{aligned} \tag{51}$$

$$\begin{aligned} c^2\Delta p_{\mathrm{d}m} + \omega^2 p_{\mathrm{d}m} &= 0 && \text{in } \Omega_f \\ \partial_n p_{\mathrm{d}m} &= \rho_f \boldsymbol{w}_m \cdot \boldsymbol{n} && \text{on } \Sigma_f \\ p_{\mathrm{d}m} &= 0 && \text{on } \Gamma_{fa} \end{aligned} \tag{52}$$

Solving the linear system (50) for given applied forces $\mathbf{f}_b$, $\mathbf{f}_s$ gives the solution of the SFSI problem in terms of the degrees of freedom $\mathbf{q}$. The fields in each domain can then be computed using the following formulae :

$$\boldsymbol{u}_\mathrm{d}(\boldsymbol{x}) = \boldsymbol{u}_{\mathrm{do}}(\boldsymbol{x}) + \sum_{m=1}^{M} \boldsymbol{u}_{\mathrm{d}m}(\boldsymbol{x}) q_m \quad , \quad p(\boldsymbol{x}) = \sum_{m=1}^{M} p_{\mathrm{d}m}(\boldsymbol{x}) q_m \tag{53}$$

Up to now this solution procedure is quite formal since expressions (47) assembling the matrices and the vectors depend on the solution of the boundary value problems (42-44). However it is worth noticing that only the tractions along the interfaces (the flux for pressure fields in the fluid) associated with these radiated fields are required when assembling these terms. Thus the Boundary Element Method is particularly well suited to approximate these fields. This method will be presented in Section 4.

The choice of the finite basis $\{\boldsymbol{w}_m\}_{m=1,M}$ is much simpler and can be based on standard FEM or modal reduction techniques as proposed in the next section.

3.5. FEM AND REDUCTION TECHNIQUES

In order to build the finite dimensional basis $\{\boldsymbol{w}_m\}$ a finite element mesh $\mathcal{M}^h$ of the structure, the soil-fluid interface and eventually the free surface of the soil is defined. It can include either 3D elements, thick or thin shell elements, beam elements... Surface meshes $\mathcal{M}_s^h$ and $\mathcal{M}_f^h$ for the boundary of the soil and the boundary of the fluid and an FE mesh $\mathcal{M}_b^h$ for the structure are deduced from this original mesh $\mathcal{M}^h$. On mesh $\mathcal{M}^h$, the standard FE shape function basis $\{\boldsymbol{w}_{Ii}^h\}$ is defined, I standing for the node number and i for the local degrees of freedom at this node.

Using such a basis, matrices $\mathbf{M}_b$ and $\mathbf{K}_b$ are the classical sparse FEM matrices of the structure. On the contrary, matrices $\mathbf{M}_f$ and $\mathbf{K}_s$ are *a priori* full complex matrices since all degrees of freedom on $\mathcal{M}_f^h$ and $\mathcal{M}_s^h$ are coupled throughout the local boundary value problems (42-44). In order to decrease the storage requirements for these matrices, the degrees of freedom associated with basis functions $\{\boldsymbol{w}_{Ii}\}$ can be sorted in order to split M_i dof associated with inner nodes inside Ω_b and M_c dof associated with boundary nodes on $\mathcal{M}_f^h$ and $\mathcal{M}_s^h$. M_c being usually much smaller than M, matrices $\mathbf{M}_f$ and $\mathbf{K}_s$ can be stored only as $M_c \times M_c$ matrices. The M_i equations associated with inner nodes can even be locally eliminated from the final system since Ω_s and Ω_f do not contribute to these equations. However performing this elimination requires the computation of a full complex impedance matrix of the structure depending on the frequency ω. Doing so may lead to a more demanding storage requirement for this matrix than for the frequency independent sparse matrices $\mathbf{M}_b$ and $\mathbf{K}_b$. Indeed, even when damping is present, the impedance matrix of Ω_b varies strongly with respect to the frequency due to local natural frequencies. Thus a very refined sampling along the frequency axis is required to approximate this variation. On the contrary $\mathbf{M}_f$ and $\mathbf{M}_s$ are usually smooth functions of ω since, because of radiation damping, unbounded domains do not exhibit strong resonances.

3.5.1. *Component Mode Synthesis*

In order to overcome the aforementioned limitations and to facilitate the coupling between FEM and BEM computed codes a modal reduction technique is preferred. Modal reduction techniques seek approximate solutions of a problem with a large number of dofs within a subspace of small dimension. Component Mode Synthesis (CMS) methods [72, 20] form a large class of reduction methods where the subspace is selected *a priori*. SFSI equation (50) is rewritten in the following synthetic form for a sufficiently refined basis $\{\boldsymbol{w}_m^h\}_{m=1,M^h}$ using standard FEM approximation :

$$\left\{\mathbf{K}^h(\omega) - \omega^2\mathbf{M}^h(\omega)\right\}\mathbf{q}^h(\omega) = \mathbf{f}^h(\omega) \tag{54}$$

The idea of a displacement based approximation (Ritz analysis) is to seek the approximate answer within the subspace spanned by the basis $\{\boldsymbol{w}_m^H\}_{m=1,M^H}$ being deduced from the basis $\{\boldsymbol{w}_m^h\}_{m=1,M^h}$ by the projection $M^h \times M^H$ matrix $\mathbf{T}$, the associated dofs $\mathbf{q}^H$ satisfying :

$$\left\{\mathbf{K}^H(\omega) - \omega^2\mathbf{M}^H(\omega)\right\}\mathbf{q}^H(\omega) = \mathbf{f}^H(\omega) \tag{55}$$

where :

$$\mathbf{K}^H = \mathbf{T}^T\mathbf{K}^h\mathbf{T} \qquad \mathbf{M}^H = \mathbf{T}^T\mathbf{M}^h\mathbf{T} \qquad \mathbf{f}^H = \mathbf{T}^T\mathbf{f}^h \tag{56}$$

The error between the refined solution $\mathbf{q}^h$ and the solution on the reduced model is then given by :

$$\mathbf{e}^h = \mathbf{q}^h - \mathbf{T}\mathbf{q}^H \tag{57}$$

The quality of the reduced basis being given by the ratio between the energy assocated with the error $\mathbf{e}^h$ and the one associated with the reduced solution $\mathbf{q}^H$ which reads at each frequency ω :

$$e(\omega) = \frac{\mathbf{e}^{h^T}\mathbf{K}_o\mathbf{e}^h}{\mathbf{q}^{H^T}\mathbf{K}_o^H\mathbf{q}^H} \tag{58}$$

with $\mathbf{K}_o$ any convenient positive definite matrix equivalent to the strain energy in the system, $\mathbf{K}_o^H$ being its projection on the reduced basis. For example the static stiffness can be chosen :

$$\mathbf{K}_o = \mathbf{K}_b + \mathbf{K}_s(0) \qquad \mathbf{K}_o^H = \mathbf{T}^T\mathbf{K}_o\mathbf{T}$$

The proposed error estimate seems useless since the error $\mathbf{e}^h$ has to be computed and then the refined solution $\mathbf{q}_h$. To avoid such computations it has to be noticed that $\mathbf{e}^h$ satisfies the following dynamic equation on the refined mesh with residual forces $\mathbf{r}^h(\omega)$ on the right hand side :

$$\begin{aligned} \left\{\mathbf{K}^h(\omega) - \omega^2\mathbf{M}^h(\omega)\right\}\mathbf{e}^h(\omega) &= \mathbf{r}^h(\omega) && (59)\\ \mathbf{r}^h(\omega) &= \mathbf{f}^h(\omega) - \left\{\mathbf{K}^h - \omega^2\mathbf{M}^h\right\}\mathbf{T}\mathbf{q}^H(\omega) && (60) \end{aligned}$$

Moreover $\mathbf{e}^h$ belongs to the subspace $V_H^\perp$ which is orthogonal to the vector space spanned by $\{\boldsymbol{w}_m^H\}$. Then assuming that on $V_H^\perp$ the dynamic stiffness is positive definite, e.g. there exist two constants c_1 and c_2 such that :

$$c_1\mathbf{q}^T\mathbf{K}_o\mathbf{q} \leq |\mathbf{q}^T(\mathbf{K}^h - \omega^2\mathbf{M}^h)\mathbf{q}| \leq c_2\mathbf{q}^T\mathbf{K}_o\mathbf{q} \qquad \forall\mathbf{q} \in V_H^\perp$$

one has the following error estimate :

$$e(\omega) \approx \frac{\mathbf{q}_r^{h\,T}\mathbf{K}_o\mathbf{q}_r^h}{\mathbf{q}^{H\,T}\mathbf{K}_o^H\mathbf{q}^H} \tag{61}$$

where $\mathbf{q}_r^h$ is the static response of the system to the residue $\mathbf{r}^h$ e.g. satisfying :

$$\mathbf{K}_o\mathbf{q}_s^h = \mathbf{r}^h(\omega)$$

The reader can refer to Appendix 10.2. to find more details of the mathematical justification of this formula and especially the fact that $\mathbf{q}_r^{h\,T}\mathbf{K}_o\mathbf{q}_r^h$ is the natural norm for the residual forces $\mathbf{r}^h(\omega)$ to be preferred to the usual euclidian norm.

While the error estimation is first used to establish convergence of the approximation, it provides a natural mechanism to correct the initial reduced basis $\{\boldsymbol{w}^H\}$ by adding displacement residuals to it, taking for example :

$$\boldsymbol{w}_{M^H+1}^H = \boldsymbol{W}^h\Re(\mathbf{q}_r^h) \qquad \boldsymbol{w}_{M^H+2}^H = \boldsymbol{W}^h\Im(\mathbf{q}_r^h) \tag{62}$$

where several strategies can be defined to restrict the number of frequencies and loading cases (ω_k, u_k) used to build the refined basis.

3.5.2. *Principal Directions*

When defining a reference problem to build a reduction basis, it often happens that its dimension is larger than really needed. A simple mechanism to select vectors within a basis is thus fundamental. A simple way to do so consists of looking for the eigenmodes of the reduced approximate model e.g. $(\mathbf{q}_l^H, \omega_l)$ satisfying :

$$\mathbf{K}_o^H\mathbf{q}_l^H = \omega_l^2\mathbf{M}_o^H\mathbf{q}_l^H \qquad \mathbf{M}_o^H = \mathbf{M}_b^H + \mathbf{M}_f^H(0) \tag{63}$$

and selecting modes satisfying $\omega_l \leq \omega_{\max}$. This reorthogonalisation of the reduced basis is particularly important when several right hand sides and thus several residual forces are treated. It turns out to be a much more efficient technique than taking the standard Singular Value Decomposition of the multiple right hand side or equivalently the Karhunen-Loeve expansion of the applied forces as proposed in Section 2.5..

4. Boundary Integral Equations and the BEM

Boundary Integral Equations are particularly adapted to solve linear boundary value problems on unbounded elastic or acoustic domains. This section is dedicated to the numerical approximation of these equations in the case of layered half-spaces. For an homogeneous acoustic half-space the image technique is well known, so the developments are restricted here to the more complex case of a layered elastic half-space. A regularized variational direct integral equation is first given together with general properties for the regularizing tensor. Then the numerical aproximation using the standard Galerkin technique will be discussed.

4.1. REGULARIZED BOUNDARY INTEGRAL EQUATION IN A LAYERED HALF-SPACE

It may be recalled that the objective is to solve in the frequency domain a set of boundary value problems (42-43) with mixed non homogeneous boundary conditions in a perturbed layered half space Ω_s. It is assumed in the following that the part of the boundary of Ω_s that does not belong to the planar free surface of the half space is bounded and denoted by Σ_s. For the sake of simplicity it is also assumed that displacement boundary conditions are applied on Σ_s even though more complex situations may arise in practical applications.

The generic BVP reads :

$$\begin{aligned} \mathbf{Div}\boldsymbol{\sigma}_s(\boldsymbol{u})+\boldsymbol{f} &= -\rho_s\omega^2\boldsymbol{u} && \text{in } \Omega_s \\ \boldsymbol{u} &= \boldsymbol{u}_o && \text{on } \Sigma_s \\ \boldsymbol{t}_s(\boldsymbol{u}) &= \boldsymbol{0} && \text{on } \Gamma_{sa} \subset S \end{aligned} \tag{64}$$

where S is the free-surface of a layered half space and where Σ_s may reach the free surface S or be partially included in it. Σ_s and Ω_s^{b}, the support of $\boldsymbol{f}$, are assumed to be bounded. The required outputs for the problem are the traction field $\boldsymbol{t} = \boldsymbol{t}_s(\boldsymbol{u})$ on Σ_s and the displacement field $\boldsymbol{u}$ in some bounded region inside Ω_s, namely Ω_s^{b}.

Classically, $\boldsymbol{t}$ is sought in a suitable functional space $\boldsymbol{V}'_\Sigma$ for traction (see Section 10.1.) and satisfies the following variational regularized direct integral equation on the boundary of Ω_s :

$$\mathcal{U}_\Sigma(\boldsymbol{t},\boldsymbol{t}') = \mathcal{T}'_\Sigma(\boldsymbol{u}_o,\boldsymbol{t}') + \mathcal{U}^{\mathrm{b}}_\Sigma(\boldsymbol{f},\boldsymbol{t}') \quad \forall \boldsymbol{t}' \in \boldsymbol{V}'_\Sigma \tag{65}$$

The bilinear forms $\mathcal{U}_\Sigma(\boldsymbol{t},\boldsymbol{t}')$, $\mathcal{T}'_\Sigma(\boldsymbol{u}_o,\boldsymbol{t}')$ and $\mathcal{U}^{\mathrm{b}}_\Sigma(\boldsymbol{f},\boldsymbol{t}')$ are defined as follows for every traction field $\boldsymbol{t}$ and $\boldsymbol{t}'$ in $\boldsymbol{V}'_\Sigma$, every displacement field $\boldsymbol{u}_o$ defined on the boundary Σ

and every body force $\boldsymbol{f}$ applied on Ω_s^{b} belonging respectively to suitable functional spaces $\boldsymbol{V}_\Sigma$ and $\boldsymbol{V}'$ defined in Section 10.1. :

$$\mathcal{U}_\Sigma(\boldsymbol{t},\boldsymbol{t}') = \int_{\Sigma_s} \boldsymbol{t}'^T(\boldsymbol{x}') \int_{\Sigma_s} \boldsymbol{U}_s^G(\boldsymbol{x},\boldsymbol{x}')\boldsymbol{t}(\boldsymbol{x})dS_{\boldsymbol{x}}dS_{\boldsymbol{x}'} \tag{66}$$

$$\mathcal{U}_\Sigma^{\mathrm{b}}(\boldsymbol{f},\boldsymbol{t}') = \int_{\Sigma_s} \boldsymbol{t}'^T(\boldsymbol{x}') \int_{\Omega_s^{\mathrm{b}}} \boldsymbol{U}_s^G(\boldsymbol{x},\boldsymbol{x}')\boldsymbol{f}(\boldsymbol{x})dS_{\boldsymbol{x}}dS_{\boldsymbol{x}'} \tag{67}$$

$$\begin{aligned}\mathcal{T}_\Sigma'(\boldsymbol{u}_o,\boldsymbol{t}') &= \int_{\Sigma_s} \boldsymbol{t}'^T(\boldsymbol{x}') \int_{\Sigma_s} \left\{\boldsymbol{T}_s^G(\boldsymbol{x},\boldsymbol{x}')\boldsymbol{u}_o(\boldsymbol{x}) - \boldsymbol{T}^o(\boldsymbol{x},\boldsymbol{x}')\boldsymbol{u}_o(\boldsymbol{x}')\right\} dS_{\boldsymbol{x}}dS_{\boldsymbol{x}'} \\ &+ \int_{\Sigma_s} \boldsymbol{t}'^T(\boldsymbol{x}')\boldsymbol{D}(\boldsymbol{x}')\boldsymbol{u}_o(\boldsymbol{x}')dS_{\boldsymbol{x}'}\end{aligned} \tag{68}$$

where $\boldsymbol{v}^T$ stands for the transpose of any vector $\boldsymbol{v}$. In these formulae $\boldsymbol{U}_s^G(\boldsymbol{x},\boldsymbol{x}')$ stands for the first Green's tensor of a layered half-space e.g. for any unit vectors $\boldsymbol{a}'$ and $\boldsymbol{a}$, $\boldsymbol{a}'^T\boldsymbol{U}_s^G(\boldsymbol{x},\boldsymbol{x}')\boldsymbol{a}$ is the displacement field at point $\boldsymbol{x}$ along direction $\boldsymbol{a}$ being generated by a point force $\boldsymbol{a}'$ applied at $\boldsymbol{x}'$. $\boldsymbol{T}_s^G$ is the second Green's tensor e.g. $\boldsymbol{a}'^T\boldsymbol{T}_s^G(\boldsymbol{x},\boldsymbol{x}')\boldsymbol{a}$ is the component of the traction fields along direction $\boldsymbol{a}$ at $\boldsymbol{x}$ applied on the boundary Σ for a point force $\boldsymbol{a}'$ applied at $\boldsymbol{x}'$. $\boldsymbol{T}^o$ is a regularization tensor that has to be determined such that for $r = ||\boldsymbol{x} - \boldsymbol{x}'|| \to 0$:

$$\lim_{r\to 0} \left(\left|\boldsymbol{T}_s^G(\boldsymbol{x},\boldsymbol{x}') - \boldsymbol{T}^o(\boldsymbol{x},\boldsymbol{x}')\right|\right) < +\infty \tag{69}$$

$$\boldsymbol{T}^o(\boldsymbol{x},\boldsymbol{x}') = \frac{\boldsymbol{G}^o}{r^2} \tag{70}$$

where $\boldsymbol{G}^o$ is a second order tensor depending only on $\boldsymbol{x}'$. Finally, $\boldsymbol{D}$ is a second order tensor defined, for any unbounded domain Ω_s, as a function of $\boldsymbol{T}^o$ as follows:

$$\boldsymbol{D}(\boldsymbol{x}') = \mathbf{I} - \int_{\Sigma_s^{\mathrm{s}}} \boldsymbol{T}^o(\boldsymbol{x},\boldsymbol{x}')dS_{\boldsymbol{x}} \tag{71}$$

where Σ_s^{s} is the image of Σ_s with respect to S, the free-surface of the half-space.

It is worth noticing that the second term on the right hand side of equation (71) vanishes when Σ_s is closed and remains at a finite distance from the free surface S. It is also the case when Σ_s is locally included in S since $\boldsymbol{T}_s^G$ vanishes. Moreover this integral is non singular as long as $\boldsymbol{x}'$ remains at a finite distance from the free surface. When $\boldsymbol{x}'$ is on the edge of Σ_s and thus belongs to the free surface, the variational approach adopted here still holds since the measure of this edge is equal to zero (see Figure 2).

Some additional mathematical properties are summarized in Appendix 10.1. and attention is now focused on the regularizing tensor and the numerical approximation.

4.2. REGULARIZING TENSORS

It will be shown in Section 6 that a regularizing tensor $\boldsymbol{T}^o$ satisfying properties (69 and 70) exists in some practical cases and especially for a stratified half space where

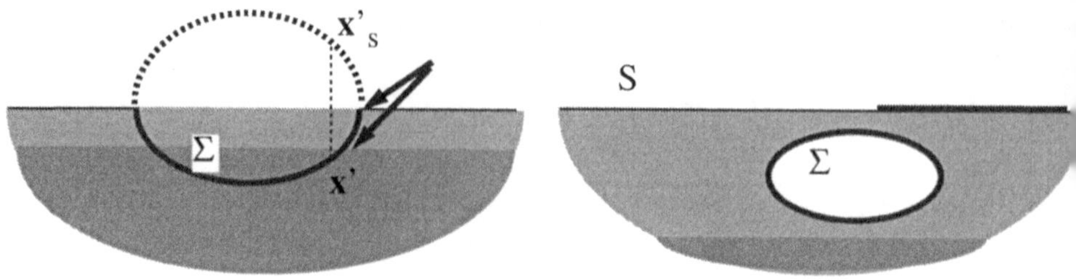

Figure 2. Two generic BEM problems requiring specific regularization on the left and standar or no regularization on the right. On the left diagram the image of the boundary is plotted wit dashed lines.

moreover the tensor $\boldsymbol{G}^o$ can be determined numerically. However such an approac is not necessary in some particular cases. First of all, when a part of Σ_s belong to the free surface $\boldsymbol{T}_s^G$ and thus $\boldsymbol{T}^o$ vanishes. Moreover when a part of Σ_s crosse a locally homogeneous region, the static second Green's tensor of an homogeneou space is a regularizing tensor. Kelvin and Boussinesq solutions can also be use on the free surface or along an interface (see [34]). However this is only a par of the solution since the Green's tensors $\boldsymbol{T}_s^G$ and $\boldsymbol{U}_s^G$ still have to be compute numerically. Then due to numerical error there is little chance that property (69 still holds. This problem becomes even more complex when hysteretic or viscou damping is included since the corresponding analytical solutions do not exist. Fo these reasons it is believed that the present approach is the most efficient bot from the theoretical and from the numerical points of view.

Moreover, it is interesting to notice that this approach differs from usual reg ularization techniques using static solutions and rigid body motions. Indeed thi formulation, fully documented in [15], is based on an invariant property of $\boldsymbol{T}^o$:

the integral of $\boldsymbol{T}^o$ on any surface remains invariant when a similarity transfor centred on $\boldsymbol{x}'$ is applied to this surface.

From this property it is proved that for a bounded domain the free term in th non regularized boundary integral equation is equal to the integral of $\boldsymbol{T}^o$ on it boundary. As a consequence the integral of $\boldsymbol{T}^o$ over a closed surface which doe not include the source point $\boldsymbol{x}'$ vanishes.

4.3. BOUNDARY ELEMENTS

The numerical approximation of the variational boundary integral equation (65 is simply obtained by looking for $\boldsymbol{t}$ in $\boldsymbol{V}'_{\Sigma h}$ a finite dimensional sub-space of $\boldsymbol{V}'$ with given basis $\{\boldsymbol{\phi}_k\}_{k=1,N_h}$ and taking for $\boldsymbol{t}'$ all elements of this basis. $\boldsymbol{V}'$ being a space of applied tractions the basis functions $\boldsymbol{\phi}_k$ do not have to satisf strong regularity conditions. In particular they can be chosen as piecewise constan vectors $\boldsymbol{e}_k$ with a local support restricted to a surface element S_k. Using thi approximation the following linear system is then obtained :

$$\mathbf{Cg} = \underline{\mathbf{u}} \quad , \quad \boldsymbol{t} = \sum_k \phi_k g_k \tag{72}$$

with $\mathbf{C}$, $\mathbf{g}$ and $\underline{\mathbf{u}}$, $N_h \times N_h$ matrices and N_h a vector.

$$[\mathbf{C}]_{kk'} = \iint_{S_{k'} \times S_k} \boldsymbol{e}_{k'}^T \boldsymbol{U}_s^G(\boldsymbol{x}, \boldsymbol{x}') \boldsymbol{e}_k dS_{\boldsymbol{x}} dS_{\boldsymbol{x}'} \tag{73}$$

$$\begin{aligned} \underline{\mathbf{u}}_{k'} = & \int_{S_{k'}} \boldsymbol{d}_{k'}(\boldsymbol{x}') \cdot \boldsymbol{u}_o(\boldsymbol{x}') dS_{\boldsymbol{x}'} \\ & + \boldsymbol{e}_{k'}^T \int_{S_{k'}} \int_{\Sigma_s \setminus S_{k'}} \boldsymbol{T}_s^G(\boldsymbol{x}, \boldsymbol{x}') \boldsymbol{u}_o(\boldsymbol{x}) dS_{\boldsymbol{x}} dS_{\boldsymbol{x}'} \\ & + \boldsymbol{e}_{k'}^T \int_{S_{k'}} \int_{S_{k'}} \left\{ \boldsymbol{T}_s^G(\boldsymbol{x}, \boldsymbol{x}') \boldsymbol{u}_o(\boldsymbol{x}) - \boldsymbol{T}^o(\boldsymbol{x}, \boldsymbol{x}') \boldsymbol{u}_o(\boldsymbol{x}') \right\} dS_{\boldsymbol{x}} dS_{\boldsymbol{x}'} \end{aligned} \tag{74}$$

$$\boldsymbol{d}_{k'}(\boldsymbol{x}') = \boldsymbol{e}_{k'}^T \left\{ \boldsymbol{I}_{\mathrm{d}} + \int_{\Sigma_s^{\mathrm{s}} \cup \Sigma_s \setminus S_{k'}} \boldsymbol{T}^o(\boldsymbol{x}, \boldsymbol{x}') dS_{\boldsymbol{x}} \right\} \tag{75}$$

All terms in this expression being at most weakly singular they can be integrated using a standard Gaussian quadrature technique. However adapted integration formulae are needed when $k = k'$ as is usually done in BEM [34]. Moreover it has to be noticed that the same integration technique does not need to be used for both integrals since the second one is usually more regular than the first one. Using only one integration point for the second integral is equivalent to the standard collocation technique.

As far as accuracy and efficiency are concerned it is more convenient to extract analytically the singular parts $\boldsymbol{U}_s^{G^{\mathrm{s}}}$ and $\boldsymbol{T}_s^{G^{\mathrm{s}}}$ of the two tensors $\boldsymbol{U}_s^G$ and $\boldsymbol{T}_s^G$. For example $\boldsymbol{T}_s^{G^{\mathrm{s}}}$ can be either equal to $\boldsymbol{T}^o$ or to the homogeneous Green's tensor when $\boldsymbol{x}'$ is located in a locally homogeneous region. With such a decomposition, the regular parts $\boldsymbol{U}_s^{G^{\mathrm{r}}}$ and $\boldsymbol{T}_s^{G^{\mathrm{r}}}$ can be computed numerically on a coarse grid. As far as Green's functions of a layered half-space are concerned the invariance of the tensor with respect to any translation along the horizontal direction and any rotation with respect to any vertical axis allows for the sampling of these tensors with respect to only three variables (z', z, r) where z and z' are respectively the vertical coordinates of $\boldsymbol{x}$ and $\boldsymbol{x}'$ ranging from 0 at the free surface to $-z_{\max}$ the maximum depth of Σ_s. r ranges from 0 to $r_{\max}$ the maximum width of Σ_s. Different grids can also be defined for near field and far field terms. Assembling techniques based on the far field grid are equivalent to clustering techniques [101, 157, 129].

4.4. COUPLING WITH OTHER NUMERICAL TECHNIQUES

The extension of the proposed formulation to cases more complex than a layered half space is straightforward as long as the decomposition of $\boldsymbol{U}_s^G$ and $\boldsymbol{T}_s^G$ into an analytical singular part and a regular part holds. For example when the regular

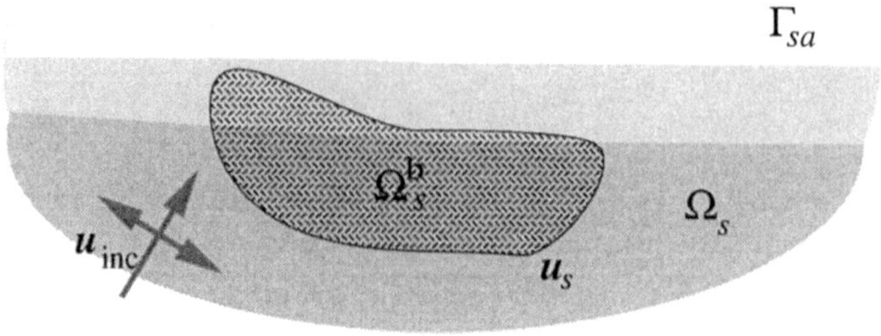

Figure 3. Model of volume coupling

part can be taken as the Green's function of a locally homogeneous or stratified half space, the regular part is nothing other than the field scattered by the non local heterogeneities and can be computed using any desired numerical technique including BEM or domain decomposition techniques [164] or high frequency approximations [27].

When such techniques cannot be applied the proposed approach still holds by using indirect integral equations instead of direct ones. Indeed in this case, the fictitious domain technique is recovered [96], $\underline{\boldsymbol{u}}$ in equation (72) being nothing but the projection of the prescribed boundary condition $\boldsymbol{u}_o$ on $\boldsymbol{V}_{\Sigma h}$ whereas $\boldsymbol{g}$ is the the Lagrange Multiplier used to enforce this boundary condition. In particular this approach can be employed when finite elements are used to compute the bilinear form $\mathcal{U}_\Sigma$. Extension to direct boundary integral equations is more difficult since in this case the bilinear form $\mathcal{T}_\Sigma$ is not clearly defined. A final extension then consists in coupling FEM and BEM inside a three-dimensional region instead of along a given interface Σ. This technique is described in Section 4.5..

4.5. FEM-BEM COUPLING INSIDE A VOLUME

When complex or stochastic analyses are to be performed it is unlikely that BEM can solve the local problem (42) and (43) since the soil shows strong heterogeneities in addition to horizontal layering. A solution to account for such cases consists in incorporating a bounded heterogeneous part of the soil in the FEM model of the structure Ω_b. However, this procedure increases the number of dofs associated with Ω_b and has also a strong negative effect on the modal reduction technique. Another approach is to account for integration cells in the BEM framework. However, strongly singular integrals for a layered elastic domain have to be evaluated since up to now no efficient regularization procedure has been derived for them.

A Dual Domain Decomposition Technique based on volume coupling is proposed hereafter [164, 165]. It follows the same lines as the one proposed in Section 3 and is fully compatible with it. The same notation is kept but, for the sake of simplicity, the fluid domain will be ignored and it will be assumed that Ω_b, denoted now by

Ω_s^{b} to avoid any confusion, is the bounded heterogeneous part of the soil as shown in Figure 3.

This assumption implies that the geometrical domains Ω_s and Ω_b overlap and more precisely that Ω_b is strictly included inside Ω_s. We are now free to decide which of the physical parameters $\boldsymbol{m}$ are attributed either to Ω_s or to Ω_s^{b}, $\boldsymbol{m}_s$ indicating the "soil" portion, whereas $\boldsymbol{m}_b$ stands for the additional contribution satisfying :

$$\boldsymbol{m} = \boldsymbol{m}_s + \boldsymbol{m}_b \qquad \text{in } \Omega_s^{\mathrm{b}} \qquad \boldsymbol{m} = \boldsymbol{m}_s \qquad \text{in } \Omega_s \setminus \Omega_s^{\mathrm{b}} \tag{76}$$

The displacement field $\boldsymbol{u}_{\mathrm{d}}$ has now to satisfy the boundary value problem:

$$\begin{array}{rcc} \mathbf{Div}\boldsymbol{\sigma}_s(\boldsymbol{u}_{\mathrm{d}}) - \boldsymbol{\phi} = -\rho_s\omega^2\boldsymbol{u}_{\mathrm{d}} & \text{in} & \Omega_s \\ \boldsymbol{t}_s(\boldsymbol{u}_{\mathrm{d}}) = \mathbf{0} & \text{on} & \Gamma_{sa} \end{array} \tag{77}$$

for a given $\boldsymbol{\phi}$ defined on Ω_s^{b}. When $\boldsymbol{\phi}$ belongs to a proper functional space $\boldsymbol{V}_s'$ (see Section 10.1.), the problem (77) is well posed as long as the physical parameters $\boldsymbol{m}_s$ remain positive and internal damping is accounted for. Moreover when the Green's functions of Ω_s are known, $\boldsymbol{u}_{\mathrm{d}}$ can be computed using the following integral equation :

$$\boldsymbol{u}_{\mathrm{d}}(\boldsymbol{x}') = -\int_{\Omega_s^{\mathrm{b}}} \boldsymbol{U}_s^G(\boldsymbol{x}, \boldsymbol{x}')\boldsymbol{\phi}(\boldsymbol{x})dV_{\boldsymbol{x}} \tag{78}$$

In order to satisfy the equilibrium equation in the soil with the parameters equal to $\boldsymbol{m}$ instead of $\boldsymbol{m}_s$, $\boldsymbol{\phi}$ inside Ω_s^{b} has to satisfy :

$$\begin{array}{rcc} \mathbf{Div}\boldsymbol{\sigma}_b(\boldsymbol{u}_b) + \boldsymbol{\phi} = -\rho_b\omega^2\boldsymbol{u}_b & \text{in} & \Omega_s^{\mathrm{b}} \\ \boldsymbol{t}_b(\boldsymbol{u}_{\mathrm{d}}) = \mathbf{0} & \text{on} & \partial\Omega_s^{\mathrm{b}} \end{array}$$

or, in a weak sense, for every $\boldsymbol{v}_b \in \boldsymbol{V}_s^{\mathrm{b}}$:

$$\int_{\Omega_s^{\mathrm{b}}} \{\boldsymbol{\sigma}_b(\boldsymbol{u}_b) : \boldsymbol{\epsilon}(\boldsymbol{v}_b) - \omega^2\rho_b\boldsymbol{u}_b \cdot \boldsymbol{v}_b\}dV \quad = \quad \int_{\Omega_s^{\mathrm{b}}} \boldsymbol{\phi} \cdot \boldsymbol{v}_b dV \tag{80}$$

Since physical parameters $\boldsymbol{m}_{\mathrm{b}}$ are not necessarily positive, equation (80) does not define a well posed problem for $\boldsymbol{u}_b$, $\boldsymbol{\phi}$ being given. Then, equation (80) should not be considered as an independant equation on $\boldsymbol{u}_b$. Indeed, in order to satisfy the original problem, $\boldsymbol{u}_b$ has to be equal to $\boldsymbol{u}_s$ over all Ω_s^{b} which gives :

$$\boldsymbol{u}_b = \boldsymbol{u}^{\mathrm{inc}} + \boldsymbol{u}_{\mathrm{d}} = \boldsymbol{u}^{\mathrm{inc}} - \boldsymbol{u}_{\boldsymbol{\phi}} \qquad \text{inside } \Omega_s^{\mathrm{b}} \tag{81}$$

or, in a weak sense, as follows for all $\boldsymbol{\phi}' \in \boldsymbol{V}_s'$:

$$\int_{\Omega_s^{\mathrm{b}}} \boldsymbol{u}_b \cdot \boldsymbol{\phi}' dV + \mathcal{U}^{\mathrm{b}}(\boldsymbol{\phi}, \boldsymbol{\phi}') = \int_{\Omega_s^{\mathrm{b}}} \boldsymbol{u}^{\mathrm{inc}} \cdot \boldsymbol{\phi}' dV \tag{82}$$

where $\mathcal{U}^{\mathrm{b}}(\boldsymbol{\phi},\boldsymbol{\phi}')$ is a coercive bilinear form on $\boldsymbol{V}'\times\boldsymbol{V}'$ (see Appendix 10.1.) defined as follows:

$$\mathcal{U}^{\mathrm{b}}(\boldsymbol{\phi},\boldsymbol{\phi}') = \int_{\Omega_s^{\mathrm{b}}}\int_{\Omega_s^{\mathrm{b}}} \boldsymbol{\phi}^T(\boldsymbol{x}')\boldsymbol{U}_s^G(\boldsymbol{x},\boldsymbol{x}')\boldsymbol{\phi}(\boldsymbol{x})dV_{\boldsymbol{x}}dV_{\boldsymbol{x}'} \tag{83}$$

Thanks to this property, it has been shown in [164] that the mixed variational problem (80-82) is well posed and can be approximated expanding $\boldsymbol{u}$ and $\boldsymbol{v}_b$ on a standard finite element basis $\{\boldsymbol{w}_m\}_{m=1,M}$ on a given mesh $\mathcal{M}$ of Ω_s^{b}, $\boldsymbol{\phi}$ and $\boldsymbol{\phi}'$ being expanded on a discrete basis $\{\boldsymbol{\phi}_k\}_{k=1,N_h}$ of $\boldsymbol{V}'$ consisting for example of constant shape functions on each element of this mesh :

$$\boldsymbol{\phi} = \sum_{k=1}^{N_h} \boldsymbol{\phi}_k g_k \qquad \boldsymbol{u}_b = \sum_{m=}^{M} \boldsymbol{w}_m q_m \tag{84}$$

This leads to the final linear system:

$$\begin{pmatrix} \mathbf{K}^{\mathrm{b}} - \omega^2\mathbf{M}^{\mathrm{b}} & \mathbf{B} \\ \mathbf{B}^T & \mathbf{U} \end{pmatrix} \begin{pmatrix} \mathbf{q} \\ \mathbf{g} \end{pmatrix} = \begin{pmatrix} \mathbf{0} \\ \mathbf{u}_{\mathrm{inc}} \end{pmatrix} \tag{85}$$

Here K^{b} and M^{b} are the $M \times M$ FEM stiffness and mass matrices of the Ω_s^{b} computed with physical parameters $\boldsymbol{m}_b$ and $\mathbf{B}$ and $\mathbf{U}$ and defined as follows :

$$[\mathbf{U}]_{kk'} = \int_{V_{k'}}\int_{V_k} \boldsymbol{e}_{k'}^T\boldsymbol{U}_s^G(\boldsymbol{x},\boldsymbol{x}')\boldsymbol{e}_k dV_{\boldsymbol{x}}dV_{\boldsymbol{x}'} \tag{86}$$

$$[\mathbf{B}]_{mk'} = \int_{\Omega_s^{\mathrm{b}}} \boldsymbol{\phi}_{k'}(\boldsymbol{x}')\boldsymbol{w}_m(\boldsymbol{x}')dV_{\boldsymbol{x}'} = V(V_k'\cap V_m)\boldsymbol{e}_{k'}^T\boldsymbol{e}_m \tag{87}$$

$$[\mathbf{u}_{\mathrm{inc}}]_m = \boldsymbol{u}^{\mathrm{inc}}(\boldsymbol{x}_m)\cdot\boldsymbol{e}_m \tag{88}$$

where V_k is the three-dimensional finite element on which $\boldsymbol{\phi}_k$ is constant and equal to $\boldsymbol{e}_k$, V_m is the support of $\boldsymbol{w}_m$, $\boldsymbol{e}_m$ being the direction of displacement vector associated to this shape function and where $V(D)$ denotes the volume of the 3D set D.

Keeping the same basis functions but satisfying the continuity equation (81) only at the nodes of the mesh gives the following system of equations for $\mathbf{g}$:

$$\{\mathbf{U} + \mathbf{U}^{\mathrm{b}T}(\mathbf{K}^{\mathrm{b}} - \omega^2\mathbf{M}^{\mathrm{b}})\mathbf{U}^{\mathrm{b}}\}\mathbf{g} = \mathbf{U}^{\mathrm{b}T}(\mathbf{K}^{\mathrm{b}} - \omega^2\mathbf{M}^{\mathrm{b}})\mathbf{u}_{\mathrm{inc}} \tag{89}$$

with $\mathbf{U}^{\mathrm{b}}$ defined as follows :

$$[\mathbf{U}^{\mathrm{b}}]_{mk'} = \int_{V_k} \boldsymbol{e}_{k'}^T\boldsymbol{U}_s^G(\boldsymbol{x}_m,\boldsymbol{x}')\boldsymbol{e}_m dV_{\boldsymbol{x}'} \tag{90}$$

where $\boldsymbol{x}_m$ is the coordinate of the node associated with FE shape function $\boldsymbol{w}_m$.

It is worth noticing that these integrals are at most weakly singular and can be computed using the same procedure as the one described for surface Boundary

Elements in Section 4.3.. Thus, thanks to the variational approach the derivation of the Green's tensor is in fact performed in equation (89) using an auxiliary FEM mesh. It has been shown in [164, 165] that this procedure gives the same results as those obtained by standard substructuring techniques in the case of an homogeneous inclusion embedded in a layered half-space with roughly the same computational effort. However in contrast with standard substructuring techniques, it can account for heterogeneous inclusions.

5. Unbounded Interfaces

Numerical techniques based on Finite Elements, Domain Decomposition or Boundary Elements are based on the assumption that the unknown fields to be approximated must have bounded supports, and in fact almost the same assumption has to be made for the applied loads. In many practical analyses such an assumption is far too restrictive even when complex Green's functions are used. Thus for very large structures such as tunnels , boreholes or sheet-piles other techniques have to be developed.

The main purpose of this section is to show that, when the entire domain or some parts of it satisfy invariance properties, the proposed numerical techniques can still be used on a reference cell as long as the proper integral transform is applied at the same time on the unknown fields and on the loads. In particular this means that restrictions on the invariance of the domain are not restrictions on the loads that can be of any type. This procedure has been already used when time-dependent problems have been transformed to the frequency domain thanks to the invariance with respect to time. Of course these techniques can be applied only for linear problems.

5.1. GENERAL SPACE-WAVENUMBER TRANSFORM

When the geometrical domain denoted by Ω, the boundary conditions and the physical parameters $\boldsymbol{m}$ are invariant for a group of isometries $\mathcal{T}$:

$$\mathcal{T} = \{\tau : \Omega \to \Omega / \tau(\Omega) = \Omega; \tau(\boldsymbol{m}) = \boldsymbol{m}\} \tag{91}$$

this domain can be formally written as :

$$\widetilde{\Omega} \times \Omega_{\mathcal{T}} \tag{92}$$

where $\widetilde{\Omega}$ is the subset of Ω from which Ω can be recovered applying all transformations τ in $\mathcal{T}$, $\Omega_{\mathcal{T}}$ being the set of indices of $\mathcal{T}$. The expression of $\widetilde{\boldsymbol{u}}$, the integral transform of any field $\boldsymbol{u}$ with respect to the space variables belonging to $\Omega_{\mathcal{T}}$ is then given by :

$$\widetilde{\boldsymbol{u}}(\widetilde{\boldsymbol{x}}, \boldsymbol{k}) = \int_{\Omega_{\mathcal{T}}} \boldsymbol{u}(\boldsymbol{x}) e^{i\langle \boldsymbol{k}, \boldsymbol{x}_{\mathcal{T}} \rangle} d\boldsymbol{x}_{\mathcal{T}} \qquad \widetilde{\boldsymbol{x}} \in \widetilde{\Omega} \tag{93}$$

$\boldsymbol{k}$, being the *wavenumber* belonging to $\Omega'_{\mathcal{T}}$, the dual of $\Omega_{\mathcal{T}}$ for the duality product $\langle ., . \rangle$ defined on the unit circle.

In particular one can define :

- The Fourier series of $\boldsymbol{u}$ for axisymmetrical domains around axis $\boldsymbol{e}_2$ takin[g] $\Omega_\mathcal{T} = \{0 \leq \theta < 2\pi\}$, $\Omega'_\mathcal{T} = \mathbb{Z}$, $\widetilde{\Omega} = \{\widetilde{\boldsymbol{x}} = x_1\boldsymbol{e}_1 + x_2\boldsymbol{e}_3\}$, $\langle \boldsymbol{k}, \boldsymbol{x}_\mathcal{T}\rangle = n\theta$:

$$\widetilde{\boldsymbol{u}}(\widetilde{\boldsymbol{x}}, \boldsymbol{k} = n) = \int_0^{2\pi} \boldsymbol{u}(\boldsymbol{x}) e^{in\theta} \qquad \widetilde{\boldsymbol{x}} \in \widetilde{\Omega} \qquad (94$$

- the Fourier transform of $\boldsymbol{u}$ with respect to the p directions $\{\boldsymbol{e}_i\}_{1\leq i\leq p\leq}$ taking $\Omega_\mathcal{T} = \Omega'_\mathcal{T} = \mathbb{R}^p$, $\boldsymbol{x}_\mathcal{T} = \sum_{i=1,p} x_i\boldsymbol{e}_i$, $\widetilde{\Omega} = \{\widetilde{\boldsymbol{x}} = \sum_{i=p+1,n} x_i\boldsymbol{e}_i, x_i \in \mathbb{R}, i = p+1, n\}$, $\langle \boldsymbol{k}, \boldsymbol{x}_\mathcal{T}\rangle = \sum_{i=1,p} k_i x_i$:

$$\widetilde{\boldsymbol{u}}(\widetilde{\boldsymbol{x}}, \boldsymbol{k}) = \int_{\mathbb{R}^p} \boldsymbol{u}(\boldsymbol{x}) e^{i\boldsymbol{k}\cdot\boldsymbol{x}_\mathcal{T}} d\boldsymbol{x}_\mathcal{T} \qquad \widetilde{\boldsymbol{x}} \in \widetilde{\Omega} \qquad (95$$

  which can be replaced by the Hankel transform when $p = 2$ by taking cy[-]clindrical coordinates for $\boldsymbol{x}_\mathcal{T}$ and $\boldsymbol{k}$ as given in Appendix 10.4..

- the Floquet transform of $\boldsymbol{u}$ for a domain periodic along $\boldsymbol{e}_2$ by taking $\Omega_\mathcal{T} = \{\boldsymbol{x}_\mathcal{T} = nL\boldsymbol{e}_2, n \in \mathbb{Z}\}$, L being the period, $\Omega'_\mathcal{T} = \{k \in [0, 2\pi/L]\}$, $\widetilde{\Omega} = \{\widetilde{\boldsymbol{x}} = \boldsymbol{x}/0 \leq \boldsymbol{x}\cdot\boldsymbol{e}_2 < L\}$, $\langle \boldsymbol{k}, \boldsymbol{x}_\mathcal{T}\rangle = knL$:

$$\widetilde{\boldsymbol{u}}(\widetilde{\boldsymbol{x}}, \boldsymbol{k} = k) = \sum_{n\in\mathbb{Z}} \boldsymbol{u}(\widetilde{\boldsymbol{x}} + nL\boldsymbol{e}_2) e^{iknL} \qquad \widetilde{\boldsymbol{x}} \in \widetilde{\Omega} \qquad (96$$

Most of these transformations are widely used in numerical modelling in order t[o] solve boundary value problems on symmetric domains especially in Finite Element[s] [37], their use for boundary integral equations being well known in physics [100]. I[n] earthquake engineering and structural dynamics the Fourier Transform has bee[n] associated with BEM for example in [151, 18] using the analytical Fourier trans[-]form of Green's function of an homogeneous elastic medium or the one of a layere[d] half-space [14, 133]. Fourier series for axisymmetrical or non-axisymmetrical load[s] on axisymmetrical domains has been used for example in [156, 84] and in [187, 80] in a substructuring technique. The main difficulty here is the accurate computatio[n] of the axisymmetrical Green's function that does not have a closed form solutio[n] for elastodynamic problems. The expansion proposed in [188] has been shown to b[e] very effective to model very long axisymmetrical boreholes [27, 65]. Axisymmetri[c] Green's functions for a layered half-space are presented in [8, 182, 39].

The Floquet Transform [90] has been recently used to model periodic domain[s] [1] and results given in [61, 62, 60] are summarised here as an example of integra[l] transform used in connection with Domain Decomposition and boundary integra[l] equations. It may be noted that when the period L tends to 0 the Fourier transfor[m] is retrieved .

5.2. INVARIANT OPERATORS

The above mentioned integral transforms expand any field $\boldsymbol{u}$ defined on a periodic domain in a finite or infinite set of fields $\widetilde{\boldsymbol{u}}$ defined on the reference cell $\widetilde{\Omega}$ that can be extended on the whole domain Ω using invariance properties. In the case of the Floquet transform $\widetilde{\boldsymbol{u}}$ satisfies :

$$\widetilde{\boldsymbol{u}}(\widetilde{\boldsymbol{x}} + L\boldsymbol{e}_2) = e^{-ikL}\widetilde{\boldsymbol{u}}(\widetilde{\boldsymbol{x}}) \tag{97}$$

Moreover an inverse transform can be defined in order to recover the original field $\boldsymbol{u}$ once the fields $\widetilde{\boldsymbol{u}}$ are known :

$$\boldsymbol{u}(\widetilde{\boldsymbol{x}} + nL\boldsymbol{e}_2) = \frac{1}{L}\int_{-\pi/L}^{\pi/L} \widetilde{\boldsymbol{u}}(\widetilde{\boldsymbol{x}}, k)e^{-inkL}dk \tag{98}$$

Then using the linearity of the equation, a boundary value problem for $\boldsymbol{u}$ with applied loads $\boldsymbol{f}$ on a periodic domain Ω can be equivalently solved on $\widetilde{\Omega}$ for all $\widetilde{\boldsymbol{u}}$ with applied loads $\widetilde{\boldsymbol{f}}$. With the invariance property (97), the problem can then be restricted to $\widetilde{\Omega}$ provided that $\widetilde{\boldsymbol{u}}$ satisfies these invariance properties on the boundary of the cell. In the periodic case and with Σ_o being the left boundary of the cell as shown in Figure 4, $\widetilde{\boldsymbol{u}}$ has to satisfy equation (97) for any point $\boldsymbol{x}$ on Σ_o. When the period tends to 0 this constraint simply becomes the classical definition of the Fourier transform of the derivative with respect to the invariant axis, $\widetilde{\cdot}$ denoting the Fourier Transform :

$$\partial_2\widetilde{\boldsymbol{u}}(\widetilde{\boldsymbol{x}}) = -ik\widetilde{\boldsymbol{u}}(\widetilde{\boldsymbol{x}}) \tag{99}$$

5.3. DOMAIN DECOMPOSITION ON INVARIANT DOMAINS

With the previously mentioned properties equations (30-41) still apply on the reference cell provided that the integral transform is applied on all variables and that generalized boundary conditions (97) are satisfied by all fields on Σ_o. These generalized boundary conditions can be transformed to standard periodicity conditions writing all fields $\widetilde{\boldsymbol{u}}$ as follows :

$$\widetilde{\boldsymbol{u}}(\widetilde{\boldsymbol{x}}) = e^{-ikx_2}\widehat{\boldsymbol{u}}(\widetilde{\boldsymbol{x}}) \tag{100}$$

where $\widehat{\boldsymbol{u}}$ satisfies standard periodicity conditions. However this transformation will change equations (30-41) when expressed as functions of $\widehat{\cdot}$ fields since the derivations along the invariant axis have to be applied both on $\widehat{\boldsymbol{u}}$ and on the exponential term. This is currently done when the Fourier Transform is applied. However it is believed that keeping $\widetilde{\cdot}$ fields as unknowns is simpler and safer. Moreover it gives the correct limit when L tends to 0 as long as constraint (97) is accounted for.

Thus the standard domain decomposition technique can be applied on the reference cell $\widetilde{\Omega}$ and local boundary value problems (42) and (43) on $\widetilde{\Omega}_s$ and (44) on

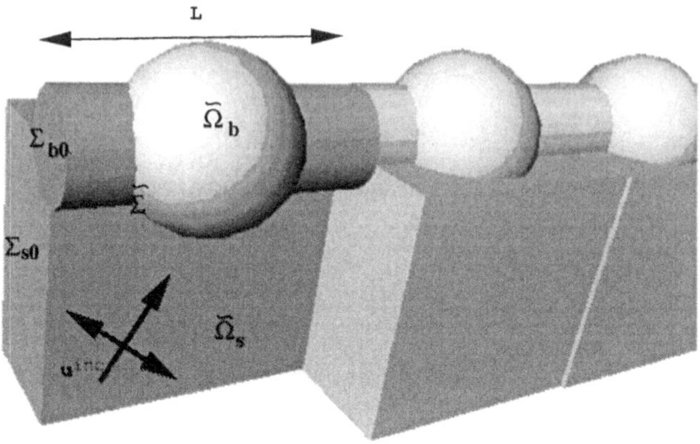

Figure 4. Model of an periodic domain

$\widetilde{\Omega}_f$ can be defined as long as generalized boundary conditions are satisfied by the local fields.

Thanks to the conditions satisfied by the coupling field, $\widetilde{\boldsymbol{u}_\Sigma}$ and the virtual field $\widetilde{\boldsymbol{v}_\Sigma}$ integrals on the boundary of the cell Σ_0 and Σ_L are of opposite signs due the orientation of the normal vector and then disappear. Finally equation (45) still holds adding $\tilde{.}$ on all fields and performing the integral (46-47) on $\widetilde{\Omega_b}$, $\widetilde{\Sigma_{sf}}$ and $\widetilde{\Sigma_{bs}}$. The nice feature is that now fields only on bounded domains or interfaces have to be approximated.

The approximation of equation (45) is certainly the most difficult step since basis functions $\widetilde{\boldsymbol{w}}_m(\widetilde{\boldsymbol{x}})$ have to satisfy the periodicity conditions (97) on Σ_0. A way to build such a basis has been proposed in [61] in the context of component mode synthesis.

The second difficult step consists in solving the local boundary value problems (42-44) for $\widetilde{\boldsymbol{u}_\mathrm{d}}$, $\widetilde{\boldsymbol{u}_\mathrm{do}}$ and $\widetilde{p_\mathrm{d}}$ using boundary integral equation and Boundary Elements, keeping in mind that these fields have to satisfy the periodicity conditions (97).

5.4. BEM ON INVARIANT DOMAINS

The key point when trying to solve a boundary value problem on a periodic domain using boundary integral equations is to get rid of integrals along the boundary of the cell. Indeed, a periodic domain which is infinite not only along the invariance direction but also along the transverse directions cannot be accounted for by standard BIE. Σ_0 belongs to the boundary of $\widetilde{\Omega}$ and thus the integral equation has to be written on an unbounded surface.

However, in boundary integral equations, boundaries where the Green's functions satisfy the imposed boundary conditions disappear from the final equation since the fields will automatically satisfy them. Then in order to deal with BIE on a bounded surface Green's functions satisfying the periodic boundary conditions have to be built.

Fortunately, the integral transforms give immediately the answer since the "invariant" is nothing but the integral transform of the original Green's function. This property is well known for axisymmetric BEM or 2D BEM and can be proved very easily in taking the integral transform of the field equations satisfied by the original Green's function. As far as periodic BIE is concerned the Floquet-Green's function simply reads [61] :

$$\widetilde{\boldsymbol{U}_s^G}(\widetilde{\boldsymbol{x}}, \widetilde{\boldsymbol{x}}') = \sum_{-\infty}^{+\infty} e^{inkL} \boldsymbol{U}_s^G(\widetilde{\boldsymbol{x}}, \widetilde{\boldsymbol{x}}' + nL\boldsymbol{e}_2) \tag{101}$$

and can be shown to satisfy (97) with respect to $\widetilde{\boldsymbol{x}}'$ and with respect to $\widetilde{\boldsymbol{x}}$ changing k by $-k$. Moreover since this expression simply consists in the superposition of several Green's function with at most one of the singular point on the reference cell, the associated integral equation can be regularized using the original regularizing tensor. Thus Section 4 holds for periodic domains adding $\widetilde{\cdot}$ on all variables and all surfaces.

The only remaining difficulty is related to the computation of the series in (101). As long as damping is present it is uniformly convergent for $||\widetilde{\boldsymbol{x}} - \widetilde{\boldsymbol{x}}'|| > \epsilon$ and strictly positive frequencies [61]. As a consequence this convergence may be very slow especially for small frequencies and small wavenumbers which becomes even worse when two-dimensional periodicity is accounted for. However it has to be said that the numerical convergence of (101) is not necessary in the Boundary Element Method since only the convergence of the integral of $\widetilde{\boldsymbol{U}_s^G}$ on the source and the receiver elements is required to assemble the final linear system [82].

When Green's functions of a layered half-space are to be accounted for, the expression (101) becomes very expensive to evaluate. Fortunately, thanks to existing relationships between Floquet transforms and Fourier transforms and as the Green's functions of a layered half-space are computed in the horizontal wavenumber domain, the following expression turns out to be much more efficient since it can be evaluated using Fast-Fourier-Transforms [41] :

$$\widetilde{\boldsymbol{U}_s^G}(k, \widetilde{\boldsymbol{x}}, \widetilde{\boldsymbol{x}}') = \sum_{n=-\infty}^{+\infty} \int_{-\infty}^{+\infty} \widehat{\widehat{\boldsymbol{U}_s^G}}(k_1, k + \frac{2n\pi}{L}, x_3, x_3) \, e^{-i\{(k+\frac{2n\pi}{L})(x_2 - x_2') + k_1(x_1 - x_1')\}} dk_1 \tag{102}$$

where $\widehat{\widehat{\boldsymbol{U}_s^G}}$ is the 2D Fourier Transform of $\boldsymbol{U}_s^G$ with respect to the x_1, and x_2 axes, the two first arguments being the wavenumbers along these two axes varying from

$-\infty$ to $+\infty$. This simple expression must not be allowed to hide the numerical difficulties associated with it and to the authors knowledge the treatment of the singular terms in (102) is still open. Finally one can remark that when L tends to 0 the series on n disappears in (102), but singular terms still have to be evaluated. In the end, taking $k = 0$ gives the 2D Green's functions of a layered half space.

5.5. NON INVARIANT UNBOUNDED INTERFACES

Invariant domains or subdomains are not so easy to find in nature and the scope of applicability of techniques described in this section may not appear to be as wide as expected even when combined with DD techniques and BEM for complex media. Nevertheless it can be argued that such techniques are able to account for the most important physical phenemena and in particular guided waves as will be shown in Section 7. However these waves are very sensitive to perturbations of the medium and the consequences of the loss of invariance have to be investigated and improvements sought. A few guidelines are given here :

5.5.1. *Statistically Homogeneous Random Medium*

Surprisingly, for strongly perturbated domains the invariant approach is still effective. Indeed for a domain showing random perturbations that are homogenous in space and the covariance depending only on the separation distance, the invariant property being satisfied in a statistical sense. Then periodic models with a period much larger than the correlation length are shown to be very good approximations of the original model as far as ensemble averages are concerned. In particular both diffusion and localisation effects occurring in a random media can be analysed as shown in Section 7.

5.5.2. *Weakly Perturbed Invariant Domains*

As long as the perturbations around the perfectely invariant case remain small iterative solutions may be considered. In particular when the problem under consideration can be described as an invariant domain with perturbations at some distance from the invariant interface the interaction problem on this interface can be decoupled from the propagation problem in the media. Similar techniques can be applied when the surface is weakly perturbed or curved. This technique has been widely used in geophysics to model seismic experiments in boreholes [18, 19, 27, 65] and some results are given in Section 7.

5.5.3. *Truncated Invariant Domains*

The perturbation approach fails when the invariant domain is truncated since the perturbation is not weak -compact in the mathematical sense- and the three dimensional problem has to be studied. Some arguments have then to be found to justify the approximation by a numerical model. Some mathematical developments have been proposed in [36] whereas some numerical experiments that suggest rules for such an approximation (see [80] for the case on borehole geophysics). From this point of view BEM appears to be much more efficient that FEM since even when the mesh is truncated, bulk waves are still propagating toward infinity while they are reflected in FEM approaches. However special attention has to be paid to guided waves as these waves are reflected when the BE mesh is truncated. Special infinite elements can be developed to filter these waves [27], an example will be shown in Section 7.

The other big issue as far as three dimensional models are concerned is the computational efficiency since the approximated model will have a large number of dofs. In particular, partial symmetries on a regular mesh can be used to reduce the computational cost when assembling the BEM system as proposed in [27]. A much more general approach is the Multi-pole Expansion [157, 129]. Moreover invariant problems can be used as efficient preconditioners in iterative solvers.

6. Green's Functions of a Layered Half-space

Analytical solutions for the Green's function of a layered half-space do not exist except in the simple case of an homogeneous half-space [22]. Numerical techniques to compute these functions have been first introduced by Thomson and Haskell [179, 105] using the propagator method in the wavenumber domain and then using an inverse Fourier Transform [40]. Their scheme is very unstable for large wave numbers and Dunkin has proposed some modifications to avoid these instabilities. In the early eighties Kennett [122] proposed a new algorithm based on the reflection-transmission coefficients which has been shown to be much more stable. More recently impedance formulations have been proposed by several authors following Kausel's first work [117] and have been extended to poroelastic media [75]. Apsel and Luco [135] were the first to use these Green's functions to solve boundary integral equations where the sources were put outside of the boundary in order to avoid computations of singular terms. In [55], we have shown that Green's functions of the layered half-space based on Kennett's algorithm and an inverse Fast Hankel Transform [48, 44, 47] may be efficiently used in a Boundary Element Method provided that singularities are efficiently accounted for using adapted regularisation procedures [12]. We recall briefly in the following that the procedure consists in finding the displacement field $\boldsymbol{u}_m$ in each layer $C_m = \mathbb{R}^2 \times]z_{m-1}, z_m[$ satisfying the following equations :

$$\begin{array}{rll} \mathbf{Div}\sigma_m(\boldsymbol{u}_m) + \boldsymbol{f} = \rho_m \omega^2 \boldsymbol{u}_m & \text{in} & C_m \\ \boldsymbol{t}_m(\boldsymbol{u}_m) = \boldsymbol{t}_{m+1}(\boldsymbol{u}_{m+1}) & \text{on} & \Sigma_m \\ \boldsymbol{u}_m = \boldsymbol{u}_{m+1} & \text{on} & \Sigma_m \\ \boldsymbol{t}_1(\boldsymbol{u}_1) = \mathbf{0} & \text{on} & \Sigma_o \end{array} \tag{103}$$

where z_m is the depth of interface Σ_m between layer C_m and C_{m+1}, $\boldsymbol{f} = \delta(\boldsymbol{x} - z_s \boldsymbol{e}_z)\boldsymbol{a}'$ the source on the z axis at depth z_s, $\boldsymbol{t}_m$ denoting the traction vector on an horizontal plane and Σ_o being the free surface. Because of the cylindrical symmetry of the layered half-space with respect to the vertical axis centred on the source location a cylindrical frame will be used in the following. Moreover defining for any vector field $\boldsymbol{v}(r,\theta,z)$ and scalar field $y(r,\theta,z)$, the following differential operators:

$$\begin{array}{rcl} \boldsymbol{v}' & = & \boldsymbol{v} - v_z\boldsymbol{e}_z \\ \mathrm{div}'\boldsymbol{v} & = & (\partial_r(rv_r) + \partial_\theta v_\theta)/r \\ \mathrm{rot}'\boldsymbol{v} & = & (\partial_r(rv_r) - \partial_\theta v_\theta)/r \\ \Delta' y & = & (\partial^2_{rr} + r^{-1}\partial_r + r^{-2}\partial^2_{\theta\theta})y \\ \mathbf{grad}' y & = & \partial_r y\boldsymbol{e}_r + \frac{1}{r}\partial_\theta y\boldsymbol{e}_\theta \end{array} \quad (104)$$

decomposition between P-SV and SH problems is obtained. Indeed, in each layer the stress-strain relationships and the dynamic equilibrium may be written as two independent first order differential systems with respect to z:

$$\partial_z \mathbf{b}_m = \mathbf{A}_m \mathbf{b}_m + \mathbf{g}_m \quad (105)$$

where $\mathbf{b}_m$, the Thomson-Haskel vector being equal to $(\boldsymbol{u}_m \cdot \boldsymbol{e}_z, \mathrm{div}'\boldsymbol{u}, \boldsymbol{t}_m \cdot \boldsymbol{e}_z, \mathrm{div}'\boldsymbol{t})$ or to $(\mathrm{rot}'\boldsymbol{u}_m, \mathrm{rot}'\boldsymbol{t}_m)$ is continuous across each interface. The source term $\mathbf{g}_m$ in (105) is equal to $(0, 0, f_z, \mathrm{div}'\boldsymbol{f})$ or to $(0, \mathrm{rot}'\boldsymbol{f})$ and is of order 0 or ± 1 with respect to θ. The operators $\mathbf{A}$ given in formula (125) and (126) depend only on operator Δ', the elastic parameters of the layer and the circular frequency ω.

Since the Bessel functions are the eigenfunctions of Δ' in a cylindrical coordinate system the Hankel transform $\mathcal{H}_n$ [189, 175] (see Appendix 10.4.) with $n \in \{0, \pm 1\}$ applied to equation (105) simply replaces this differential operator by $-k^2$, k being the horizontal wavenumber. The equation (105) becomes:

$$\partial_z \widetilde{\mathbf{b}}_m = \widetilde{\mathbf{A}}_m \widetilde{\mathbf{b}}_m + \widetilde{\mathbf{g}}_m \quad (106)$$

where $\widetilde{\mathbf{b}}_m = (\widetilde{\mathbf{u}}^T, \widetilde{\mathbf{t}}^T)^T$, $\widetilde{\mathbf{A}}_m$ and $\widetilde{\mathbf{g}}_m$ are given in Appendix 10.3..

6.1. SOLUTION IN THE SLOWNESS SPACE

In the wavenumber domain system (106) can be solved by diagonalizing matrix $\mathbf{A}$ whose eigenvalues in the P-SV case are the vertical wave numbers $\pm i\gamma_\alpha$, $\alpha = p, s$ the general solution in each layer L being :

$$\widetilde{\mathbf{b}}_m(z) = \mathbf{M}_m \mathbf{D}_m(z) \mathbf{c}_m \quad (107)$$

where $\mathbf{M}_m$ is the matrix of the eigenmodes of $\mathbf{A}$, $\mathbf{D}_m(z)$ the diagonal matrix representing the downward and upward propagation, $\mathbf{c}_m$ being the degrees of freedom in each layer. These dofs in each layer are related by the following boundary interface and radiation conditions :

$$\widetilde{\mathbf{t}}_1(z_o) = \mathbf{0} \qquad \widetilde{\mathbf{b}}_m(z_m) = \widetilde{\mathbf{b}}_{m+1}(z_m) \qquad \mathbf{c}_M = (.,0,.,0)^T \quad (108)$$

The corresponding system of equations is efficiently solved using the reflection transmission coefficient technique which appears to be nothing but a standard Gaussian elimination on the total system (see [55]). Once degrees of freedom $\mathbf{c}_m$ are computed for each layer, displacements and tractions along an horizontal interface and the other components of the stress tensor are then computed using an inverse Hankel Transform and the following formulae :

$$
\begin{array}{rcl}
\Delta \boldsymbol{u}' & = & \mathbf{grad}'\mathrm{div}'\boldsymbol{u} - \mathbf{rot}(\boldsymbol{e}_z \mathrm{rot}'\boldsymbol{u}) \\
\Delta \boldsymbol{t}' & = & \mathbf{grad}'\mathrm{div}'\boldsymbol{t} - \mathbf{rot}(\boldsymbol{e}_z \mathrm{rot}'\boldsymbol{t}) \\
\sigma_{\theta\theta} & = & (1-2\kappa^2)\{\sigma_{zz} + 2\mu \mathrm{div}'\boldsymbol{u}\} + 2\mu/r(\partial_\theta u_\theta + u_r) \\
\sigma_{rr} & = & (1-2\kappa^2)\{\sigma_{zz} - (3\lambda + 4\mu)\mathrm{div}'\boldsymbol{u}\} - 2\mu/r(\partial_\theta u_\theta + u_r) \\
\sigma_{r\theta} & = & \mu \mathrm{rot}'\boldsymbol{u} - 2\mu/r(u_\theta - \partial_\theta u_r)
\end{array}
\tag{109}
$$

Final expressions for $\boldsymbol{u}$ and $\boldsymbol{\sigma}$ are given in Appendix 10.5..

6.2. FAST INVERSE HANKEL TRANSFORM

The inverse Hankel transform can be efficiently performed using the following formula as proposed in [44] :

$$
\begin{array}{rcl}
\mathcal{H}_n(f)(r) & = & (-i)^n/\pi \int_0^{2\pi} \widehat{g}(r\cos\phi)\cos(n\phi)d\phi \\
g(k) & = & f(k)k \qquad \text{for } k \geq 0 \\
g(k) & = & 0 \qquad \text{for } k < 0
\end{array}
\tag{110}
$$

$\widehat{\ }$ denoting the Fourier transform which is performed using a classical FFT algorithm. Integration in (110) is then computed using a standard integration formula.

6.3. SINGULARITIES

In order to use such Green's functions in a standard BEM, singularities of the displacement and of the traction vectors have to be carefully dealt with. It turns out that the local singularities around the point source in the physical space have the following form in the horizontal wavenumber domain [55]:

$$
\begin{array}{rcl}
\widetilde{\boldsymbol{u}} & = & e^{-k|z-z_s|}(|z-z_s|\mathbf{0}(1) + \mathbf{0}(1/k)) \qquad k \to 0 \qquad z \to z_s \\
\widetilde{\boldsymbol{t}} & = & ke^{-k|z-z_s|}(|z-z_s|\mathbf{0}(1) + \mathbf{0}(1/k)) \qquad k \to 0 \qquad z \to z_s
\end{array}
\tag{111}
$$

$\mathrm{rot}'\boldsymbol{u}$ and $\mathrm{div}'\boldsymbol{u}$ having the same singularities as the tractions. From the mathematical point of view this shows that due to the exponential term the inverse Hankel transform is uniformly convergent for $z \neq z_s$ which is not the case when $z = z_s$. From the numerical point of view this means that when z is close to z_s some special techniques have to be used. The basic idea is to remove this singular behaviour when k becomes large and then to take its inverse transform analytically. Indeed it appears that the corresponding integrals have analytical expressions given by :

$$
\begin{array}{rcl}
I_{mn}(z,r) & = & \int_0^{+\infty} e^{-kz} J_m(kr) k^{n-1} dk = \frac{h_{nm}(\varphi)}{R^n} \\
R & = & \sqrt{r^2+z^2} \qquad \cos\varphi = z/R
\end{array}
\tag{112}
$$

These terms being computed analytically and the regular part being computed using the Fast Hankel Transform the resulting Green's functions can be used in a BE program. Moreover one can remark that analytical expressions of the regularizing tensor $\boldsymbol{T}^o$ and $\boldsymbol{U}^o$ are available which satisfy property (70).

7. Applications

This section illustrates some applications to real case studies of the theoretical and numerical framework. The domain decomposition technique described above for Soil-Fluid-Structure Interactions is first used in Section 7.1. to analyse the dynamic behaviour of an arch dam. The power of reduction techniques for Soil-Structure Interaction analysis is proposed in Section 7.2. applied to nuclear reactor buildings. Section 7.3. shows the efficiency of FE-BE coupling on a volume to deal with random properties of the soil and Section 7.4. describes the spectral approach when analysing the seismic safety of periodic sheet-piles.

In Section 7.5. these techniques are applied to seismological issues and in particular topographic effects whereas Section 7.6. is devoted to the coupling between site effect and SSI in dense urban areas. In particular it is shown that periodic models of a city along the two horizontal directions are able to give very valuable insight into the interactions between many buildings randomly distributed inside the city. The importance of new phenomena such as diffusion and localisation are pointed out.

Finally Section 7.7. is devoted to borehole geophysics analyses where the same methods apply even when the domain under consideration is very slender but not invariant.

7.1. SOIL-FLUID-STRUCTURE INTERACTION

The domain decomposition approach proposed in Section 3 has been applied to a wide range of analyses concerning the seismic safety of large structures such as dams. Figure 5 shows the impact of the Soil-Fluid-Structure Interaction on the resonant modes of a 200 meters high arch dam built in a narrow valley. BEM has been used for the rock and for the reservoir. In both cases the Green's function of an homogeneous elastic and acoustic domain are considered. The BEM mesh has been truncated at some distance from the dam (typically twice its width). The dam has been modeled using standard thick shell finite elements and a reduced model consisting of 15 eigenmodes of the dam and 20 static modes for the foundation have been used. The seismic loading consists in a set of plane P and SV waves with different incidences and azimuths. More complete results can be found in [70, 46] and since another chapter of the book is dedicated to SFSI for arch dams, the reader is referred to it.

7.2. MODAL REDUCTION FOR SSI

The proposed methodologies are illustrated in the case of the seismic response of the building shown in Figure 6. More complete results can be found in [123].

For structures with large foundations, the number of dofs on the interface is quite large and BEM computational costs increase rapidly with the number of these dofs. Thus, there is a strong interest in not only reducing internal structure dofs but also interface dofs.

Eigenmodes	S	SSI	FSI0	SSI	SFSI
1	2.75	2.7	2.4	2.4	2.35
SFSI 1	*	*	*	*	2.7
2	3.6	3.5	3.3	3.2	3.0
SFSI 2	*	*	*	*	3.3
3	4.5	4.3	4.1	4.0	3.8
4	5.7	5.5	5.1	5.2	4.8
5	5.8	5.6	5.2	5.2	4.9
6	5.9	5.8	5.2	5.2	4.9
SFSI 3	*	*	*	6.3	5.7
7	7.6	7.5	6.6	6.6	6.4
8	7.9	7.6	7.1	7.3	7.1
9	8.2	8.	7.4	7.3	7.2

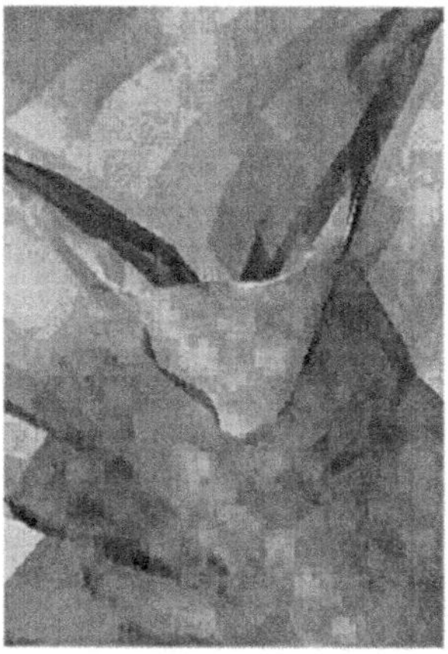

Figure 5. Resonance frequency of an arch dam depending on its environment; S - Rigid foundation; SSI - Soil-Structure Interaction with an empty reservoir; FSI0 - Fluid-Structure Interaction with an incompressible fluid; FSI - Fluid-Structure Interaction with a compressible fluid, SFSI - Soil-Fluid-Structure Interaction with a compressible fluid (from [70] http://lilas.mss.ecp.fr/DVLPT/Logiciels/MISS/gallery.html)

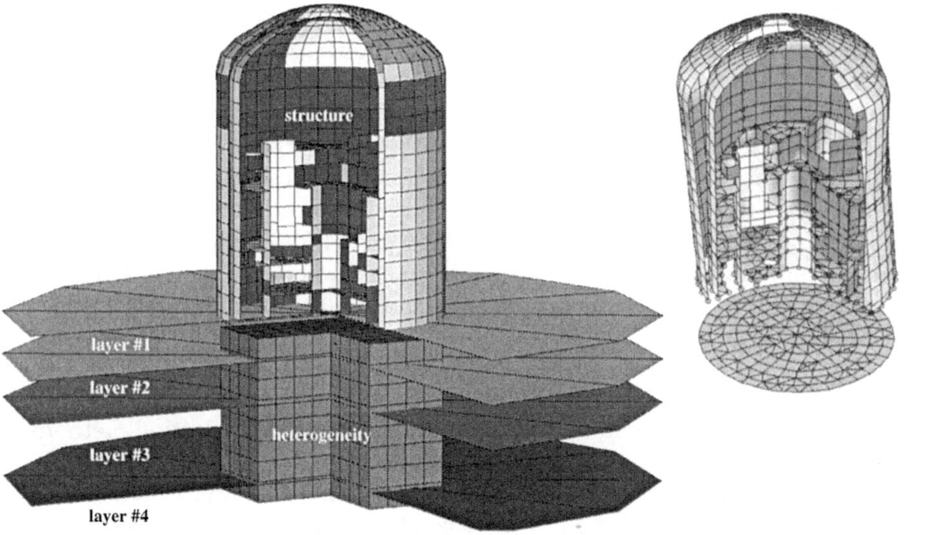

Figure 6. (left) Example of a pressurized water reactor building on a stratified soil with random heterogeneity, (right) First fixed interface mode for the model. The soil-structure interface is lowered for the plot.

Here, the soil impedance is obtained from the computer code MISS3D [57] by means of a BEM for a layered soil [12, 55]. The building model, a Craig-

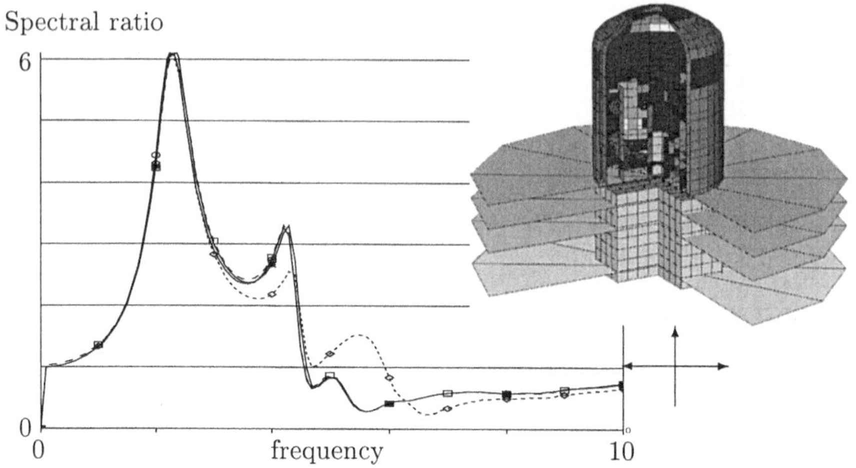

Figure 7. Frequency response of a nuclear reactor building resting of a layered half space to a vertically incident S wave (from [68]).

Bampton reduction of the initial building model, is exported from the computer code *code_aster* [78]. This model has 558 interface dofs, and 133 fixed interface modes. The loaded modes used in the following sections are computed on the basis spanned by the fixed interface modes. This is clearly an approximation but it has a marginal impact on the results shown here. On the contrary assuming a rigid foundation is not a good approximation as shown in Figure 6 especially when the second and the third resonances of the structure are considered.

The reference elastic stiffness $\mathbf{K}_0$ is produced by combining the elastic model of the building and a layer of ground springs, whose values are found by taking along each translation direction the mean of the real part of the soil impedance at 1 Hz (approximately $2.5e9N/m$).

7.2.1. *Selecting Dynamic Interface Modes*

In this Section, we seek to prove the validity of interface selection methods. To do so, four cases are compared.

The reduction basis for model FI combines fixed interface modes and a variable number of modes of the model condensed onto the soil-structure interface. Keeping all the interface modes would be equivalent to using the Craig-Bampton method [72]. Note that these modes can be approximated by building a mass matrix defined on the interface only, which may be more efficient numerically than condensing the model.

The reduction basis for model LO keeps modes of the model associated with the reference stiffness $\mathbf{K}_0$ and thus corresponds to a loaded interface CMS method. The drawback of this method is that there is nolonger a decoupling between internal and interface deformations. This might be inefficient for a structure with many internal resonances.

For both models FI and LO, we compare a base version and a version where the static response $\mathbf{K}_0^{-1}\mathbf{f}$ to the spatial distribution of loads is linked to the incoming waves computed at the low end of the frequency range. Keeping the real and imaginary parts for the three types of waves, leads to adding six interface modes. Better strategies for building a static correction are considered in Section 7.2.2.

Figure 8 shows the strain energy error. The value shown is the maximum error for 15 frequencies between 1 and 15 Hz and 3 loading cases: SV, SH and P with vertical incidences.

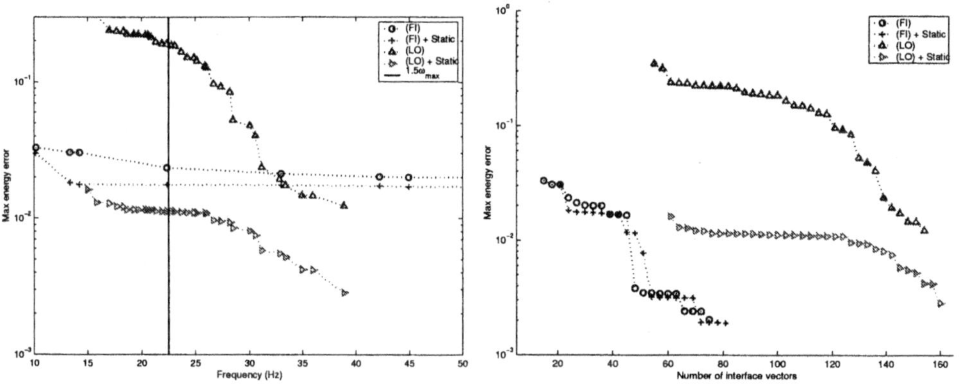

Figure 8. Maximum of energy error (left) as a function of frequency of last retained mode and (right) as a function of number of modes with non zero displacement on interface -133 fixed interface modes are accounted for in the FI model- (from [123]).

It comes out clearly that the best choice will depend on the relative cost of evaluating the soil impedance using the BEM and evaluating the response. The use of static correction terms, for the spatial distribution of the inputs, is a very useful safeguard and should not be omitted. The use of the $1.5\omega_{\max}$ threshold is safe if such static correction is included.

Convergence often shows long series of modes with little variation. Evaluating the error by comparing a nominal model and a somewhat richer (where a number of additional interface modes have been included) can thus give very misleading results.

7.2.2. *Selecting Input Shapes for Static Correction*

An important specific property of seismic loads is that the spatial distribution of loads varies with frequency. Doing a proper static correction, thus requires an appropriate treatment of these variations.

The simple approach, used in Section 7.2.1., is to take the loads at a restricte number of frequencies

$$\mathbf{b}_I = \{\Re(\mathbf{f}_s(\omega_1)), \Im(\mathbf{f}_s(\omega_1)) \;\; ... \;\; \Im(\mathbf{f}_s(\omega_N))\} \qquad (113$$

and include the static correction $\mathbf{K}_o^{-1}\mathbf{b}_I$ in the interface basis (typically afte orthonormalization).

Based on the discussion in Section 3.5.2., a first extension is to take a fairly larg number of frequencies to produce $\mathbf{b}_I$, compute the singular value decomposition o these loads, and keep the vectors associated with the largest singular values. Th SVD can be applied to $\mathbf{b}_I$ or to $\mathbf{K}_o^{-1}\mathbf{b}_I$. Figure 9 shows that SVD slopes obtaine when considering each wave type separately or simultaneously is almost the sam so that this is an open choice. When considering a large number of loading cases (wave types and 7 incidences), the overall slope of the singular values is obviousl smaller but the number of vectors found for a threshold of 10^{-2} or 10^{-3} is stil quite small. One should also note that singular values of $\mathbf{K}_o^{-1}\mathbf{b}_I$ decrease muc more rapidly than those of $\mathbf{b}_I$. This illustrates the smoothing effect of computin the static response.

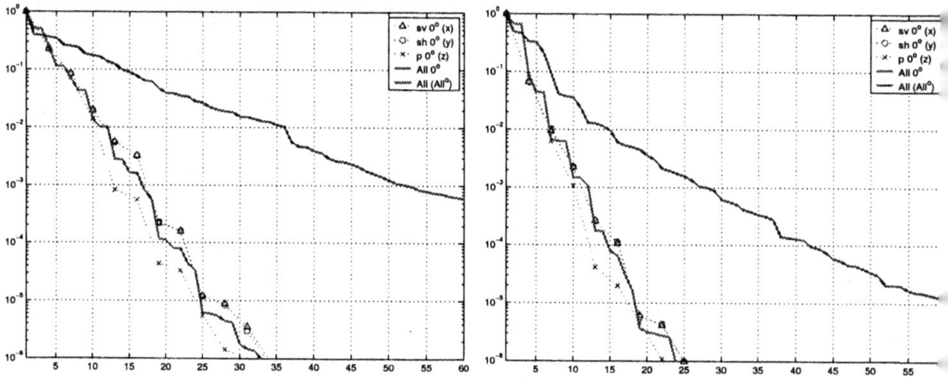

Figure 9. Singular values of the decomposition of $\mathbf{f_I}$ (left) and $\mathbf{K_o}^{-1}\mathbf{f}_I$ (right) for SV waves a 0°, SH at 0°, P at 0°, all waves at 0°, all waves at 7 different incidences (from [123]).

7.3. SSI ON A RANDOM SOIL

The substructuring method on a volume interface is applied to a realistic exampl of a nuclear power plant resting on a circular spread footing of radius $R \approx 25\,m$ (see Figure 6 for the geometry and mesh layout). We restrict our analysis to th case of random Lamé moduli for the soil. This means that the elasticity tensor i the soil medium occupying Ω_s^b is isotropic and depends only on two parameters λ and μ_b written in the perturbation zone Ω_s^b as:

$$\lambda(\boldsymbol{x}) = \underline{\lambda}(z) + \lambda(\boldsymbol{x}); \quad \mu(\boldsymbol{x}) = \underline{\mu}(z) + \mu(\boldsymbol{x}) \qquad (114$$

where $\underline{\lambda}$ and $\underline{\mu}$ are Lamé moduli for the soil mean stratification, function of the depth $z = \boldsymbol{x} \cdot \boldsymbol{e}_3$. The mass density of soils is known to vary smoothly with depth, so its perturbations are neglected in this case. The deterministic free field considered is a vertically incident plane SH-wave. The structural displacements on a fixed basis are expanded using the first 30 eigenmodes, with constant critical damping rates $\zeta_A = 7\%$. The stratified soil medium (without heterogeneity) has 3 horizontal layers and the bedrock (layer #4 in Figure 6) whose characteristics are summarized in Table 1. The corresponding Green's tensors $\boldsymbol{U}_s^G$ and $\boldsymbol{T}_s^G$ of

TABLE 1. Characteristics of the stratified soil: thickness, Lamé parameters and mass density (the hysteretic damping rate is 5%).

	h (m)	$\underline{\lambda}$ (MPa)	$\underline{\mu}$ (MPa)	$\underline{\rho}$ (kg/m^3)
layer #1	10.5	$1.16\,10^4$	$0.37\,10^4$	2400
layer #2	10.5	$1.95\,10^4$	$0.76\,10^4$	2500
layer #3	19.5	$2.12\,10^4$	$0.60\,10^4$	2400
bedrock	∞	$1.62\,10^4$	$0.99\,10^4$	2500

the stratified half-space are numerically computed as proposed in Section 6. The random soil heterogeneity occupies a cubic block $40 \times 40 \times 40\,m^3$ and its auto-correlation function is taken in the form:

$$[\mathbf{R}(\boldsymbol{x}, \boldsymbol{x}')]_{pq} = \sigma_p \sigma_q \exp\left(-\frac{\|\boldsymbol{x} - \boldsymbol{x}'\|}{\ell}\right) \tag{115}$$

where $p, q \in \{1, 2\}$; σ_1 and σ_2 are the standard deviations of Lamé's moduli and l is the correlation length. The plot on the left of Figure 10 displays the log-scaled amplitude of the horizontal displacement $u_x = \boldsymbol{u}_b \cdot \boldsymbol{e}_1$ on top of the structure normalized by the incident wave amplitude when: (i) the correlation length ℓ for the soil is set to $50\,m$ and the standard deviations $\sigma = \sigma_1 = \sigma_2$ of Lamé's moduli are of the order of 25%; (ii) the correlation length is $\ell = 20\,m$ and the standard deviations of Lamé's moduli are of the order of 50%. In the first case 500 trials and 5 modes were used in the Karhunen-Loeve expansion, the relative error ε obtained being $\varepsilon = 11\%$; in the second case 1000 trials and 20 modes were used, the relative error being $\varepsilon = 8\%$. Mean values and confidence intervals are plotted, the latter being defined by the upper and lower envelopes such that $P(|u_x - E < u_x > | < U_c) \geq p_c$ with a confidence level $p_c = 95\%$. The upper and lower envelopes are produced using Chebychev's inequality [127] which results in $U_c(\boldsymbol{w})$ such that:

$$p_c = 1 - \frac{V_{u_x}}{U_c^2} \tag{116}$$

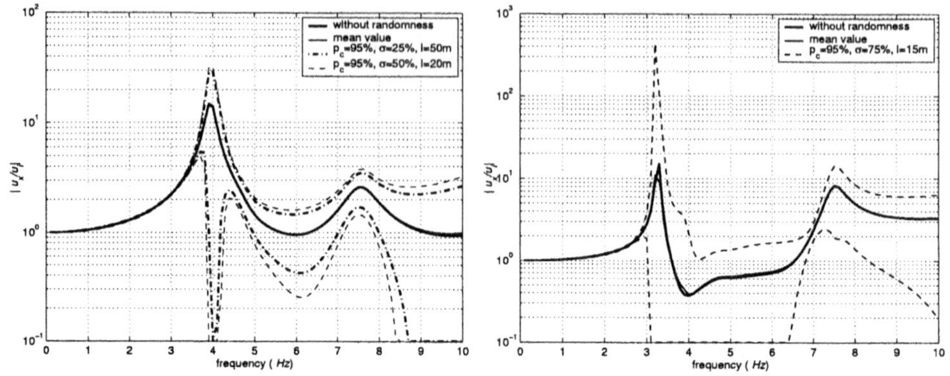

Figure 10. Normalized structural displacement under a vertically incident SH-wave for randomly perturbed soil stiffness, (right) 3 stiff layers stratification: response without randomness (thick solid line), mean response (thin solid line), and confidence intervals for (i) $\ell = 50\,m$ and $\sigma = 25\%$ (thick dashed lines), and for (ii) $\ell = 20\,m$ and $\sigma = 50\%$ (thin dashed lines), (right) 1 soft layer stratification: response without randomness (thick solid line), mean response (thin solid line), and confidence interval for $\ell = 15\,m$ and $\sigma = 75\%$ (thin dashed lines).

where $V_{u_x} = E < |u_x - \underline{u_x}|^2 >$. The mean values for both cases are seen to be almost undistinguishable from the amplitude observed in the deterministic case; however the confidence regions are, as expected, quite different. From an engineering point of view, these results show that a reduction of the structural vibration amplitude up to 5% with $\sigma = 25\%$, and up to 10% with $\sigma = 50\%$, can be obtained with a confidence of 95% when the soil uncertainty underneath the foundation is accounted for. We also note that, for the present configuration, the fundamental eigenfrequency of the coupled soil-structure system is almost unchanged by the presence of random heterogeneities.

The soil described in Table 1 above is quite stiff as compared to usual soils. Other calculations have been performed with a softer soil having only one horizontal layer above the bedrock. The thickness of this layer is $40.5\,m$, its density is $\underline{\rho} = 2400\,kg/m^3$, its Young's modulus is $\underline{E} = 5 \cdot 10^3\,MPa$ and its Poisson's coefficient is $\underline{\nu} = 0.38$ (corresponding to $\underline{\lambda} = 5.74 \cdot 10^3\,MPa$ and $\underline{\mu} = 1.81 \cdot 10^3\,MPa$); the characteristics of the bedrock are unchanged from the previous example. The right plot in Figure 10 is the same as the left one, except that now $\ell = 15\,m$ and $\sigma = 75\%$. 3000 trials were used for these parameters and $K = 40$ modes for the Karhunen-Loeve expansion of the perturbations of Lamé's moduli. In this example, Soil-Structure Interaction effects are more pronounced than in the previous case because of the softening of the soil. The dispersion of the structural response for the same incident wave as in the previous case is increased, but qualitatively similar conclusions for the reduction factors can be drawn: a reduction of the structural vibration amplitude up to 15% can be achieved with a confidence of 95%. The fundamental eigenfrequency of the coupled system is almost unchanged as well. However slight discrepancies can already be observed for the higher mode at roughly 7.5 Hz. This trend agrees with the results obtained in the frame of the more general theory of symmetric positive-definite random matrices [169]. As regards reduction of structural vibration levels, comparable results, both qualitatively and quantitatively, were obtained by Toubalem *et al.* [181] by a simplified approach.

7.4. SFSI FOR PERIODIC SHEET-PILES

We present here an analysis for the dynamic behaviour of diaphragm and quay walls. These retaining walls (see Figure 11) reaching up to 1 kilometre long, are made of identical panels of about six metres long connected by joints. The panels are anchored using tiebacks. Their static behaviour is relatively well-known, which is not the case concerning their dynamic behaviour. In fact when an earthquake occurs, waves propagate in many directions. For inclined incident waves or surface incident waves, the panels do not vibrate in phase creating differential displacements which may lead to water infiltration and other dangerous phenomena. The presence of joints and anchors as well as the 3D characteristics of the loadings require a full 3D analysis. Regarding the dimensions of such structures a Finite Element model would lead to a huge number of degrees of freedom. A Boundary Element Method is more suitable since only the interfaces between domains need to be meshed. Nevertheless for an industrial case and even using BEM and a high performance supercomputer the numerical model can account only for six panels using a full 3D approach; which is clearly not enough to account for interaction effects between panels.

The periodic approach proposed in this paper is particularly suited for these analyses as quay walls are periodic. The generic cell consists of one panel with its anchors, the corresponding slice of soil and eventually the fluid as shown in Figure 11. The panel is a $0.85 \times 6 \times 22.5m^3$ concrete thick plate. Three inclined anchoring beds of respective length equal to $18, 24, 28m$ from top to bottom, reinforce its static stability. They are anchored on half their length and connected to the panel at $3.5, 8.5, 12.5m$ from the top with an angle of about 30 degrees. The panel lies in a linear elastic stratified halfspace made up of two layers referred to by numbers 1 and 2 with the following physical parameters $c_{s1} = 170m/s$, $c_{s2} = 240m/s$, $c_{p1} = c_{p2} = 1500m/s$, $\rho_1 = \rho_2 = 2000kgm^{-3}$, $\beta_1 = \beta_2 = 5\%$, where c_s and c_p are respectively the shear and compressional wave velocities, ρ being the mass density and β the hysteretical damping ratio. The loading is a plane shear wave propagating either vertically or inclined with an angle of 30 degrees with respect to the vertical axis. The polarisation is normal to the quay wall along the $\boldsymbol{e}_1$ direction. In order to make a comparative study, several models with and without the anchors and the water have been analysed (see Figure 11).

- Case AW denotes the case with anchors and water,
- case AN with anchors and without water,
- case FW stands for free of anchors with water,
- case FN standing for free of anchors without water.

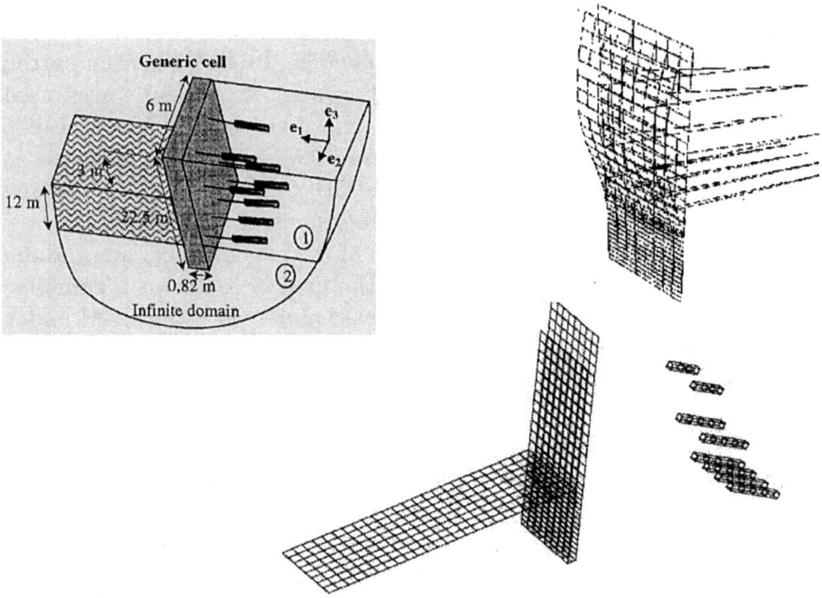

Figure 11. The generic cell of the quay wall (top left), the BEM mesh for the soil including anchors (bottom), the forth eigenmode of the panel (top right).

As shown in Section 3.5. a Finite Element analysis is required for the structure (the free panel in cases FW and FN and with the anchors in cases AW and AN) in order to determine the displacement field decomposition (see equations (49)). The 20 first eigenmodes of a free panel have been selected. This choice is made according either to an *a priori* criterium choosing the last natural frequency larger than 2.5 times the maximum frequency ($[0.1Hz, 12.5Hz]$ in our study) or an *a posteriori* criterium: verifying that the last participation factors q_m are small.

The BEM mesh for the soil is shown in Figure 11 and includes neither the free surface of the soil nor the interface between the two layers since Green's functions of a layered half-space are used. It then consists of the interface with the panel and the soil-fluid interface. The mesh of this last interface has been truncated $30m$ away from the panel. The BEM mesh for the fluid consists of the soil-fluid interface shown in Figure 11 and the interface with the panel (not shown in Figure 11). The free surface of the fluid is not meshed since the Green's functions of an acoustic half-space are used for the water.

A preliminary verification as been performed concerning the number of cells (threshold $n_{\max}$) to be taken into account in formula (101) when computing the numerical periodic Green's function. Indeed since in this case we are using the Green's functions of a stratified half-space we were not able to use the convergence results given in [62]. The conclusion was that up to 17 panels ($2n_{\max} + 1$) have to be taken into account to reach convergence.

Results in the frequency domain show that the most active participation factors are those corresponding to a flexural mode of the plate in the $\boldsymbol{e}_1$ direction which is the polarisation of the incident field. Moreover the moduli of these participation factors decrease when the mode number increases justifying the truncation of the modal basis above 20 modes.

Displacement on the top and bottom of the panel as a function of the frequency and for the cases considered are shown in Figure 12. These FRF are normalized using the displacement at the top of the layer which explains the reductions in amplitudes of the bottom displacement curves. We notice that the $\boldsymbol{e}_1$ direction is the most active one because it is the loading excitation direction. The water level and the anchors appear to play a relatively marginal role in the dynamics of this system. Moreover dynamic amplification factors remain low. More interesting are the relative displacements between two nearby panels for an inclined incident wave. Time results in Figure 13 show that these displacements remain relatively small and are somehow proportional to the velocity on top of the panel, the proportionality factor being roughtly equal to $\sin\theta_i c_{s1}/L$, where θ_i is the incident angle and L the size of width of the panel. It can be concluded that the centres of the panels follow the displacements of the incident field. This conclusion is confirmed by the fact that no significant dynamic effect has been observed for this structure.

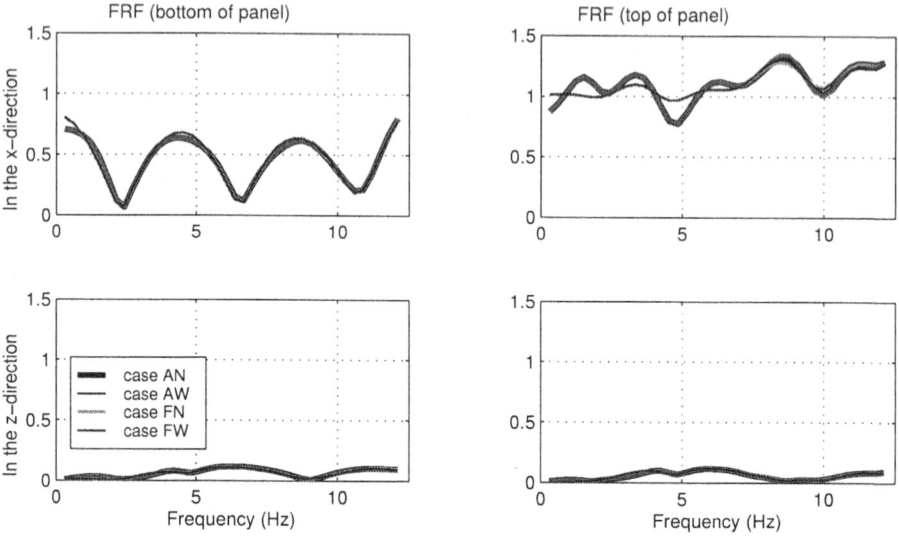

Figure 12. Displacement FRF on top and bottom of the panel in the normal $\boldsymbol{e}_1$ (x) and th
vertical $\boldsymbol{e}_3$ (z) direction.

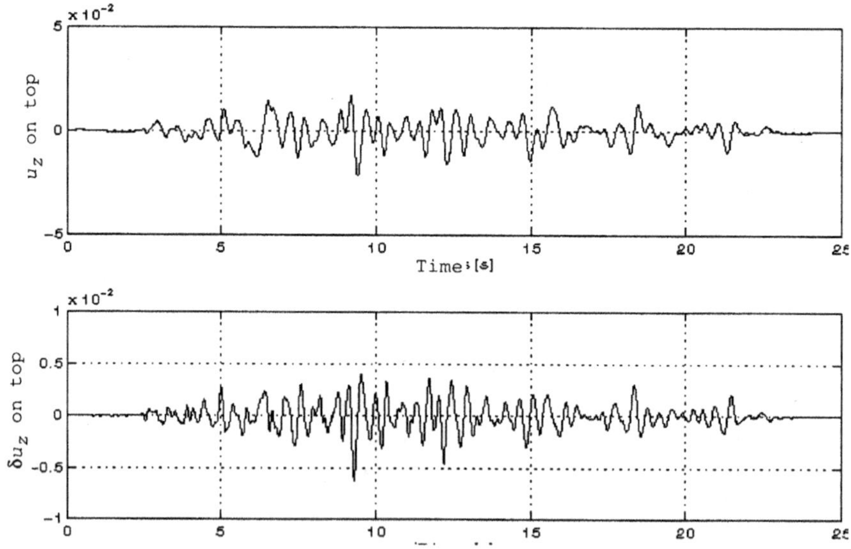

Figure 13. Displacements on top of the panel (top plot) and relative displacements on top c
the joint between two panels (bottom plot) as a function of time (from [82])

7.5. TOPOGRAPHIC SITE EFFECTS USING SSI FRAMEWORK

The proposed methodology can be applied to geological structures as well as man-made structures. One of the possible applications is the analysis of site or topography effects as shown in [55] and in Figure 14 showing the frequency response of a hemispherical mountain on a homogeneous half-space and both having the same physical properties.

Denoting the mountain by Ω_b this problem corresponds to a SSI analysis with a very flexible foundation. Two subdomains have been defined both of them being computed using a BEM with the full-space Green's functions for the mountain and the half-space Green's functions for the underlying bedrock. The mesh consists in two parts, Γ_{ba} the free surface of the montain approximated using 300 elements and Σ_{bs} the interface between the bedrock and the mountain including 420 elements. No mesh is required for the free surface of the half-space since half-space Green's functions are used. The mesh appearing in Figure 14 is included for visualisation purposes.

The major interest of using the Green's function of a layered half-space is clearly visible on time domain results given in Figure 15. Indeed Rayleigh waves generated by the interaction between the incident plane wave and the mountain are free to propagate away from the mountain and no spurious reflections of these waves appear. Using the Green's functions of the full space with a meshed free-surface would have given rise to such phenomena unless this mesh is extended far away from the mountain. In this latter case the computational effort would be prohibitive.

From the physical point of view it may be noticed in the frequency response that somehow the mountain behaves like a building in classical SSI and shows resonance frequencies in the low frequency range. Of course, due to the aspect ratio of this mountain and to the weak impedance contrast at the bottom of the mountain, the amplification factor is much smaller than in the case of a building and remains below 3. Another phenomenon that does not currently appear in SSI is the amplification on top of the mountain for the entire frequency range as shown in Figure 15. Indeed this phenomenon is due to the constructive interference between the two Rayleigh waves generated at the bottom of the mountain and reaching the top at the same time whatever their frequency content is.

This example not only shows the ability of the proposed numerical techniques to handle 3D topography effect as shown in another chapter of this book, it also provides interpretations of these phenomena in terms of classical structural dynamic concepts.

7.6. THE CITY-SITE EFFECT

The displacement fields scattered by a set of randomly distributed buildings over a soft layer under a vertically incident S wave in the bedrock is characterized using a 2D-periodic model. The distribution of the 324 buildings as a function of the number of storeys is given on top-left of Figure 17. For each box in this figure,

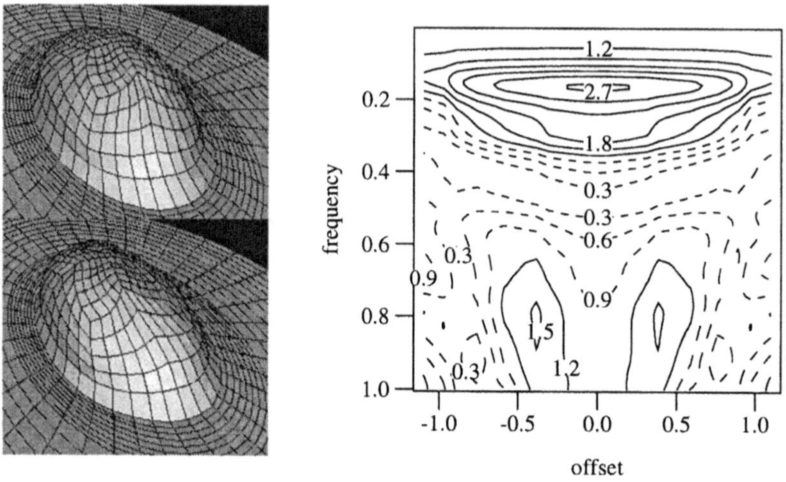

Figure 14. Response of a $2 \times 1 \times 0.5km$ hemispherical mountain on an homogeneous half-space ($c_p = 2c_s = 1km/s$) for a vertically incident SH wave polarized along the the major axis: deformation of the mountain at a given time step, frequency response on top of the minor axis (frequency in Hertz) (from [56, 59], http://lilas.mss.ecp.fr/DVLPT/Logiciels/MISS/gallery.html).

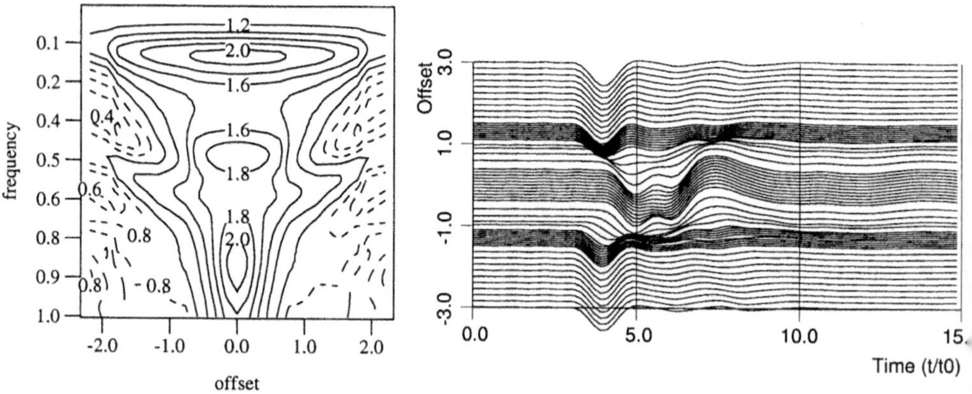

Figure 15. Response of a $2 \times 1 \times 0.5$km hemispherical mountain on an homogeneous half-space ($c_p = 2c_s = 1$km/s) for a vertically incident S wave polarized along the minor axis: frequency response on top of the minor axis (frequency in Hertz), time response for a Ricker input signal centred at 0.65Hz (from [56, 59])

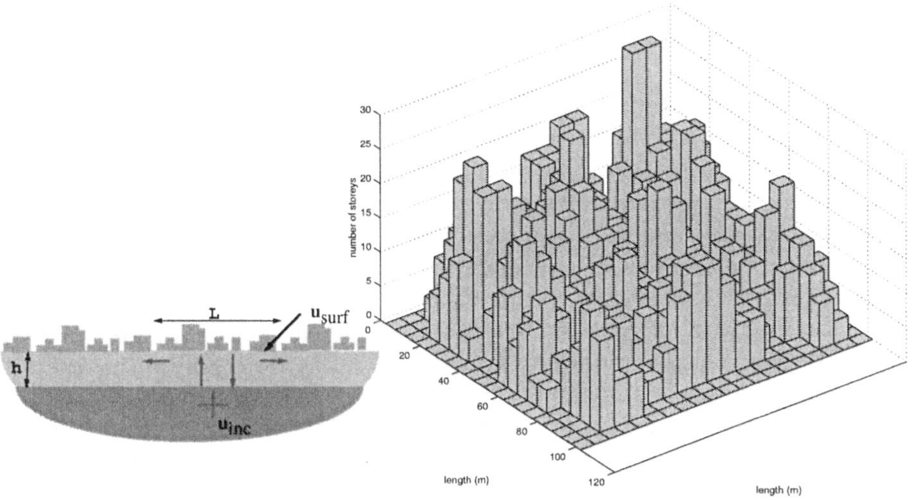

Figure 16. Interaction between a geological site made of one soft layer of thickness H lying on a homogeneous half-space with a random city modelled by an $L \times L$ cell of $n = 324$ buildings periodically reproduced along the two horizontal axes ($h = 30m$, $L = 100m$, $c_s = 30, 300m/s$, $c_p = 1500, 3000m/s$, from [60]).

buildings have exactly the same dynamic behaviour, all buildings having $5m \times 5m$ spread footings. Although not realistic this assumption will allow us to perform some averaging on these samples. The only differences between two buildings in this are their relative positions with respect to other buildings. These buildings are randomly distributed on a $100m \times 100m$ square cell as shown in Figure 17, this cell being periodically reproduced along the two horizontal axis. Convergence in series (101) has been reached for 27×27 cells. The small size of the individual foundations allows us to perform the computations up to $1Hz$ (at this frequency there are 6 points per wavelength). The frequency increment has been set to $0.01Hz$ which is consistant with the width of the expected resonance peak (the damping in the buildings is equal to 7%).

7.6.1. *Spectral Ratios*

On the top-right of Figure 17 the spectral ratios between the top layer response at each point of the reference cell and the expected free field at a free bedrock have been plotted and on the bottom-right of this figure the same results normalized with respect to the free field on top of the layer are given. On the first plot the first two resonances of the layer at $0.25Hz$ and $0.75Hz$ can clearly be identified. Up to the first resonance a very weak dispersion is observed around the response of a single layer without any building on it. Indeed the building distribution shows that none of them have their natural frequencies in this range (the natural frequency in the horizontal direction is set to $10/N_{storeys}$). Between the first and the second resonance frequency of the layer the individual responses become more and more scattered, this scattering increasing almost linearly with the frequency. Apart from this scattering, the mean value follows the reference curve. On the contrary when the second resonance of the layer is reached a significant reduction of the mean value is observed. This reduction is due to some interaction between the building response and the resonance of the layer. After the second resonance one still observes a small reduction in amplitude and the scattering keeps on growing. From these results it can be concluded that in terms of amplification the buildings do not strongly modify the site response. Small reductions may even be expected around the resonance frequency of the layer. The most significant phenomenon is the linear increase of the amplitude of scattered field with the frequency. Before studying in detail this scattering effect the building response is first analysed.

On the bottom-left of Figure 17 the spectral ratios between the top of the building responses and the free field on top of the layer have been plotted. In order to distinguish the contribution of each type of building the statistical analysis for each sample given on the top-left plot has been performed. On the same graph the top to bottom spectral responses of each building in the cell have been plotted and no dispersion on these results (thin lines) is observed. As for the site response it may be noticed that the scattering increases strongly with the frequency. In particular 10 storey buildings having their natural frequency at $1Hz$ show a very scattered response ranging from 4 to 9 compared to 7 which is the top to bottom amplification. The mean response of higher buildings remains close to the top to bottom response, amplification ranging from 6 to 9. The case of 15 storey buildings is a little peculiar as these buildings have their natural frequency around $0.75Hz$ which is also the second resonance frequency of the layer. Indeed a reduction of the amplification and a small frequency shift towards the low frequencies is observed. Nevertheless one can conclude that in this case the cross building interaction does not strongly modify the building response in terms of amplification. Of course these conclusions should be confirmed for other cases with more realistic building models.

As a conclusion of this study it has been shown [60] that the scattered field induced by a random building distribution shows the following properties :

- the scattered field is negligible when the frequency is lower than a cutoff frequency equal to the lowest natural frequency of the buildings ($0.33Hz$ in our case),
- below this cutoff frequency the correlation length and the correlation frequency do not depend on the frequency,
- above this cutoff frequency the correlation length is decreasing with the inverse of the frequency,
- this correlation length is of the order of one quarter of the wavelength of the S wave in the layer,
- the relaxation time shows mainly the same behaviour as the correlation length and is of the order of the relaxation time of the building.

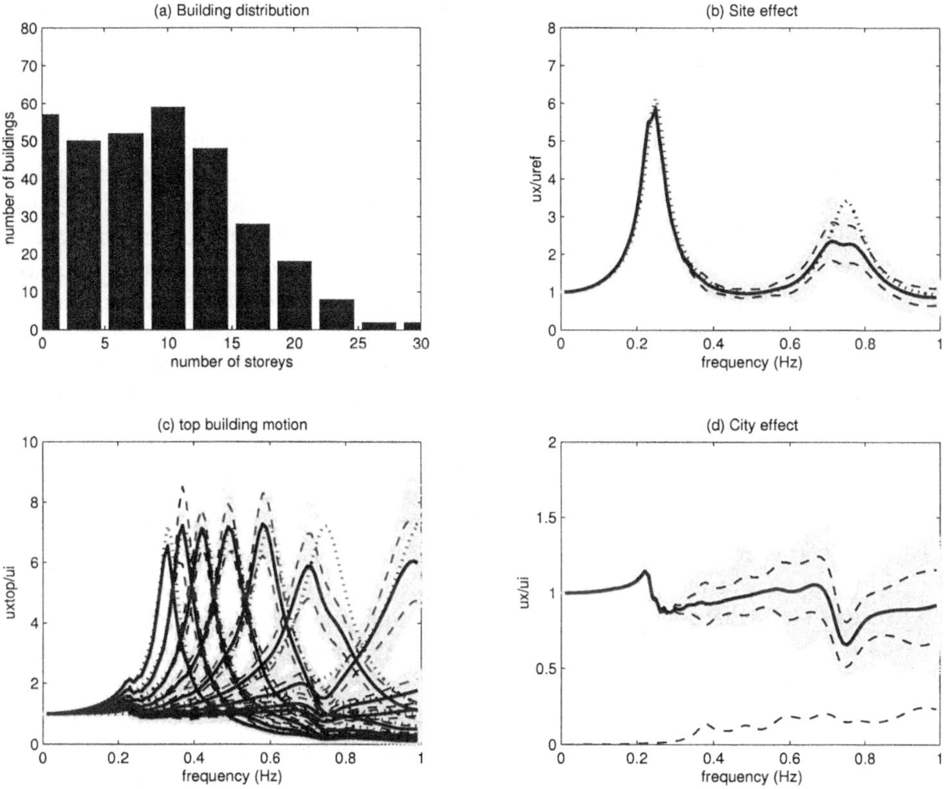

Figure 17. The seismic response of 324 buildings distributed randomly on top of a layered half-space (individual results are indicated by gray areas, thick solid lines indicate the mean values and thick dashed lines the mean values $\pm$ the standard deviation) : (top-left) the building distribution, (top-right) spectral ratios between the response on top of the layer and on top of a free bedrock (the thin dotted line indicates the response without the buildings), (bottom-right) spectral ratios between the response on top of the layer and the free-field at the same place and the standard deviation of the scattered field, (bottom-left) spectral ratios between the response at the top of the building and the response at the top of the layer in the free field conditions (the thin lines with dots indicate the top to bottom spectral ratio of the buildings).

Then the coherency of our numerical results follows roughly the classical exponen tial law :

$$|\gamma(r,\omega)| = e^{-(a+b\omega)r/c_s} \quad (117$$

with $a \approx 0.5$ and $b \approx 0.3$ which is in accordance with experimental results giver in [106] for a thiner and softer layer but for higher frequencies.

7.7. SSI IN BOREHOLE GEOPHYSICS

Fluid-soil interactions play a major role in borehole geophysics since they generat guided waves or "tube waves" along the borehole that strongly modify both th transmission of the energy inside the geological formation and the pressure fiel recorded in the receiver borehole. These phenomena have been extensively studie using the proposed numerical techniques and particularly the reflection of thes waves by heterogeneities of the borehole or the geological formation (see Figur 18).

The major difficulty comes from the ratios between typical length-scales ap pearing in the problem : namely the length of the borehole of the order of 1km the radius of about 10cm, the thickness of the tube which is a few centimetres the wavelength from 0.1 to 10 metres.

Axisymmetrical [16, 80] and translationally invariant models [176, 18, 19] hav been used to treat such problems. The emission from the borehole and the recep tion in the borehole have been uncoupled from the propagation in the geologica formation in order to cope with 3D problems arising from a non horizontal inter face at some distance from the borehole (see Figure 19 and [65]). In this particula case the Green's functions of the geological formation have been computed usin ray techniques.

In order to be able to account for a long enough borehole in the axisymmetri cal model special numerical tools have been developed. First special integratio schemes have been used to account for axisymmetrical boundary elements ver close to the vertical axis. A reduction factor of about 60 has been obtained by ac counting for the partial invariance of the tube in the assembling process provide that a regular mesh is used. Finally, special radiation conditions have been define at the end of the mesh in order to prevent the reflection of the guided wave. Th efficiency of this condition can be seen on either Figure 18 or 19 (see [27] for mor details).

8. Conclusion

As a conclusion of the chapter we hope to have been able to convince the reade that the general theoretical and numerical approaches proposed in Sections 2 to 5 lead to a very wide spectrum of applications as presented in Section 7. Thus th price paid in generalizing the standard SSI framework is largely overcome by th value of the results obtained in many different analyses. Being able to analyse a the same time SSI and site effects is of primary importance to assess the safety o

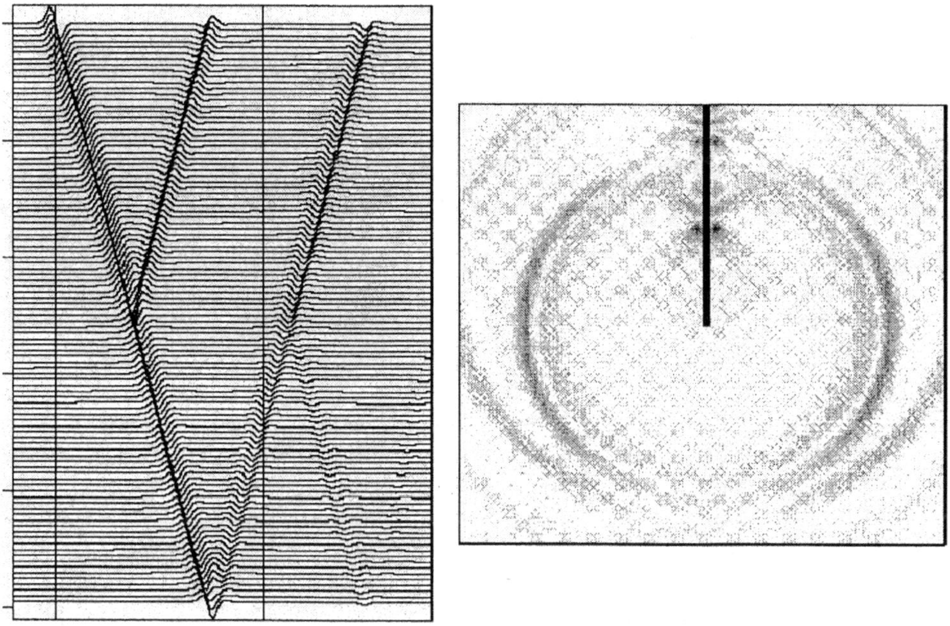

Figure 18. Fields emitted in a seismic experiment inside a borehole meshed with about 200 boundary elements: (left) the pressure field in the fluid as a function of time and distance along a borehole crossing an interface, (right) the displacement field in an homogeneous geological formation at a given time step (from [27]).

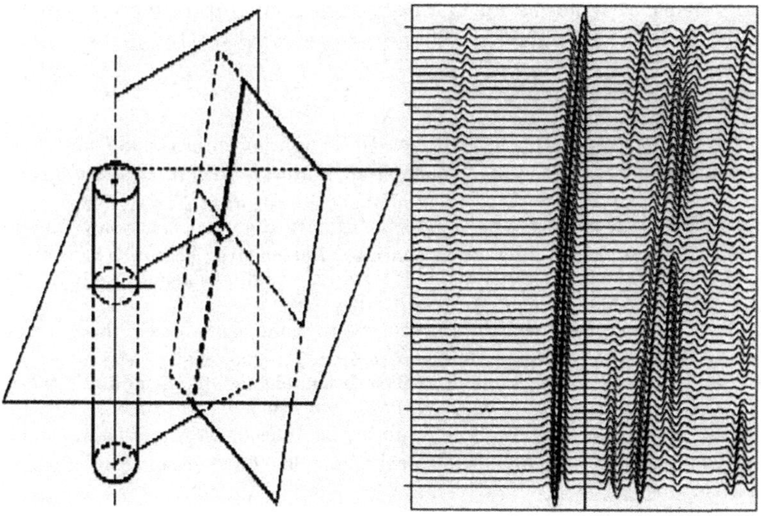

Figure 19. Pressure field in the borehole after reflection on an non horizontal interface : (left) layout, (right) pressure field as a function of time and distance along the borehole (from [27]).

large structures such as dams and to understand the seismic behaviour of a whole city resting on soft deposits.

Compared to a standard Finite Element technique that would discretise equivalently the entire domain under study, the hierarchical approach proposed here appears to be very valuable for at least two reasons. First it re-uses standard tools available in the structural dynamics community and combines them with propagation methods widely used in the geophysics community, taking advantage of both. It then can easily model basic physical phenomena occuring in practical situations and in particular resonant behaviour of the structures and propagation in an invariant propagation medium. Far field interactions are thus efficiently modelled.

Extension to heterogeneous and random media using either volume coupling or periodic models is the third main interest of this formulation. Indeed on the first hand it provides engineers with error-bounds on the results when the domain under study is known only in a statistical sense. On the other hand it can account for very important phenomena such as diffusion and localisation of waves in random media. Even though these phenomena are now recognized to play a major role in the generation and the propagation of seismic waves [43], they are currently ignored in usual practice.

The main drawback of the proposed approach is the linear assumption that seems to underly most of this work. In many cases, and in particular in the domain decomposition technique, this assumption can be relaxed. However space limitations have not allowed us to present these developments and the reader is referred to [67] for a time-frequency algorithm accounting for non linear contact conditions at the soil-structure interface or to [68] for the volume coupling between FD-BEM and TD-FEM whereas for strongly non-linear biphase constitutive laws of the soil, full TD-FEM has been preferred as proposed in [11, 17].

9. References

[1] T. Abboud, V. Mathis, and J.-C. Nedelec. Diffraction of an electromagnetic travelling wave by a periodic structure. In G. Cohen et al., editors, *Third international conference on Mathematical and Numerical Aspects of Wave Propagation*. SIAM, 1995.

[2] N. Abrahamson, J. Schneider, and J. Stepp. Empirical spatial coherency functions for application to soil-structure interaction analyses. *Earthquake Spectra*, 7(1):1–27, 1991.

[3] S. Ahmad and P. Banerjee. Multi-domain BEM for two-dimensional problems of elastodynamics. *Int. J. Num. Meth. Eng.*, 26:891–911, 1988.

[4] S. Ahmad and P. Banerjee. Inelastic transient elastodynamic analysis of three-dimensional problems by BEM. *Int. J. Num. Meth. in Eng.*, 29:371–390, 1990.

[5] S. Ahmad and P. K. Banerjee. Time-domain transient elastodynamic analysis of 3-d solids by BEM. *Int. J. Num. Meth. in Eng.*, 26:1709–1728, 1988.

[6] H. Antes and O. V. Erstoff. Analysis of absorption effects on the dynamic response of dam reservoir systems by boundary element methods. *Earthquake. Eng. Struct. Dyn.*, 15:1023–1036, 1987.

[7] H. Antes and O. V. Erstoff. Dynamic response of 2D and 3D block foundations on a halfspace with inclusions. *Soil Dyn. Earthquake Engrg*, 13:305–311, 1994.

[8] R. Apsel and J. Luco. Torsional response of a rigid embedded foundation. *J. Engrg. Mech. Div. ASCE*, 102:957–970, 1976.

[9] A. Askar and A. Cakmak. Seismic waves in random media. *Prob. Engrg. Mech.*, 3(3):124–129, 1988.

[10] D. Aubry, A. Baroni, and D. Clouteau. Seismic borehole modeling with a mixed boundary and finite-element method. In *Proc. SEG Los Angeles*, volume PP4.4, pages 751–754, 1994.

[11] D. Aubry, A. Baroni, D. Clouteau, A. Fodil, and A. Modaressi. Modeling of ballast behaviour in railway track. In *Int. Conf. on soil mechanics*, 1999.

[12] D. Aubry and D. Clouteau. A regularized boundary element method for stratified media. In G. Cohen and al., editors, *Math. and num. aspects of wave propagation, Proc. 1st Int. Conf.*, pages 660–668. SIAM, 1991.

[13] D. Aubry and D. Clouteau. *A subdomain approach to dynamic soil-structure interaction*, pages 251–272. Ouest Editions/AFPS, Nantes, 1992.

[14] D. Aubry, D. Clouteau, and G. Bonnet. Modelling of wave propagation due to fixed or mobile dynamic sources. In N. Chouw and G. Schmid, editors, *Workshop Wave '94, Wave propagation and Reduction of Vibrations*, pages 109–121. Berg-verlag, 1994.

[15] D. Aubry, D. Clouteau, J. Crepel, and F. Gillon. Approche par sous-domaines et lments finis de frontire des effets de site tridimensionnels. Rapport final M.R.T, appel d'offre 1988/ Gnie Civil 88.F.0863, Ecole Centrale de Paris, 1991.

[16] D. Aubry, D. Clouteau, J. Dompierre, and J. Svay. The seismic borehole problem, a mixed bem-fem modal approach in the frequency domain. In *Proc. Numerical Methods in Engineering'92*, Brussels, 1992.

[17] D. Aubry, D. Clouteau, and A. Modaressi. Interaction dynamique sol-structure. In *Proc. jour. Calcul tridimensionnel en geotechnique.* Presse des Ponts et Chausses, 1998.

[18] D. Aubry, D. Clouteau, and J. Svay. The seismic horizontal borehole problem, a mixed bem-fem modal approach in the frequency domain. In *2nd int. Conf. on Mathematical and Numerical Aspects of Wave Propagation*, Philadelphia, 1993. SIAM.

[19] D. Aubry, D. Clouteau, and J. Svay. Tube waves in horizontal boreholes. In *Proc. SEG Los Angeles*, volume BG1.8, pages 28–31, 1994.

[20] E. Balmès. Optimal Ritz vectors for component mode synthesis using the singular value decomposition. *AIAA Journal*, 34(6):1256–1260, 1996.

[21] A. Bamberger and T. Ha Duong. Formulation variationnelle espace-temps pour le calcul par potentiel retardé de la diffraction d'une onde acoustique (i). *Math. Meth. in Appl. Sc.*, 8:405–435, 1986.

[22] P. Banerjee and S. M. Manoon. A fundamental solution due to a periodic point force in the interior of an elastic half-space. *Earth Eng. Struct. Dyn.*, 19:91–105, 1990.

[23] P. Banerjee and R. Shaw. *Developments in Boundary Element Methods.* Elsevier Applied Science, Barking, Essex, U.K., 1979.

[24] P. Bard. Diffracted waves and displacement field over two-dimensional elevated topographies. *Geophys. J; R. Asrt. Soc.*, 71:731–760, 1982.

[25] P. Bard and M. Bouchon. The two dimensional resonance of sediment-filled valley. *Bull. Seism. Soc. Am.*, 75:517–541, 1985.

[26] P.-Y. Bard and J. Riepl-Thomas. *Wave propagation in complex geological structures and local effects on strong ground motion*, pages 38–95. Number ISBN 1-85312-744-2 in Advances in Earthquake Engineering. WIT Press, 1999.

[27] A. Baroni. *Modélisation du couplage sol-fluide pour la sismique entre puits.* PhD thesis, Ecole Centrale de Paris, Châtenay-Malabry, november 1996.

[28] T. Belytschko, C. H. S., and Y. Y. Lu. A variationally coupled finite element-boundary element method. *Comp. Struct.*, 33:17–20, 1989.

[29] J. Bérenger. A perfectly matched layer for the absorption of electromagnetic wave. *J. Comp. Phys.*, 114, 1994.

[30] C. Bernardi, N. Debit, and Y. Maday. Coupling finite element and spectral methods: First results. *Math. Comp.*, 189:21–39, 1990.

[31] D. E. Beskos. Boundary element methods in dynamic analysis. *Appl. Mech. Rev.*, 40:1–23, 1987.

[32] D. E. Beskos. Boundary element methods in dynamic analysis, part. II (1986–1996). *Appl. Mech. Rev.*, 50:149–197, 1997.

[33] P. Bisch, editor. *11th European conference on earthquake engineering*, Rotterdam, September 1998. Balkema.

[34] M. Bonnet. *Boundary integral equations methods for Solids and Fluids*. John Wiley and sons, 1999.

[35] M. Bonnet, G. Maier, and C. Polizzotto. Symmetric Galerkin boundary element method. *Appl. Mech. Rev.*, 1997.

[36] A. Bonnet-BenDhia and A. Tillequin. A generalized mode matching method for the junction of open waveguides. In A. Bermudez et al, editor, *Fifth international conference on Mathematical and Numerical Aspects of Wave Propagation*, pages 399–402. SIAM, 2000.

[37] A. Bossavit. Symmetry, groups and boundary value problems : a progressive introduction to non commutative harmonic analysis of partial differential equations in domains with geometrical symmetry. *Comp. Meth. in Appl. Mech. Engng.*, 56:167–215, 1986.

[38] M. Bouchon. A simple complete numerical solution to the problem of diffraction of SH waves by an irregular surface. *J. Acous. Soc. Am.*, 77:1–5, 1985.

[39] M. Bouchon. A numerical simulation of the acoustic and elastic wavefields radiated by a source on a fluid-filled borehole embedded in a layered medium. *Geophysics*, 58:475–481, 1993.

[40] M. Bouchon and K. Aki. Discrete wavenumber representation of seismic source wave field. *Bull. Seism. Soc. Am.*, 67:259–277, 1977.

[41] E. O. Brigham. *The Fast Fourier transform*. Prentice-Hall, 1974.

[42] C.A.Brebbia, L. Wrobel, and J. Telles. *Boundary element techniques. Theory and application in engineering*. Springer-Verlag, 1984.

[43] M. Campillo, L. Margerin, and K. Aki. *Seismology*, volume 144 of *Springer Tracts in Modern Physics*, pages 87–99. Springer, 1998.

[44] S. M. Candel. Simultaneous calculation of Fourier-Bessel transforms up to order n. *J. of Computational Physics*, 44:243–261, 1981.

[45] C. Canuto, M. Hussaini, A. Quarteroni, and T. A. Zang. *Spectral methods in fluid dynamics*. Springer Verlag, New York, 1988.

[46] X. Chang and D. Clouteau. Rponse sismique de barrages vote sous l'interaction sol-fluide-structure : influence de la rigidit du sol et des champs incidents. Rapport interne, Ecole Centrale de Paris, 1995.

[47] F. Chapel. Boundary element method applied to linear soil-structure interaction on a heterogeneous soil. *Earthquake Engrg. Struct. Dyn.*, 15:815–829, 1987.

[48] F. Chapel and C. Tsakaladis. Computation of the Green's functions of elastodynamics for a layered half space through a Hankel transform, applications to foundation vibration and seismology. In I. Kawamoto, editor, *Numerical Methods in Geomechanics*, pages 1311–1318. Balkema, 1985.

[49] K. Chen and T. Soong. Covariance properties of waves propagating in a random medium. *J. Acoust. Soc. Am.*, 49(5):1639–1642, 1971.

[50] L. Chernov. *Wave propagation in a random medium*. McGraw-Hill, New York, 1960.

[51] L. Chu, A. Askar, and A. Cakmak. Earthquake waves in a random medium. *Int. J. Num. Methods Geomech.*, 5:79–96, 1981.

[52] C. Chua, T. Balendra, and K. Lo. Groundborne vibrations due to trains in tunnels. *Earthquake Engineering and Structural Dynamics*, 21:445–460, 1992.

[53] C. Chua, K. Lo, and T.Balendra. Building response due to subway train traffic. *J. of Geotechnical Engineering*, 121(11), 1995.

[54] R. Clough and J. Penzien. *Dynamics of structures*. McGraw-Hill, New York, 1975.

[55] D. Clouteau. *Propagation d'ondes dans des milieux héérogènes, Application à la tenue d'ouvrages sous séismes*. PhD thesis, Ecole Centrale de Paris, 1990.

[56] D. Clouteau. Diffraction d'ondes lastiques sur une montagne, comparaison 2d et 3d. Rapport interne, Ecole Centrale de Paris, 1994.

[57] D. Clouteau. *Miss: Manuel scientifique, rev. 6.2.* Laboratoire MSSMat, Ecole Centrale de Paris, Châtenay-Malabry, 1999.

[58] D. Clouteau and D. Aubry. 3d seismic soil-fluid-structure interaction for arch dams including source and site effects. In *Proc. of Eurodyn93 Int. Conf.*, pages 1217–1224, Rotterdam, 1993. Balkema.

[59] D. Clouteau and D. Aubry. Site effects on 3d elevated topography. In A. Cakmak and C. Brebbia, editors, *Soil Dynamics and Earthquake Engineering VII*, pages 291–298, Southampton, 1995. Computational Mechanics Publications.

[60] D. Clouteau and D. Aubry. Modifications of the ground motion in dense urban areas. *Journal of Computational Acoustics*, 9(4):1–17, 2001.

[61] D. Clouteau, D. Aubry, M. Elhabre, and E. Savin. *Periodic and stochastic BEM for large structures embedded in an elastic half-space*, pages 91–102. Chapman & Hall, CRC Press, London, 1999.

[62] D. Clouteau, D. Aubry, and M. L. Elhabre. Periodic bem and fem-bem coupling : application to seismic behaviour of very long structures. *Comp. Mech.*, 25(6):567–577, 2000.

[63] D. Clouteau, D. Aubry, and E. Savin. Influence of free field variability on soil–structure interaction. In Bisch [33].

[64] D. Clouteau, D. Aubry, and B. Tardieu. Influences of site effects and incident fields on the seismic response of an arch dam. In G. Duma, editor, *Earthquake Engineering, Proc. 10th European conf., 28 August – 2 September 1994, Wien, Austria*, pages 1837–1842, Rotterdam, 1994. Balkema.

[65] D. Clouteau, A. Baroni, and D. Aubry. Boundary integrals and ray method coupling for seismic borehole modeling. In J. A. DeSanto, editor, *Fourth international conference on Mathematical and Numerical Aspects of Wave Propagation*, page 768. SIAM, 1998.

[66] D. Clouteau, G. Degrande, and G. Lombaert. Some theoretical and numerical tools to model traffic induced vibrations. In S. Chouw, editor, *WAVE2000 - Wave Propagation, Moving Load, Vibration Reduction*, pages 13–28, Rotterdam, December 2000. Balkema.

[67] D. Clouteau and G. Devesa. Dcollement des fondations sous chargement sismique : mthodes temporelle et temps/frquence. In J. B. et al., editor, *Actes 5ème Colloque National en Calcul des Structures*, volume 2, pages 1065–1073, Teknea, Toulouse, 2001.

[68] D. Clouteau, G. Devesa, and G. Jacquart. Couplage fem-bem : applications pratiques en génie parasismique. In Ladevèze et al. [128], pages 215–220.

[69] D. Clouteau and E. Savin. Influence of soil variability on soil–structure interaction. In Bisch [33].

[70] D. Clouteau and B. Tardieu. Calculs sismiques tridimensionnels d'un barrage vote : effets de site et interaction sol-fluide-structure. In *Proc. 2me Colloque national de gnie parasismique.* APFS, 1993.

[71] D. M. Cole, D. D. Kosloff, and J. B. Minster. A numerical boundary integral equation method for elastodynamics i. *Bull. Seism. Soc. Am.*, 68:1331–1357, 1978.

[72] R. Craig and M. Bampton. Coupling of substructures for dynamic analysis. *A.I.A.A. J. Aircraft*, 6(7):1313–1319, 1968.

[73] F. de Barros and J. Luco. Identification of foundation impedance functions and soil properties from vibration tests of the Hualien containment model. *Soil Dyn. Earthquake Engrg.*, 14:229–248, 1995.

[74] A. de La Bourdonnaye, C. Farhat, A. Macedo, F. Magoulès, and F. Roux. A non-overlapping domain decomposition method for the exterior Helmholtz problem. *Contempory Mathematics*, 218:42–66, 1998.

[75] G. Degrande, G. D. Roeck, P. V. den Broeck, and D. Smeulders. Wave propagation in layered dry, saturated and unsaturated poroelastic media. *International Journal of Solids and Structures*, 35(34-35), 1998.

[76] P. V. den Broeck and G. D. Roeck. The vertical receptance of track including soil-structure. In L. Fryba and J. Naprstek, editors, *Proceedings of the 4th European Conference on Structural Dynamics: Eurodyn '99*, pages 837–842, Rotterdam, June 1999. A.A. Balkema.

[77] A. Der Kiureghian. A coherency model for spatially varying ground motion. *Earthquak Engng. and Struct. Dyn.*, 25:99–111, 1996.
[78] G. Devesa. Oprateurs code_aster lire_miss_3d impr_miss_3d : documentation d'utilisatio code_aster. Technical Report U4.78.01 et U4.28.01, E.d.F/D.E.R, 1998.
[79] J. Dominguez and O. Maseo. Earthquake analysis of arch dams I: Dam foundation inter action. *J. Engrg. Mech.*, 119(3):496–512, 1993.
[80] J. Dompierre. *Modlisation de la sismique de puit en axisymtrique*. PhD thesis, Ecol Centrale de Paris, 1994.
[81] M. Dravinski and T. Mossessian. Scattering of plane harmonic p, sv and rayleigh wave by dipping layers of arbitrary shape. *Bull. Seism. Soc. Am.*, 77:212–235, 1987.
[82] M. Elhabre. *Interaction sismique Sol-Fluide-Structure*. PhD thesis, Ecole Centrale d Paris, 2000.
[83] M.-L. Elhabre, D. Clouteau, and D. Aubry. Seismic behaviour of diaphragm walls. I Bisch [33].
[84] J. M. Emperador and J. Dominguez. Dynamic response of axisymmetric embbedde foundations. *Earth. Eng. Struct. Dyn.*, 18:1105–1117, 1989.
[85] B. Engquist and A. Majda. Radiation boundary conditions for acoustic and elastic wav calculations. *Comm. Pure Appl. Math.*, 32:313–357, 1979.
[86] S. K. S. et al. The Mexico earthquake of September 19, 1985 – a study of amplificatio of seismic waves in the valley of Mexico with respect to a hill zone site. *Earth. Spectra* 4(4):653–673, 1988.
[87] E. Faccioli, F. Maggio, A. Quarteroni, and A. Tagliani. Spectral-domain decompositio for the solution of acoustic and elastic wave equations. *Geophysics*, 61:1160–1174, 1996.
[88] C. Farhat and F. Roux. A method for finite element tearing and interconnecting and it parallel solution algorithm. *Int. J. Num. Meth. in Eng.*, 32:1205–1227, 1991.
[89] J. Favre. Errors in geotechnics and their impact on safety. *Computers & Structures* 67:37–45, 1998.
[90] M. G. Floquet. Sur les quations diffrentielles linaires coefficients priodiques. *Ann. d l'Ecole Normale*, 12, 1883.
[91] E. Fukuyama and R. Madariaga. Rupture dynamics of a planar fault in a 3d elasti medium: rate- and slip-weakening friction. *Bull. Seism. Soc. Am.*, 88:1–17, 1998.
[92] L. Geli, P. Bard, and B. Jullien. The effect of topography on earthquake ground motion a review and new results. *Bull. Seism. Soc. Am.*, 78:42–63, 1988.
[93] R. Ghanem and R. Kruger. Numerical solution of spectral stochatsic finite element sys tems. *Comp. Methods Appl. Mech. Engrg.*, 129:289–303, 1996.
[94] R. Ghanem and P. Spanos. *Stochastic finite elements: a spectral approach*. Springe Verlag, New York, 1991.
[95] Ghanem R.G. Ingredients for a general purpose stochastic finite elements implementatio *Comp. Methods Appl. Mech. Engrg.*, 168:19–34, 1999.
[96] R. Glowinski, T. Pan, and J. Periaux. A fictitious domain method for Dirichlet probler and applications. *Comp. Mech. in Appl. Mech. and Eng.*, pages 283–303, 1994.
[97] Grabe J. Spatial variation of soil stiffness : spectral density approach. *Soil Dyn. Earth quake Engrg.*, 13:25–29, 1994.
[98] H. Grundmann, M. Lieb, and E. Trommer. The response of a layered half-space to traffi loads moving along its surface. *Archive of Applied Mechanics*, 69, 1999.
[99] J. Gutierrez and A. Chopra. A substructure method for earthquake analysis of structure including structure-soil interaction. *Earthquake Engineering and Structural Dynamic* 6:51–69, 1978.
[100] J. Guy and B. Mangeot. Use of group theory in various integral equations. *SIAM Appl. Math.*, 40:390–399, 1981.
[101] W. Hackbusch and Z. Nowak. On the fast matrix multiplication in the BEM by pan clustering. *Numer. Math.*, 54:463–491, 1989.
[102] H. Hao and S. Zhang. Spatial ground motion effect on relative displacement of adjace building structures. *Earthquake Engrg. Struct. Dyn.*, 28:333–349, 1999.

[103] R. S. Harichandran and E. Vanmarke. Stochastic variation of earthquake ground motion in space and time. *ASCE J. Engrg. Mech.*, 112(2):154–174, 1986.

[104] R. S. Harichandran. Estimating the spatial variation of earthquake ground motion from dense array recordings. *Structural Safety*, 10:219–233, 1991.

[105] N. A. Haskell. The dispersion of surface waves on multilayered media. *Bull. seism. Soc. Am.*, 43:17–34, 1953.

[106] M. Hoshiya and K. Ishii. Evaluation of kinematic interaction of soil-foundation systems by a stochastic model. *Soil Dyn. Earthquake Engrg.*, 2(3):128–134, 1983.

[107] T. Hughes. *The Finite Element Method, Linear Static and Dynamic Finite Element Analysis.* Prentice-Hall International, 1987.

[108] J. Hujeux. *Une loi de comportement pour le chargement cyclique des sols*, pages 287–302. Presses Ponts et Chausses, Paris, 1985.

[109] H. Hunt. Stochastic modelling of traffic-induced ground vibration. *J. of Sound and Vibration*, 144(1), 1991.

[110] D. Inman. *Engineering Vibration.* Prentice-Hall, Englewood Cliffs, N.J., 1994.

[111] A. Ishimaru. *Wave propagation and scattering in random media.* Oxford University Press, 1978.

[112] F. Janod and O. Coutant. Seismic response of three-dimensional topographies using a time-domain Boundary Element Method. *Geophys. J. Int.*, 142:603–614, 2000.

[113] C. Johnson and J. Nédélec. On the coupling of boundary integral and finite element methods. *Math. Comp.*, 35:1036–1079, 1980.

[114] M. Kahan, R. Gibert, and P. Bard. Influence of seismic waves spatial variability on bridges : a sensitivity approach. *Earthquake Engrg. Struct. Dyn.*, 25:795–814, 1996.

[115] D. L. Karabalis. A simplified 3D time-domain BEM for dynamic soil-structure interaction problems. *Engng. Anal. with Bound. Elem.*, 8:139–145, 1991.

[116] E. Kausel and A. Pais. Stochastic deconvolution of earthquake motions. *ASCE J. Eng. Mech.*, 113(2):266–277, 1987.

[117] E. Kausel and R. Peek. Dynamic loads in the interior of a layered stratum : an explicit solution. *Bull. Seism. Soc. Am.*, 72:1459–1481, 1982.

[118] E. Kausel and J. Roesset. Dynamic stiffness of circular foundations. *J. eng. Div. ACSE*, CI:771–785, 1975.

[119] E. Kausel and J. Roesset. Stiffness matrices for layered soils. *Bull. Seism. Soc. Am.*, 71(6):1743–1761, 1981.

[120] H. Kawase. Time-domain response of semi-circular canyon for incident P, SV and Rayleigh waves calculated by discrete wavenumber Boundary Element Method. *Bull. Seism. Soc. Am.*, 78:1415–1437, 1988.

[121] H. Kawase and K. Aki. A study on the response of soft basin for incident S, P and Rayleigh waves with special reference to long duration observed in Mexico city. *Bull. Seism. Soc. Am.*, 79:1361–1382, 1989.

[122] B. L. N. Kennett. *Seismic wave propagation in stratified media.* Cambridge University Press, 1983.

[123] G. Kergourlay, E. Balms, and D. Clouteau. Model reduction for efficient FEM/BEM coupling. In *ISMA, Leuven*, 2000.

[124] W. Kohler, G. Papanicolaou, and B. White. Reflection of waves generated by a point source over a randomly layered medium. *Wave Motion*, 13:53–87, 1991.

[125] W. Kohler, G. Papanicolaou, and B. White. Localization and mode conversion for elastic waves in randomly layered media I. *Wave Motion*, 23:1–22, 1996.

[126] D. P. N. Kontoni and D. E. Beskos. Application of DR-BEM to inelastic dynamic problems. In C. A. Brebbia, J. Dominguez, and F. Paris, editors, *Boundary Elements XIV.* Computational Mechanics Publications, 1992.

[127] P. Krée & C. Soize. *Mathematics of random phenomena.* MIA, Reidel Publishing, Dordrecht, 1986.

[128] P. Ladevèze, D. G. Degeorges, and M. Raous, editors. *Actes* 4ème *Colloque National en Calcul des Structures*, volume 1, Teknea, Toulouse, 1999.

[129] C. Lage and C. Schwab. Wavelet Galerkin algorithms for boundary integral equations. *J. Sci. Comput.*, 20(6):2195–2222, 1999.

[130] P. Lewicki, R. Burridge, and G. Papanicolaou. Pulse stabilisation in a strongly heterogeneous layered medium. *Wave Motion*, 20:177–195, 1994.

[131] W. Liu, T. Belytschko, and A. Mani. Random field finite elements. *Int. J. Num. Methods Engrg.*, 23:1831–1845, 1986.

[132] C. Loh, J. Penzien, and Y. Tsai. Engineering analysis of SMART-1 array accelerograms. *Earthquake Engrg. Struct. Dyn.*, 10:575–591, 1982.

[133] G. Lombaert, G. Degrande, and D. Clouteau. Deterministic modelling of free field traffic induced vibrations. *Soil Dyn. and earth. eng.*, 19(7):473–488, 2000.

[134] J. E. Luco. Linear soil-structure interaction. Technical Report UCRL-15272, Lawrence Livermore National Laboratory, 1980.

[135] J. E. Luco and R. J. Apsel. On the Green's functions for a layered half-space, Part I and II. *Bull. Seism. Soc. Am.*, 73:909–950, 1983.

[136] G. Maier, M. Diligenti, and A. Carini. A variational approach to boundary element elastodynamic analysis and extension to multidomain problems. *Comp. Meth. in Appl. Mech. Engng.*, 92:193–213, 1991.

[137] G. Manolis. Shear wave propagation in a stochastic halfplane using Green's function. In C.A. Brebbia et al., editor, *Boundary Elements XIV, vol. 2*, pages 123–134, Computational Mechanics Publications, Southampton, 1992.

[138] G. Manolis and R. Shaw. Green's function for the vector wave equation in a mildly heterogeneous continuum. *Wave Motion*, 24:59–83, 1996.

[139] G. D. Manolis. A comparative study on three boundary element method approaches to problems in elastodynamics. *Int. J. Num. Meth. in Eng.*, 19:73–91, 1983.

[140] G. D. Manolis and R. Shaw. Harmonic wave propagation through viscoelastic heterogeneous media exhibiting mild stochasticity – I. Fundamental solutions. *Soil Dyn. Earthquake Engrg.*, 15:119–127, 1996.

[141] H. Matthies and C. Bucher. Finite elements for stochastic media problems. *Comp. Methods Appl. Mech. Engrg.*, 168:3–17, 1999.

[142] R. McNeal. A hybrid method of component mode synthesis. *Computers and structures*, 1(4):581–601, 1971.

[143] L. Meirovitch. *Elements of Vibration Analysis*. McGraw-Hill Inc., New York, N.Y., 1989.

[144] S. Muscat. L'effet ville. Master's thesis, Ecole Centrale de Nantes, 1995.

[145] J. Nédélec and J. Planchard. Une méthode d'éléments finis de pour le résolution numérique d'un problème extérieur dans $\mathbb{R}^3$. *R.A.I.R.O.*, 7:105–129, 1973.

[146] R. Paolucci. Soil-structure interaction effects on an instrumented building in Mexico city. *European Earthquake Engineering*, 3:33–44, 1993.

[147] M. Papadrakakis and A. Kotsopulos. Parallel solution of stochastic finite element methods using Monte-Carlo simulation. In S. Idelsohn, E. Oñate & E. Dvorkin, editor, *Computational mechanics. New trends and applications*, Barcelona, 1998.

[148] G. Papanicolaou. Mathematical problems in geophysical wave propagation. In *Documenta Mathematica, proc. Int. Congress of Mathematicians, Extra Volume ICM 98*, volume I, pages 241–265, Berlin, 1998.

[149] A. Pecker and J. Teyssandier. Seismic design for the foundation of the rion-antirion bridge. In *Proc. Institution of Civil Engineers, Geotechnical Engineering*, volume 131, pages 4–11, 1998.

[150] H. A. Pedersen, B. L. Brun, D. Hatzfeld, M. Campillo, and P.-Y. Bard. Ground motion amplitude across ridges. *Bull. Seism. Soc. Am.*, 85:1786–1800, 1994.

[151] H. A. Pedersen, F. J. Sanchez-Sesma, and M. Campillo. Three- dimensional scattering by two-dimensional topographies. *Bull. Seism. Soc. Am.*, 84:1169–1183, 1994.

[152] A. Peirce and E. Siebrits. Stability analysis and design of time-stepping schemes for general elastodynamic boundary element models. *Int. J. Num. Meth. in Eng.*, 40:319–342, 1997.

[153] R. Popescu, J. Prévost, and G. Deodatis. Effects of spatial variability on soil liquefaction : some design recommendations. *Géotechnique*, 47(5):1019–1036, 1997.

[154] J. H. Prevost. Dynaflow - a finite element analysis program for the static and transient response of linear and non-linear two and three-dimensional systems. Technical report, Dept. of Civil Eng., Princeton University, 1981.

[155] E. Reinoso and M. Ordaz. Spectral ratios for Mexico City from free-field recordings. *Earth. Spectra*, 15(2):273–295, 1999.

[156] J. Rizzo and D. J. Shippy. A boundary integral approach to potential and elasticity problems for axisymmetric bodies with arbitrary boundary conditions. *Mechanics Research Communications*, 6(2):99–103, 1979.

[157] V. Rokhlin. Rapid solution of integral equations of classical potential theory. *J. Comp. Phys.*, 60:187–207, 1985.

[158] R. Y. Rubinstein. *Simulation and the Monte-Carlo method.* John Wiley & Sons, New York, 1981.

[159] F. Sanchez-Sesma. Diffraction of elastic waves by three- dimensional surface irregularities. *Bull. Seism. Soc. Am.*, 73:1621–1636, 1983.

[160] F. Sanchez-Sesma and I. A. Esquivel. Ground motion on alluvial valleys under incident plane SH waves. *Bull. Seism. Soc. Am.*, 69:1107–1120, 1979.

[161] F. Sanchez-Sesma, L. Perez-Rocha, and S. Chavez-Terez. Diffraction of elastic waves by three-dimensional surface irregularities. part ii. *Bull. seism. Soc. Am.*, 79:101–112, 1989.

[162] F. J. Sanchez-Sesma and M. Campillo. Diffraction of P, SV and Rayleigh waves by topographic features : a boundary integral formulation. *Bull. Seism. Soc. Am.*, 81:2234–2253, 1991.

[163] S. Savidis and C. Vrettos. Dynamic soil-structure interaction for foundations on nonhomogeneous soils. In G. Duma, editor, *Earthquake Engineering, Proc. 10th European conf., 28 August – 2 September 1994, Wien, Austria*, pages 599–609, Balkema, Rotterdam, 1994.

[164] E. Savin. *Effet de la variabilité du sol et du champ incident en interaction sismique sol-structure.* PhD thesis, Ecole Centrale de Paris, 1999.

[165] E. Savin and D. Clouteau. Coupling a bounded domain and an unbounded heterogeneous domain for elastic wave propagation in three-dimensional random media. *Int. J. Num. Meth. in Eng.*, In press, 2001.

[166] E. Savin, D. Clouteau, and D. Aubry. Modélisation numérique stochastique de l'interaction dynamique sol-structure. In Ladevèze et al. [128], pages 221–226.

[167] M. Schanz and H. Antes. A new visco- and elastodynamic time domain boundary element formulation. *Comp. Mech.*, 20:452–459, 1997.

[168] M. Shinozuka and G. Deodatis. Response variability of stochastic finite element systems. *ASCE J. Engrg. Mech.*, 114(3):499–519, 1988.

[169] C. Soize. A nonparametric model of random uncertainties for reduced matrix models in structural dynamics. *Probabilistic Engineering Mechanics*, 15(3):277–294, 2000.

[170] C. Song and J. P. Wolf. The scaled boundary-finite element method – alias consistent infinitesimal finite-element cell method – for elastodynamics. *Comp. Meth. in Appl. Mech. Engng.*, 147:329–355, 1997.

[171] M. Soulié, P. Montes, and V. Silvestri. Modelling spatial variability of soil parameters. *Can. Geotech. J.*, 27:617–630, 1990.

[172] P. D. Spanos and R. Ghanem. *Stochastic finite elements: a spectral approach.* Springer–Verlag, 1991.

[173] A. Stamos and D. Beskos. Dynamic analysis of large 3D underground structures by the BEM. *Earthquake Engineering and Structural Dynamics*, pages 1–18, 1995.

[174] A. Stamos and D. Beskos. 3-D seismic response analysis of long lined tunnels in half-space. *Soil Dynamics and Earthquake Engineering*, 15, 1996.

[175] B. W. Suter. Foundations of Hankel transform algorithms. *Quart. of App. Math.*, XLIX(2):267–279, 1991.

[176] J. Svay. *Modlisation de la sismique de puits en puits-horizontal.* PhD thesis, Ecole Centrale de Paris, 1995.

[177] P. L. Tallec. *Domain decomposition methods in computational mechanics.* Computational Mechanics Advances. North-Holland, 1994.

[178] I. Tarman. An improved dynamical approximation to Boussinesq equation using Karhunen-Loeve basis. *Comp. Methods Appl. Mech. Engrg.*, 144:153–162, 1997.
[179] W. T. Thomson. Transmission of elastic waves through a stratified solid medium. *J. Appl. Phys*, 21:89–93, 1950.
[180] M. Toksöz, A. Dainty, and E. Charrette. Spatial variation of ground motion due to lateral heterogeneity. *Structural Safety*, 10:53–77, 1991.
[181] F. Toubalem , B. Zeldin, F. Thouverez & P. Labbé. Vertical excitation of stochastic soil-structure interaction systems. *ASCE J. Geotech. Geoenvir. Engrg.*, 125(5):349–356, 1999.
[182] C. Tsakalidis. *Diffraction d'ondes sismiques sur les structures sur pieux et fonctions de Green du sol multicouche.* PhD thesis, Ecole Centrale de Paris, Châtenay-Malabry, july 1985.
[183] E. Vanmarcke. Probabilistic modeling of soil profiles. *ASCE J. Geotech. Engrg.*, 103(GT11):1227–1246, 1977.
[184] R. F. Vogt, J. P. Wolf, and H. Bachmann. Wave scattering by a canyon of arbitrary shape in a layered half-space. *Earth. Eng. Struct. Dyn.*, 16:803–812, 1988.
[185] O. Von Estorff, A. L. Pais, and E. Kausel. Some observations on time-domain and frequency domain boundary elements. *Int. J. Num. Meth. in Eng.*, 29:785–800, 1990.
[186] C. Y. Wang and J. D. Achenbach. Elastodynamic fundamental solution for anisotropic solids. *Geophy. J. Int.*, 118:384–392, 1994.
[187] H. Wang and P. K. Banerjee. Multi-domain general axisymmetric stress analysis by BEM. *Int. J. Num. Meth. in Eng.*, 30:115–131, 1990.
[188] W. X. Wang. Expansion of Green's function in a cylindrical coordinate system. *Quaterly of Appl. Math.*, 48:499–501, 1990.
[189] G. Watson. *A treatise on the theory of Bessel functions.* Cambridge University Press, 1966.
[190] G. Wilson, H. Saurenman, and J. Nelson. Control of groundborne noise and vibration. *J. of Sound and Vibration*, 87(2), 1983.
[191] A. Wirgin and P. Bard. Effects of buildings on the duration and amplitude of ground motion in Mexico City. *Bull. Seism. Soc. Am.*, 86:914–920, 1996.
[192] J. Wolf. *Dynamic soil-structure interaction.* Prentice-Hall, Englewood Cliffs, 1985.
[193] J. Wolf and G. Darbre. Dynamic stiffness matrix of soil by the boundary - element method : conceptual aspects. *Earth. Eng. and Struc. Dyn.*, 12:385–400, 1984.
[194] H. L. Wong and J. E. Luco. Dynamic response of rigid foundations of arbitrary shape. *Earth. Eng. Struct. Dyn.*, 4:579–587, 1976.
[195] F. Yamazaki, M. Shinozuka, and G. Dasgupta. Neumann expansion for stochastic finite element analysis. *ASCE J. Engrg. Mech.*, 114(8):1335–1354, 1988.
[196] A. Zerva. On the spatial variation of seismic ground motions and its effects on lifelines *Eng. Struc.*, 16:534–546, 1994.
[197] A. Zerva and O. Zhang. Correlation patterns in characteristics of spatially variable seismic ground motions. *Earthquake Engrg. Struct. Dyn.*, 26:19–39, 1997.
[198] O. Zienkiewicz. *The finite element method in engineering science.* McGraw Hill, 1979.

10. Appendix : Mathematical Results and Formulae

10.1. MATHEMATICAL PROPERTIES OF THE VARIATIONAL BIE

Compared to standard collocation approaches, the variational BIE (65) given in Section 4.1. has the main advantage to provide us with uniqueness results under some regularity conditions on Σ_s. Let $\boldsymbol{V}$ be the space of displacement fields having a finite energy on $\mathbb{R}^{3+}$ and $\boldsymbol{V}_\Sigma$ the functional space containing the traces of such displacement fields on Σ. Provided with these functional spaces and the classical

L^2 norm on either $\mathbb{R}^{3+}$ and Σ, $\boldsymbol{V}'_\Sigma$ is the standard dual space containing surface traction $\boldsymbol{t}$ having a finite virtual work when applied to these displacements fields. This property gives the natural norm in these dual spaces defined as follows as long as Σ_s is regular enough :

$$||\boldsymbol{t}|| \quad = \quad \sup_{\boldsymbol{v} \in \boldsymbol{V}} \frac{\int_{\Sigma_s} \boldsymbol{t} \cdot \overline{\boldsymbol{v}} dS}{||\boldsymbol{v}||} \tag{118}$$

Where $||\boldsymbol{v}||$ is the square root of the energy of $\boldsymbol{v}$ on $\mathbb{R}^{3+}$. Then defining $\boldsymbol{u_t}$ as the displacement fields generated inside $\mathbb{R}^{3+}$ by forces $\boldsymbol{t}$ applied on Σ_s e.g. satisfying for all $\boldsymbol{v} \in \boldsymbol{V}$:

$$\int_{\mathbb{R}^{3+}} \left\{ \boldsymbol{\sigma}(\boldsymbol{u_t}) : \boldsymbol{\epsilon}(\overline{\boldsymbol{v}}) - \rho\omega^2 \boldsymbol{u_t} \cdot \overline{\boldsymbol{v}} \right\} dV = \int_{\Sigma_s} \boldsymbol{t} \cdot \overline{\boldsymbol{v}} dS \tag{119}$$

it can be noticed that $||\boldsymbol{t}||$ is equivalent to the energy norm of $\boldsymbol{u_t}$ as long as damping is accounted for. As a consequence the bilinear form $\mathcal{U}_\Sigma$ appears to be coercive in $\boldsymbol{V}'_\Sigma$ since :

$$\overline{\mathcal{U}_\Sigma(\boldsymbol{t}, \overline{\boldsymbol{t}})} = \int_{\mathbb{R}^{3+}} \left\{ \boldsymbol{\sigma}(\boldsymbol{u_t}) : \overline{\boldsymbol{\epsilon}(\boldsymbol{u_t})} - \rho\omega^2 \boldsymbol{u_t} \cdot \overline{\boldsymbol{u_t}} \right\} dV \tag{120}$$

The main advantage of the proposed formulation is that this result originally shown in [145] for the Laplace problem still holds when the Green's function does not have an analytical expression.

10.1.1. *Coupling on a Volume*

A similar procedure can now be applied to the volume coupling proposed in Section 4.5.. Indeed $\boldsymbol{V}_s$ being the space of fields having a finite energy on Ω_s and $\boldsymbol{V}'_s$ being its dual space equipped with the following natural norm defined for any $\boldsymbol{\phi}$ in $\boldsymbol{V}'$ by :

$$||\boldsymbol{\phi}|| \quad = \quad \sup_{\boldsymbol{v} \in \boldsymbol{V}_s} \frac{\int_{\Omega_s} \boldsymbol{\phi} \cdot \overline{\boldsymbol{v}} dV}{||\boldsymbol{v}||} \tag{121}$$

This norm is equivalent to the elastic energy in Ω_s of $\boldsymbol{u_\phi}$, the displacement field generated by the applied loads $\boldsymbol{\phi}$ on Ω_s^{b} e.g. satisfying for any $\boldsymbol{v}$ in $\boldsymbol{V}_s$:

$$\int_{\Omega_s} \left\{ \boldsymbol{\sigma}(\boldsymbol{u_\phi}) : \boldsymbol{\epsilon}(\overline{\boldsymbol{v}}) - \rho\omega^2 \boldsymbol{u_\phi} \cdot \overline{\boldsymbol{v}} \right\} dV = \int_{\Omega_s} \boldsymbol{\phi} \cdot \overline{\boldsymbol{v}} dS \tag{122}$$

Provided with some regularity assumption on Ω_s^{b} one can then conclude the coercivity of the bilinear form $\mathcal{U}^{\mathrm{b}}$ defined in (83) since :

$$\overline{\mathcal{U}^{\mathrm{b}}(\boldsymbol{\phi}, \overline{\boldsymbol{\phi}})} = \int_{\Omega_s} \left\{ \boldsymbol{\sigma}(\boldsymbol{u_\phi}) : \overline{\boldsymbol{\epsilon}(\boldsymbol{u_\phi})} - \rho\omega^2 \boldsymbol{u_\phi} \cdot \overline{\boldsymbol{u_\phi}} \right\} dV \tag{123}$$

10.2. PROPER NORM FOR RESIDUAL FORCES

The strategy developed in Appendix 10.1. to define a proper norm in the space of boundary traction $\boldsymbol{V}_\Sigma$ can be applied to define a proper norm for the residual forces when looking for a reduced model as proposed in Section 3.5.1.. Indeed, it can be noticed that the inertial term in definition (122) is not compulsary in the definition of the norm and the strain energy is a norm in itself as long as rigid body motions are excluded. This property is used to define a proper norm for the residual forces $\boldsymbol{r}$ defined in Section 3.5.1.. Indeed defining $\boldsymbol{u}_{\boldsymbol{r}}^o$ as the static displacement field on Ω_b due to applied forces $\boldsymbol{r}$ e.g. satisfying for all $\boldsymbol{v}_b \in \boldsymbol{V}_b$:

$$\int_{\Omega_b} \boldsymbol{\sigma}_b(\boldsymbol{u}_{\boldsymbol{r}}^o) : \boldsymbol{\epsilon}(\overline{\boldsymbol{v}}_b) dV = \int_{\Omega_b} \boldsymbol{r} \cdot \overline{\boldsymbol{v}_b} dS \tag{124}$$

one gets the norm of $\boldsymbol{r}$ in terms of the strain energy of the static displacement field $\boldsymbol{u}_{\boldsymbol{r}}^o$. Of course the exact field $\boldsymbol{u}_{\boldsymbol{r}}^o$ is not known, but a good approximation of this field can be computed on a refined mesh giving a good estimate of the norm of the residual forces, $\boldsymbol{r}$, and then on the norm of the error on the reduced model

10.3. MATRICES FOR THE REFLECTION-TRANSMISSION SCHEME

$$\mathbf{A}^{P-SV} = \begin{pmatrix} 0 & 2\kappa^2 - 1 & \mu^{-1}\kappa^2 & 0 \\ -\Delta' & 0 & 0 & \mu^{-1} \\ -\rho\omega^2 & 0 & 0 & -1 \\ 0 & 4\mu\Delta'(\kappa^2-1) - \rho\omega^2 & \Delta'(2\kappa^2 - 1) & 0 \end{pmatrix} \tag{125}$$

$$\mathbf{A}^{SH} = \begin{pmatrix} 0 & \mu^{-1} \\ -\mu\Delta' - \rho\omega^2 & 0 \end{pmatrix} \qquad \kappa^2 = \frac{\mu}{\lambda + 2\mu} \tag{126}$$

$$\widetilde{\mathbf{b}}_m^{P-SV} = \begin{pmatrix} u_z \\ \mathrm{div}'\boldsymbol{u}/(ik) \\ \sigma_{zz}/(i\omega) \\ \mathrm{div}'\boldsymbol{t}/(-k\omega) \end{pmatrix} \qquad \widetilde{\mathbf{g}}_m^{P-SV} = \frac{\delta(z - z_s)}{i\omega} \begin{pmatrix} 0 \\ 0 \\ a_z \\ a_x \cos\theta + a_y \sin\theta \end{pmatrix} \tag{127}$$

$$\widetilde{\mathbf{b}}_m^{SH} = \begin{pmatrix} \mathrm{rot}'\boldsymbol{u}/(ik) \\ \mathrm{rot}'\boldsymbol{t}/(-k\omega) \end{pmatrix} \qquad \widetilde{\mathbf{g}}_m^{SH} = \frac{\delta(z - z_s)}{i\omega} \begin{pmatrix} 0 \\ a_y \cos\theta - a_x \sin\theta \end{pmatrix} \tag{128}$$

$$\widetilde{\mathbf{A}}^{P-SV} = ik_s \begin{pmatrix} 0 & p(2\kappa^2 - 1) & \frac{\kappa^2}{\rho c_s} & 0 \\ -p & 0 & 0 & \frac{1}{\rho c_s} \\ \rho c_s & 0 & 0 & -p \\ 0 & \rho c_s(1 - 4p^2(1-\kappa^2)) & p(2\kappa^2 - 1) & 0 \end{pmatrix} \tag{129}$$

$$\widetilde{\mathbf{A}}^{SH} = ik_s \begin{pmatrix} 0 & \frac{1}{\rho c_s} \\ \rho c_s(1 - p^2) & 0 \end{pmatrix} \qquad p = \frac{k}{k_s} \qquad k_\alpha = \frac{\omega}{c_\alpha} \tag{130}$$

$$\gamma_s = \sqrt{1 - p^2} \qquad \gamma_p = \sqrt{\kappa^2 - p^2} \qquad \tilde{z} = k_s z \tag{131}$$

$$\begin{array}{rcl} \mathbf{D}_m^{P-SV}(z) & = & \mathbf{Diag}(e^{i\gamma_p \tilde{z}}, e^{i\gamma_s \tilde{z}}, e^{-i\gamma_p \tilde{z}}, e^{-i\gamma_s \tilde{z}}) \\ \mathbf{D}_m^{SH}(z) & = & \mathbf{Diag}(e^{i\gamma_s \tilde{z}}, e^{-i\gamma_s \tilde{z}}) \end{array} \tag{132}$$

$$\mathbf{M}_m^{P-SV}(z) = k_s \begin{pmatrix} \gamma_p & p & -\gamma_p & p \\ p & -\gamma_s & p & i\gamma_s \\ \rho c_s(2p^2-1) & -2\rho c_s p \gamma_s & \rho c_s(2p^2-1) & 2\rho c_s p \gamma_s \\ -2\rho c_s p \gamma_p & \rho c_s(1-2p^2) & 2\rho c_s p \gamma_p & \rho c_s(1-2p^2) \end{pmatrix} \tag{133}$$

$$\mathbf{M}_m^{SH}(z) = ik_s \begin{pmatrix} 1 & 1 \\ \rho c_s \gamma_s & -\rho c_s \gamma_s \end{pmatrix} \tag{134}$$

10.4. HANKEL TRANSFORM

The Hankel transform of order n denoted by $\mathcal{H}_n$ of any function $f(r)$ is equal to its inverse transform and is defined as follows :

$$\mathcal{H}_n(f)(\xi) = \int_0^\infty f(r) J_n(\xi r) r dr \tag{135}$$

10.5. RECONSTRUCTION FORMULAE

For $\boldsymbol{a} = \mathbf{0}$:

$$\begin{array}{rll} \boldsymbol{u} & = \int_0^\infty kdk & \{\widetilde{u}_1 J_o(kr)\boldsymbol{e}_z - i\widetilde{u}_2 J_1(kr)\boldsymbol{e}_r\} \\ \operatorname{div}'\boldsymbol{u} & = \int_0^\infty k^2 dk & \widetilde{u}_1 J_o(kr) \\ \boldsymbol{t} & = \omega \int_0^\infty kdk & \{i\widetilde{t}_1 J_o(kr)\boldsymbol{e}_z + \widetilde{t}_2 J_1(kr)\boldsymbol{e}_r\} \end{array} \tag{136}$$

For $\boldsymbol{a}' = \boldsymbol{e}_x$:

$$\begin{array}{rllr} \boldsymbol{u} & = \int_0^\infty kdk & \{\widetilde{u}_1 J_1(kr)\boldsymbol{e}_z - i\widetilde{u}_2(J_o - J_2)\boldsymbol{e}_r\} & \cos\theta \\ & & -i\widetilde{u}^{SH}(J_o + J_2)\boldsymbol{e}_\theta & \sin\theta \\ \operatorname{div}'\boldsymbol{u} & = \int_0^\infty k^2 dk & 2i\widetilde{u}_2 J_1(kr) & \cos\theta \\ \operatorname{rot}'\boldsymbol{u} & = \int_0^\infty k^2 dk & 2i\widetilde{u}^{SH} J_1(kr) & \sin\theta \\ \boldsymbol{t} & = \int_0^\infty kdk & \{\widetilde{t}_1 J_1(kr)\boldsymbol{e}_z - i\widetilde{t}_2(J_o - J_2)\boldsymbol{e}_r\} & \cos\theta \\ & & -i\widetilde{t}^{SH}(J_o + J_2)\boldsymbol{e}_\theta & \sin\theta \end{array} \tag{137}$$

CHAPTER 3.
THE SEMI-ANALYTICAL FUNDAMENTAL-SOLUTION-LESS SCALED BOUNDARY FINITE-ELEMENT METHOD TO MODEL UNBOUNDED SOIL

JOHN P. WOLF and CHONGMIN SONG†
Institute of Hydraulics and Energy, Department of Civil Engineering, Swiss Federal Institute of Technology Lausanne, CH-1015 Lausanne, Switzerland

1. Introduction

To analyse dynamic soil-structure interaction as in other areas of solid mechanics two well-known computational procedures are dominant, the boundary element method and the finite element method. Both exhibit their own specific features, advantages and disadvantages.

Using a fundamental solution permits the *boundary element method* to reduce the dimension of the spatial discretisation by one, as only the boundary is discretised. Another striking feature of the boundary element method is that the radiation condition at infinity is satisfied exactly when modelling unbounded soil. However, the fundamental solution yielding singular integrals can be very complicated or is not even available for general anisotropic materials in dynamics. In contrast, the *finite element method*, which does not require a fundamental solution, is more versatile, but requires the spatial discretization of the domain. In addition, when modelling unbounded soil, the radiation condition can, in general, only be satisfied approximately.

As will be demonstrated, the novel *scaled boundary finite-element method* is a fundamental-solution-less boundary-element method based on finite elements, which combines the advantages of the boundary element and finite element methods. Appealing features of its own are also present. In addition, the advantages of the analytical and numerical approaches are visible.

A brief historical review follows. Key references are specified where additional citations to the original work can be found. The scaled boundary finite-element

† Present address: School of Civil and Environmental Engineering, University of New South Wales, Sydney, NSW-2052, Australia

W.S. Hall and G. Oliveto (eds.), Boundary Element Methods For Soil-Structure Interaction, 127–173.

was originally developed to model unbounded media in elastodynamics [1]. It was called the consistent infinitesimal finite-element cell method reflecting the mechanically based derivation analogous to the early work in finite elements. It is based on the assemblage of an infinitesimal finite element cell and on similarity. Performing the limit of the infinitesimal width of the finite element cell analytically yields the consistent infinitesimal finite-element cell equation in the dynamic stiffness of an unbounded medium. For a thorough description of the consistent infinitesimal finite-element cell method with many applications the reader is referred to the book [2]. Two- and three-dimensional scalar and vector wave equations in the time and frequency domains as well as statics and incompressible material are addressed. The diffusion equation and the extension to bounded media are also covered.

A free computer program *SIMILAR* including its source code can be downloaded from *ftp://ftp.wiley.co.uk/pub/books/wolf/* and *http://lchpc25.epfl.ch/*.

Recently, the procedure has been re-derived starting from the governing partial differential equations. These are transformed from the Cartesian coordinate system to the so-called scaled boundary coordinates (radial and circumferential coordinates). In this scaled-boundary-transformation-based derivation, the numerical weighted residual technique of finite elements is applied in the circumferential directions, yielding ordinary differential equations in the radial direction which are then solved analytically. This derivation is mathematically more appealing and is consistent with today's finite element technology. The work was published in specialised journals in various areas as the development proceeded ([3], [4], [5]). Solution procedures are also discussed [6], [7] and [8]. An overview paper is available [9] as well as two tutorial articles [10], [11]. Specifically, the seismic soil-structure interaction problem is examined [12] and the scaled boundary finite-element method is put into context with other analysis methods [13]. Recent work on the analysis of the far field response of unbounded soil is also described [14], [15]. Stress singularities in fracture mechanics are examined in statics [16] and in the time domain [17]. Recent work addresses poroelastic saturated soil (Biots theory) [18]. Stress recovery and error estimation [19] as well as adaptive procedures [20] are also developed.

The goal of this chapter is to present the state of the art of the scaled boundary finite-element method for the dynamic analysis of unbounded media. To be specific, the modelling of the unbounded (infinite or semi-infinite) soil is addressed, which is essential to calculate dynamic soil-structure interaction. The static case is not examined and body loads applied to the unbounded soil are excluded. Only the concepts, key expressions and fundamental equations of the derivations and of the corresponding solution procedures are reviewed. For a detailed discussion with the definition of the nomenclature and for other aspects of the scaled boundary finite-element method, the reader should consult the cited references. Recent research which has not been published yet is however included here. Selective examples

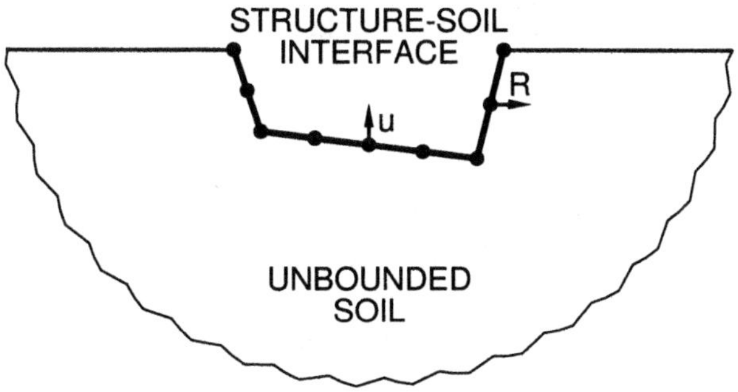

Figure 1. Interaction force-displacement relationship on discretised structure-soil interface of unbounded soil

demonstrating the versatility and accuracy are included. Conclusions address the advantages but also the restrictions of the scaled boundary finite-element method.

The paper is organised as follows. In Section 2, the dynamic unbounded soil-structure interaction problem is stated and the quantities which have to be calculated are identified. Section 3 addresses the salient concept of the approximate solution of the scaled boundary finite-element solution as compared to that of the standard finite element method. Two derivations are sketched in the following two sections. In Section 4, the scaled-boundary-transformation-based derivation with all key equations and in Section 5 the mechanically based derivation are outlined. Section 6 addresses the analytical solution in the radial direction in the frequency domain and Section 7 the corresponding numerical solution in the frequency and time domains. Section 8 discusses various extensions. Section 9 is devoted to selective numerical examples. Section 10 mentions certain results for a bounded medium. Finally, Section 11 contains concluding remarks, addressing implementation, advantages but also restrictive properties.

2. Objective of Dynamic Soil-Structure Interaction Analysis [2]

In a typical dynamic soil-structure interaction analysis based on the substructure method, the actual structure and the neighbouring soil, if present, which are irregular and can exhibit nonlinear behaviour, are modelled with finite elements. This introduces degrees of freedom within the structure and on the structure-soil interface. The dynamic behaviour of the other substructure, the linear unbounded soil, is described by the interaction force-displacement relationship in the degrees of freedom on the structure-soil interface (Figure 1)

In the frequency domain the amplitudes of the displacements $\{u(\omega)\}$ are related to those of the interaction forces $\{R(\omega)\}$ by the *dynamic stiffness matrix* $[S^{\infty}(\omega)]$ (superscript ∞ for unbounded) as follows

$$\{R(\omega)\} = [S^{\infty}(\omega)]\{u(\omega)\} \tag{1}$$

In the time domain the convolution integral applies

$$\{R(t)\} = \int_0^t [S^\infty(t-\tau)]\{u(\tau)\}\mathrm{d}\tau \quad (2$$

with the *unit-impulse response matrix* $[S^\infty(t)]$. Alternatively, the response matri to a unit impulse of accelerations $[M^\infty(t)]$ can be introduced leading to

$$\{R(t)\} = \int_0^t [M^\infty(t-\tau)]\{\ddot{u}(\tau)\}\mathrm{d}\tau \quad (3$$

The symmetric fully-coupled $[S^\infty(\omega)]$, $[S^\infty(t)]$ or $[M^\infty(t)]$ matrices are calcu lated with the scaled boundary finite-element method.

The interaction force-displacement relationship defines the contribution of th unbounded soil to the governing equations of motion of dynamic soil-structur interaction. If the coupled system is excited by seismic motion, the loading is cal culated using $[S^\infty(\omega)]$ in the frequency domain and the response (displacement and forces) at the nodes which will be on the structure-soil interface of the so called free field. Thus, the calculation of $[S^\infty(\omega)]$ is of paramount importanc and is, as far as the soil is concerned, sufficient to determine results within th structure.

In general, the response (displacements and stresses) within the unbounde soil is also of interest. This should include the far field, i.e. the unbounded soil a a large distance from the structure-soil interface.

Figure 1 represents a typical foundation embedded in a half space. Note tha the spatial discretisation is restricted to the structure-soil interface, e.g. no node are introduced on the free surface. This represents a significant advantage of th scaled boundary finite-element method compared to the boundary element metho using the fundamental solution of the full space. (The same also applies for certai interfaces between different materials).

The presence of a free surface in the unbounded soil is a dominant feature i modelling soil-structure interaction.

More general configurations are shown in Figure 2. A cavity embedded in full space (Figure 2a) could for example represent a tunnel. The entire structure soil interface S is spatially discretised. For a foundation embedded in a (distorted half space (Figure 2b) the so-called side faces A (free and fixed surfaces) are no discretized. The nodes are limited to the structure-soil interface S.

3. Salient Concept

For most practical cases, the shape of the boundary (structure-soil interface, fre surface), the variations of the material properties and the boundary condition preclude an analytical solution of the governing partial differential equations o

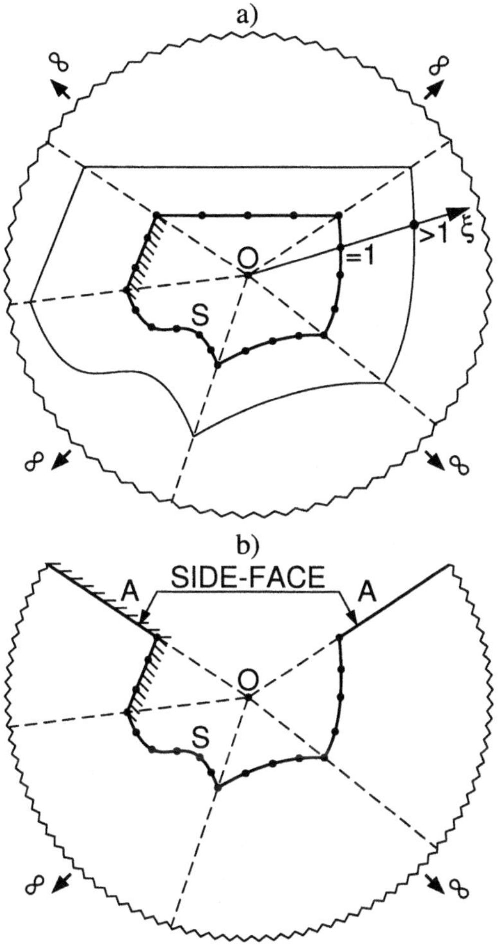

Figure 2. Structure-soil interface of unbounded soil: a) Cavity (tunnel) embedded in full space, b) Foundation embedded in half space with free and fixed side faces without spatial discretisation

elastodynamics. However, if the physical problem is governed by ordinary differential equations, classical mathematical techniques can, in important cases, lead to an exact analytical solution in the single independent variable. In certain cases symmetry exists, leading to a one dimensional problem. The governing ordinary differential equations, for instance in the radial coordinate, can then be solved exactly while the problem in three dimensions cannot be addressed analytically. To make use of these advantages also for the general case without any symmetry, a coordinate system consisting of the radial direction ξ and two local circumferential directions (parallel to the boundary and to the structure-soil interface) η, ζ is introduced (Figure 3). The governing partial differential equations are transformed from the Cartesian coordinates $\hat{x}$, $\hat{y}$, $\hat{z}$ to this coordinate system, called the *scaled boundary coordinates*. In the circumferential directions the boundary

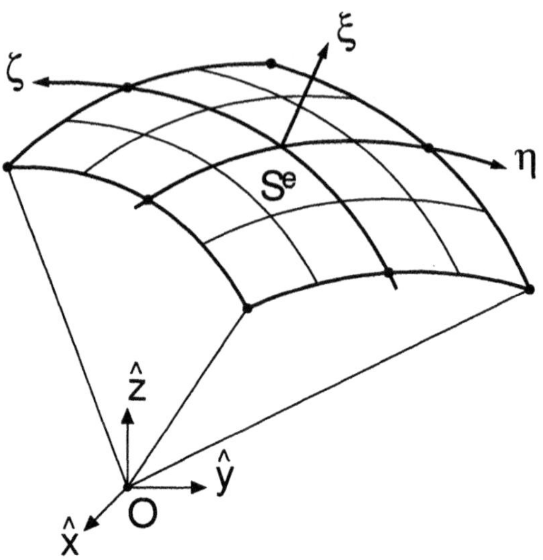

Figure 3. Scaled boundary transformation of geometry of surface finite element

is discretized with surface finite elements, reducing the governing partial differential equations to ordinary differential equations in the radial coordinate. The coefficients of the ordinary differential equations are determined by the finite element approximation in the circumferential directions. The ordinary differential equations are then solved analytically in the radial direction. Thus a novel computational procedure, called the *scaled boundary finite-element method*, is developed which *combines the advantages of the analytical and numerical approaches*. The method is a *semi-analytical* procedure for solving partial differential equations. In the circumferential directions (parallel to the boundary), where the behaviour is, in general, smooth, the weighted-residual approximation of finite elements applies, leading to convergence in the finite element sense. For the unbounded soil, the radial coordinate points away from the boundary, the structure-soil interface, towards infinity, where the boundary conditions at infinity (radiation condition) can be incorporated exactly in the analytical solution.

In more detail, the origin of the new coordinate system, called the *scaling centre* O, is chosen in a zone from which the total structure-soil interface must be visible. (For the sake of simplicity, O coincides with the origin of the Cartesian coordinate system in Figure 3). For the unbounded soil the scaling centre is located outside the domain (Figure 2a). As a special case the scaling centre O can be chosen on the extension of the boundary (see Figure 2b). In this case the total boundary is decomposed into two parts: that part of the boundary with its extension passing through the scaling centre denoted as the side face A and the remaining part S (Figure 2b). Only the structure-soil interface S is discretized with (doubly curved) surface finite elements. A typical finite element is shown in Figure 3. Denoting points on the structure-soil interface with x, y, z, the geometry is described in the local coordinate system η, ζ

$$x(\eta,\zeta) = [N(\eta,\zeta)]\{x\} \tag{4a}$$
$$y(\eta,\zeta) = [N(\eta,\zeta)]\{y\} \tag{4b}$$
$$z(\eta,\zeta) = [N(\eta,\zeta)]\{z\} \tag{4c}$$

with the mapping functions $[N(\eta,\zeta)]$ and the coordinates $\{x\}$, $\{y\}$, $\{z\}$. The three dimensional domain of the unbounded soil is defined by scaling the boundary of the nodes on the structure-soil interface with the dimensionless radial coordinate ξ measured from the scaling centre

$$\hat{x}(\xi,\eta,\zeta) = \xi x(\eta,\zeta) \tag{5a}$$
$$\hat{y}(\xi,\eta,\zeta) = \xi y(\eta,\zeta) \tag{5b}$$
$$\hat{z}(\xi,\eta,\zeta) = \xi z(\eta,\zeta) \tag{5c}$$

with $\xi = 1$ on the boundary and $\xi = 0$ in the scaling centre. For an unbounded medium $1 \leq \xi < \infty$ applies. The new coordinate system is defined by ξ and the two circumferential coordinates η, ζ. In the scaled boundary transformation $\hat{x}$, $\hat{y}$, $\hat{z}$ are replaced by ξ, η, ζ. Equation (5) describing scaling of the boundary has lead to the name of the method.

The components of the displacements, strains, stresses etc. are still defined in the Cartesian coordinate system but their position is specified in the scaled boundary coordinate system. The displacement amplitudes of the finite element on the structure-soil interface ($\xi = 1$) are interpolated using shape functions $[N(\eta,\zeta)]$. The discretization is thus restricted to this boundary. It is postulated that the same shape functions apply with the displacement amplitudes $\{u(\xi)\}$ for all surfaces with a constant ξ

$$\{u(\xi,\eta,\zeta)\} = [N(\eta,\zeta)]\{u(\xi)\} \tag{6}$$

These displacement amplitudes along the line defined by the scaling centre and the node on the structure-soil interface $\{u(\xi)\}$ are analytical functions of the radial coordinate ξ only, which, as will be demonstrated, are calculated analytically from the corresponding ordinary differential equations. In the circumferential directions, the displacements are then interpolated using the same functions used to describe the geometry of the structure-soil interface (equation(4)) and the scaling equations (equation (5)). The approximate solution determined from equation (6) could be interpreted as applying in a generalised manner the procedure of separation of variables, interpolating in the circumferential directions η, ζ the discrete values of $\{u(\xi)\}$. Equation (6) together with the definition of the scaled boundary transformation (equations (4) and (5)) forms the basis of the scaled boundary finite-element method. The difference from the standard finite element method with the displacement amplitudes in distinct nodes $\{u\}$ located throughout the unbounded soil

$$\{u(\xi,\eta,\zeta)\} = [N(\xi,\eta,\zeta)]\{u\} \tag{7}$$

is clearly visible. The so-called scaled boundary-transformation-based derivation which directly substitutes equation (6) into the governing partial differential equations is discussed in Section 4. Alternatively, the same equations follow in the mechanically based derivation in Section 5, where equation (6) is replaced by similarity in the assembly process. Note that equation (5) expresses similarity with the scaling centre coinciding with the similarity centre and ξ representing the similarity factor. The surfaces for a constant ξ are similar to the structure-soil interface. It is only due to this transformation that for a fixed η, ζ the displacements along a radial line are a function of ξ only, as expressed in equation (6).

4. Scaled-Boundary-Transformation-Based Derivation [3]

This systematic derivation involving a transformation to the scaled boundary coordinates and application of the weighted residual technique in the circumferential directions is sketched as follows.

4.1. GOVERNING EQUATIONS OF ELASTODYNAMICS

The differential equation of motion in the frequency domain expressed in displacement amplitudes $\{u\}$ in Cartesian coordinates $\hat{x}$, $\hat{y}$, $\hat{z}$, are formulated for vanishing body loads as

$$[L]^T\{\sigma\} + \omega^2\rho\{u\} = 0 \tag{8}$$

with the mass density ρ. The stress amplitudes $\{\sigma\}$ are equal to

$$\{\sigma\} = [D][L]\{u\} \tag{9}$$

with the elasticity matrix $[D]$. $[L]$ represents the differential operator

$$[L] = \begin{bmatrix} \frac{\partial}{\partial\hat{x}} & & \\ & \frac{\partial}{\partial\hat{y}} & \\ & & \frac{\partial}{\partial\hat{z}} \\ & \frac{\partial}{\partial\hat{z}} & \frac{\partial}{\partial\hat{y}} \\ \frac{\partial}{\partial\hat{z}} & & \frac{\partial}{\partial\hat{x}} \\ \frac{\partial}{\partial\hat{y}} & \frac{\partial}{\partial\hat{x}} & \end{bmatrix} \tag{10}$$

Applying the scaled boundary transformation of the geometry (equations (4) and (5)), standard procedures permit the differential operator in equation (8) to be written in the coordinate system ξ, η, ζ as

$$[L(\eta,\zeta)] = [b^1(\eta,\zeta)]\frac{\partial}{\partial\xi} + \frac{1}{\xi}\left([b^2(\eta,\zeta)]\frac{\partial}{\partial\eta} + [b^3(\eta,\zeta)]\frac{\partial}{\partial\zeta}\right) \tag{11}$$

where $[b^1(\eta,\zeta)]$, $[b^2(\eta,\zeta)]$, $[b^3(\eta,\zeta)]$ depend only on the geometry of the structure-soil interface.

4.2. BOUNDARY DISCRETISATION WITH FINITE ELEMENTS

As already discussed in Section 3, the displacement amplitudes of all surfaces with a constant ξ are specified as in equation (6). The weighting function $\{w(\xi,\eta,\zeta)\}$ is chosen consistently as

$$\{w(\xi,\eta,\zeta)\} = [N(\eta,\zeta)]\{w(\xi)\} \tag{12}$$

The weighted residual method is applied to equation (8) with $[L]$ in equation (11). Integration by parts in the circumferential directions η, ζ is performed, and the integrand of the integral over ξ is then set equal to zero, which corresponds to enforcing the equation of motion equation (8) with equation (11) exactly in the ξ direction, yielding the scaled boundary finite-element equation in displacement

$$\begin{aligned}[E^0]\xi^2\{u(\xi)\},_{\xi\xi} &+ ((s-1)[E^0] - [E^1] + [E^1]^T)\xi\{u(\xi)\},_{\xi} \\ &+ ((s-2)[E^1]^T - [E^2])\{u(\xi)\} + \omega^2[M^0]\xi^2\{u(\xi)\} = 0 \end{aligned} \tag{13}$$

with the spatial dimension s ($= 2$ or $= 3$). The coefficient matrices for $s = 3$ are

$$[E^0] = \int_S [B^1]^T[D][B^1]|J|\mathrm{d}\eta\mathrm{d}\zeta \tag{14a}$$

$$[E^1] = \int_S [B^2]^T[D][B^1]|J|\mathrm{d}\eta\mathrm{d}\zeta \tag{14b}$$

$$[E^2] = \int_S [B^2]^T[D][B^2]|J|\mathrm{d}\eta\mathrm{d}\zeta \tag{14c}$$

$$[M^0] = \int_S [N(\eta,\zeta)]^T\rho[N(\eta,\zeta)]|J|\mathrm{d}\eta\mathrm{d}\zeta \tag{15}$$

$[B^1]$, $[B^2]$, $|J|$ depend only on the geometry of the structure-soil interface. The coefficient matrices $[E^0]$, $[E^1]$, $[E^2]$, $[M^0]$ are independent of ξ. Integrations over the surface finite element and assembly similar to those in the standard finite element method are involved.

The *scaled boundary finite-element equation for displacement* formulated in the frequency domain is a system of *linear second-order ordinary differential equations* for the displacement amplitude $\{u(\xi)\}$ in the dimensionless radial coordinate ξ as the independent variable.

Introducing $\bar{\xi} = \omega\xi$, equation (13) can be rewritten as

$$[E^0]\bar{\xi}^2\{u(\bar{\xi})\},_{\bar{\xi}\bar{\xi}} + ((s-1)[E^0] - [E^1] + [E^1]^T)\bar{\xi}\{u(\bar{\xi})\},_{\bar{\xi}} + ((s-2)[E^1]^T - [E^2])\{u(\bar{\xi})\} + [M^0]\bar{\xi}^2\{u(\bar{\xi})\} = 0 \quad (16)$$

An independent variable, $\bar{\xi}$, which is the product of frequency ω and (dimensionless) radial coordinate ξ, appears. The equation could also be formulated with the same coefficients choosing the dimensionless frequency corresponding to the radial coordinate r ($= \xi r_0$, with the radial coordinate of the structure-soil interface r_0) as the independent variable

$$a = \frac{\omega\xi r_0}{c_s} \quad (17)$$

The shear wave velocity is denoted by c_s.

4.3. DYNAMIC STIFFNESS MATRIX

The amplitudes of the internal nodal forces $\{Q(\xi)\}$ which are equal to the stress amplitudes multiplied by the shape functions integrated over a surface with constant ξ are expressed as

$$\{Q(\xi)\} = \xi^{s-2}([E^0]\xi\{u(\xi)\},_{\xi} + [E^1]^T\{u(\xi)\}) \quad (18)$$

The dynamic stiffness matrix for the unbounded soil is defined for vanishing body load on the negative face as

$$-\{Q(\xi)\} = [S^\infty(\omega, \xi)]\{u(\xi)\} \quad (19)$$

yielding

$$[S^\infty(\omega, \xi)]\{u(\xi)\} = -\xi^{s-2}([E^0]\xi\{u(\xi)\},_{\xi} + [E^1]^T\{u(\xi)\}) \quad (20)$$

or

$$\xi^{2-s}[S^\infty(\omega, \xi)]\{u(\bar{\xi})\}a = -[E^0]\bar{\xi}\{u(\bar{\xi})\},_{\bar{\xi}} - [E^1]^T\{u(\bar{\xi})\} \quad (21)$$

It follows from equation (21) that $\xi^{2-s}[S^\infty(\omega, \xi)]$ is a function of $\bar{\xi}$ (or alternatively of a, equation (17)). For $s = 3$, the dynamic stiffness matrix $[S^\infty(\omega, \xi)]$ is a function of ξ and $\bar{\xi}$, i.e. of ξ and ω (or ξ and a). Equation (20) (and essentially also equation (21)) represents the *interaction force-displacement relationship* at a surface with a constant ξ. On the structure-soil interface $\xi = 1$, $[S^\infty(\omega, \xi = 1)]$ $=[S^\infty(\omega)]$ applies. In this case, $\{R(\omega)\} = -\{Q(\omega)\}$ (equation (1)), the negative sign arising from the fact that the structure-soil interface is a negative face (the sign of $\{Q\}$ is the same as for stress).

Combining equations (16) and (21) yields the *scaled boundary finite-element equation in dynamic stiffness*

$$(\xi^{2-s}[S^{\infty}(\omega,\xi)] + [E^1])[E^0]^{-1}(\xi^{2-s}[S^{\infty}(\omega,\xi)] + [E^1]^T) \\ - (s-2)\xi^{2-s}[S^{\infty}(\omega,\xi)] - \bar{\xi}(\xi^{2-s}[S^{\infty}(\omega,\xi)]),_{\bar{\xi}} - [E^2] + \bar{\xi}^2[M^0] = 0 \quad (22)$$

This is a system of *nonlinear first order ordinary differential equations* in $\xi^{2-s}[S^{\infty}(\omega,\xi)$ with $\bar{\xi}$ (or alternatively with a) as the independent variable. For the structure-soil interface ($\xi = 1$),

$$([S^{\infty}(\omega)] + [E^1])[E^0]^{-1}([S^{\infty}(\omega)] + [E^1]^T) \\ - (s-2)[S^{\infty}(\omega)] - \omega[S^{\infty}(\omega)],_{\omega} - [E^2] + \omega^2[M^0] = 0 \quad (23)$$

results. Alternatively, the dimensionless frequency corresponding to the structure-soil interface

$$a_0 = \frac{\omega r_0}{c_s} \quad (24)$$

can be introduced as the independent variable instead of ω. The coefficients in equation (23) are not changed.

4.4. HIGH FREQUENCY ASYMPTOTIC EXPANSION OF DYNAMIC STIFFNESS MATRIX [4]

As will be discussed in Sections 6 and 7, an asymptotic expansion of the dynamic stiffness matrix for high frequency permits the radiation condition to be satisfied rigorously and provides a starting value at a high but finite frequency for the numerical solution of the scaled boundary finite-element equation in dynamic stiffness for decreasing ω.

The dynamic stiffness matrix $[S^{\infty}(\omega,\xi)]$ at high frequency is expanded in a polynomial of $\mathrm{i}\bar{\xi}$ in descending order starting at one

$$\xi^{2-s}[S^{\infty}(\omega,\xi)] \approx \mathrm{i}\bar{\xi}[C_{\infty}] + [K_{\infty}] + \sum_{j=1}^{m} \frac{1}{(\mathrm{i}\bar{\xi})^j}[A_j] \quad (25)$$

The first two terms on the right-hand side represent the constant dashpot matrix $[C_{\infty}]$ and the constant spring matrix $[K_{\infty}]$ (subscript ∞ for $\omega \to \infty$). Substituting equation (25) in equation (22) and setting the coefficients of the terms in descending order of the power of $\mathrm{i}\bar{\xi}$ equal to zero determines *analytically* the unknown matrices in equation (25) sequentially. $[C_{\infty}]$ follows from

$$[C_{\infty}] = \left([\Phi]^{-1}\right)^T \lceil \Lambda \rfloor [\Phi]^{-1} \quad (26)$$

with $\lceil\Lambda\rfloor$ representing the positive root of $\lceil\Lambda^2\rfloor$ of the eigenvalue problem

$$[M^0][\Phi] = [E^0][\Phi]\left\lceil\Lambda^2\right\rfloor \tag{27}$$

This choice guarantees that $[C_\infty]$ is positive definite and thus the unbounded soil acts as a sink and not as a source of energy (*radiation condition*).

$[K_\infty]$ is determined from

$$[K_\infty] = \left([\Phi]^{-1}\right)^T [k_\infty][\Phi]^{-1} \tag{28}$$

with the (linear) Lyapunov equation

$$\lceil\Lambda\rfloor [k_\infty] + [k_\infty] \lceil\Lambda\rfloor = -\lceil\Lambda\rfloor [e^1]^T - [e^1] \lceil\Lambda\rfloor + (s-1) \lceil\Lambda\rfloor \tag{29}$$

$[A_j]$ follows from a similar Lyapunov equation.

For the structure-soil interface ($\xi = 1$) the high frequency asymptotic expansion (equation (25)) is

$$[S^\infty(\omega)] \approx \mathrm{i}\omega[C_\infty] + [K_\infty] + \sum_{j=1}^{m} \frac{1}{(\mathrm{i}\omega)^j}[A_j] \tag{30}$$

The high frequency asymptotic expansion of the dynamic stiffness matrix $[S^\infty(\omega)]$ in the frequency domain corresponds to the early time asymptotic expansion of the unit impulse response matrix $[S^\infty(t)]$ in the time domain (equation (2)). This is a consequence of the initial value theorem. The inverse Fourier transform of equation (30) is

$$[S^\infty(t)] \approx [C_\infty]\dot{\delta}(t) + [K_\infty]\delta(t) + \sum_{j=1}^{m} \frac{1}{(j-1)!}[A_j]t^{j-1}\mathrm{H}(t) \tag{31}$$

in which $\mathrm{H}(t)$ is the Heaviside step function.

It is interesting to note that the dashpot matrix $[C_\infty]$ can be calculated without solving an eigenvalue problem.

The response of the unbounded soil at the initial time can be calculated directly. The direction perpendicular to the infinitesimal area $\mathrm{d}A$ of the structure-soil interface is addressed. The unbounded soil is initially at rest. After applying the load per unit area p during the first infinitesimal time $\mathrm{d}t$ the wave front is at the distance $c_p\mathrm{d}t$ (dilatational-wave velocity c_p) and the domain of influence equals $c_p\mathrm{d}t\mathrm{d}A$. The law of conservation of momentum is formulated for the first infinitesimal time $\mathrm{d}t$. The initial momentum vanishes. The momentum (mass times velocity) at $\mathrm{d}t$ equals $\rho c_p\mathrm{d}t\mathrm{d}A\dot{u}$ (mass density ρ, velocity $\dot{u}$ at $\mathrm{d}t$). The law is written as

$$p\mathrm{d}A\mathrm{d}t = \rho c_p\mathrm{d}t\mathrm{d}A\dot{u} \qquad \text{for} \quad t = 0 \tag{32}$$

which yields

$$p = \rho c_p \dot{u} \qquad \text{for} \quad t = 0 \tag{33}$$

The initial response perpendicular to the structure-soil interface of the unbounded soil is thus modelled by a dashpot with the coefficient per unit area ρc_p which is called the *impedance*. Analogously, the initial response in the tangential directions is described by dashpots with the coefficients ρc_s.

Applying virtual-work concepts, the interaction forces can be expressed by integrating the load per unit area p. The coefficient matrix of the interaction force-nodal velocity relationship is equal to $[C_\infty]$

$$\{R\} = [C_\infty]\{\dot{u}\} \qquad \text{for} \quad t = 0 \tag{34}$$

In the far field (i.e. for a large $\bar{\xi}$), the dynamic stiffness matrix $[S^\infty(\omega, \xi)]$ in equation (21) can be replaced by the high frequency asymptotic expansion by including the first two terms only (equation(25)), resulting in the following system of linear first order ordinary differential equations in $\{u(\bar{\xi})\}$ with $\bar{\xi}$ as the independent variable

$$[E^0]\bar{\xi}\{u(\bar{\xi})\},_{\bar{\xi}} + ([E^1]^T + i\bar{\xi}[C_\infty] + [K_\infty])\{u(\bar{\xi})\} = 0 \tag{35}$$

4.5. MATERIAL DAMPING

All equations discussed above apply to a linear elastic material. They can s-traightforwardly be extended to a viscoelastic material using the correspondence principle. For demonstration the constant hysteretic material model is addressed with the same damping coefficient for shear and dilatational waves.

The correspondence principle states that the solution for hysteretic material follows from the elastic results by replacing the real Lamé constants with the corresponding complex ones (multiplication by $1 + 2i\zeta$ with the hysteretic damping ratio ζ). This yields

$$\begin{aligned} [E^{0*}] &= (1 + 2i\zeta)[E^0] \\ [E^{1*}] &= (1 + 2i\zeta)[E^1] \\ [E^{2*}] &= (1 + 2i\zeta)[E^2] \end{aligned} \tag{36}$$

The scaled boundary finite-element equation in dynamic stiffness, equation (22), becomes

$$\begin{aligned} &(\xi^{2-s}[S^\infty(\omega, \xi)] + [E^{1*}])[E^{0*}]^{-1}(\xi^{2-s}[S^\infty(\omega, \xi)] + [E^{1*}]^T) \\ &- (s-2)\xi^{2-s}[S^\infty(\omega, \xi)] - \bar{\xi}(\xi^{2-s}[S^\infty(\omega, \xi)]),_{\bar{\xi}} - [E^{2*}] + \bar{\xi}^2[M^0] = 0 \end{aligned} \tag{37}$$

with the dynamic stiffness matrix of the hysteretic material $[S^{\infty}(\omega, \xi)]$. The high frequency asymptotic expansion, equation (25), is written as

$$\xi^{2-s}[S^{\infty}(\omega, \xi)] \approx \mathrm{i}\bar{\xi}[C^{*}_{\infty}] + [K^{*}_{\infty}] + \sum_{j=1}^{m} \frac{1}{(\mathrm{i}\bar{\xi})^{j}}[A^{*}_{j}] \tag{38}$$

where

$$\begin{aligned} [C^{*}_{\infty}] &= (1 + 2\mathrm{i}\zeta)^{1/2}[C_{\infty}] \\ [K^{*}_{\infty}] &= (1 + 2\mathrm{i}\zeta)[K_{\infty}] \\ [A^{*}_{j}] &= (1 + 2\mathrm{i}\zeta)^{1+j/2}[A_{j}] \end{aligned} \tag{39}$$

The interaction force-displacement relationship (equation (21)) reads

$$\xi^{2-s}[S^{\infty}(\omega, \xi)]\{u(\bar{\xi})\} = -[E^{0*}]\bar{\xi}\{u(\bar{\xi})\},_{\bar{\xi}} - [E^{1*}]^{T}\{u(\bar{\xi})\} \tag{40}$$

4.6. UNIT IMPULSE RESPONSE MATRIX [2]

The scaled boundary finite-element equation in unit impulse response follows from the inverse Fourier transformation of the corresponding relationship in dynamic stiffness (equation (23)). To be able to perform this transformation, $[S^{\infty}(\omega)]$ is decomposed into the singular part, i.e. the value for $\omega \to \infty$, and the remaining regular part $[S^{\infty}_{r}(\omega)]$. In the light of equation (30)

$$[S^{\infty}(\omega)] = \mathrm{i}\omega[C_{\infty}] + [K_{\infty}] + [S^{\infty}_{r}(\omega)] \tag{41}$$

applies with the inverse Fourier transform

$$[S^{\infty}(t)] = [C_{\infty}]\dot{\delta}(t) + [K_{\infty}]\delta(t) + [S^{\infty}_{r}(t)] \tag{42}$$

Substituting equation (42) in the interaction force-displacement relationship yields

$$\{R(t)\} = [C_{\infty}]\{\dot{u}(t)\} + [K_{\infty}]\{u(t)\} + \int_{0}^{t} [S^{\infty}_{r}(t-\tau)]\{u(\tau)\}\mathrm{d}\tau \tag{43}$$

with the first two terms on the right-hand side representing the instantaneous response and the third term the lingering response.

Substituting equation (41) into equation (23), performing the inverse Fourier transformation and using equation (42) leads to an integral equation for $[S^{\infty}_{r}(t)]$, the scaled boundary finite-element equation in unit impulse response [4].

Alternatively, the response matrix to a unit impulse of accelerations $[M^{\infty}(t)]$ can be calculated (equation (3)). In the frequency domain, the corresponding matrix equals

$$[M^{\infty}(\omega)] = \frac{1}{(\mathrm{i}\omega)^{2}}[S^{\infty}(\omega)] \tag{44}$$

Substituting equation (44) into the scaled boundary finite-element equation in dynamic stiffness (equation (23)) yields

$$[M^{\infty}(\omega)][E^0]^{-1}[M^{\infty}(\omega)] + [E^1][E^0]^{-1}\frac{[M^{\infty}(\omega)]}{(\mathrm{i}\omega)^2} + \frac{[M^{\infty}(\omega)]}{(\mathrm{i}\omega)^2}[E^0]^{-1}[E^1]^T - s\frac{[M^{\infty}(\omega)]}{(\mathrm{i}\omega)^2} + \frac{1}{\omega}[M^{\infty}(\omega)]_{,\omega} - \frac{1}{(\mathrm{i}\omega)^4}([E^2] - [E^1][E^0]^{-1}[E^1]^T) - \frac{1}{(\mathrm{i}\omega)^2}[M^0] = 0 \quad (45)$$

Its inverse Fourier transformation leads to the integral equation

$$\int_0^t [M^{\infty}(t-\tau)][E^0]^{-1}[M^{\infty}(\tau)]\mathrm{d}\tau + \left([E^1][E^0]^{-1} - \frac{s+1}{2}[I]\right)\int_0^t\int_0^{\tau}[M^{\infty}(\tau')]\mathrm{d}\tau'\mathrm{d}\tau + \int_0^t\int_0^{\tau}[M^{\infty}(\tau')]\mathrm{d}\tau'\mathrm{d}\tau\left([E^0]^{-1}[E^1]^T - \frac{s+1}{2}[I]\right) + t\int_0^t[M^{\infty}(\tau)]\mathrm{d}\tau - \frac{t^3}{6}([E^2] - [E^1][E^0]^{-1}[E^1]^T)\mathrm{H}(t) - t[M^0]\mathrm{H}(t) = 0 \quad (46)$$

5. Mechanically Based Derivation [2]

The scaled boundary finite-element equations can also be derived using elementary concepts of mechanics familiar to engineers. In a nutshell, the procedure is based on the assemblage of an infinitesimal finite element cell (Figure 4b) and on similarity (Figure 4), followed by performing the limit of the cell width analytically.

The derivation of the scaled boundary finite-element equation in dynamic stiffness (equation (23)) is discussed as an example. In the final result only the structure-soil interface is discretised (Figure 4a). In the derivation a fictitious similar interface at an infinitesimal distance measured in the direction of the radial direction is introduced with the similarity centre O which is the same as the scaling centre in the scaled-boundary-transformation-based derivation. The similar interfaces are defined by the characteristic length r, which corresponds to the radial coordinate: r_i for the structure-soil interface (i for interior, corresponds to r_0 in equation (24)), r_e for the fictitious interface (e for exterior). r_e follows from r_i as

$$r_e = (1+w)r_i \quad (47)$$

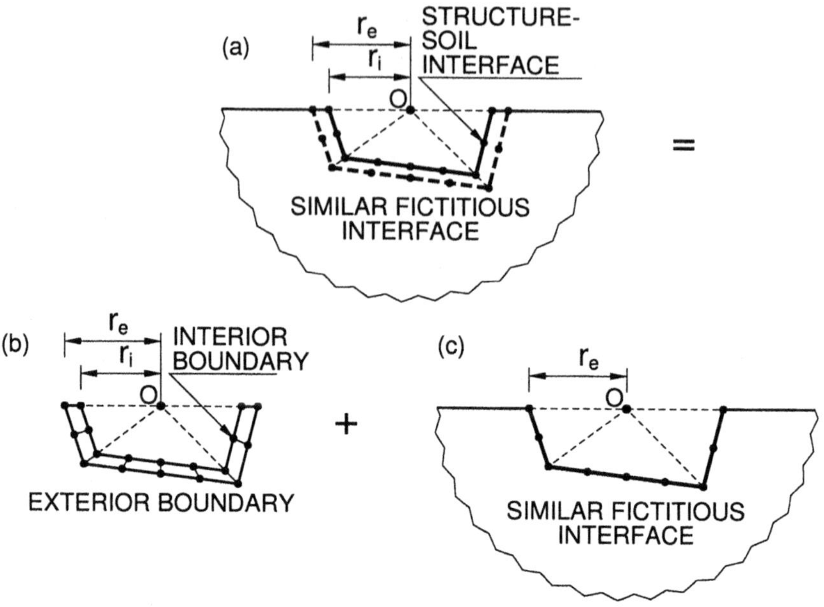

Figure 4. Concept of scaled boundary finite-element method for unbounded soil with infinitesim cell width leading to finite element discretisation of structure-soil interface only

with the infinitesimal dimensionless length w. Using a dimensionless analysis, i can be shown that the dynamic stiffness matrices of the unbounded soil at thes two interfaces are related as follows

$$r[S^{\infty}(\omega, r)]_{,r} = (s-2)[S^{\infty}(\omega, r)] + \omega[S^{\infty}(\omega, r)]_{,\omega} \quad (48$$

This equation permits the partial derivation in the radial direction (moving fron the structure-soil interface to the similar fictitious interface (Figure 4c)) to b expressed as a function of the partial derivative in frequency.

The domain between the structure-soil interface and the fictitious interface i a cell of infinitesimal width which is discretized with finite elements (Figure 4b Its interior and exterior boundaries coincide with the structure-soil interface an the fictitious interface, respectively. The arrangement of the nodes on the tw boundaries must satisfy similarity. w in equation (47) is thus called the *infinites imal dimensionless cell width.* Adding the infinitesimal finite element cell to th unbounded soil defined by the fictitious interface (Figure 4c) results in the un bounded soil defined by the structure-soil interface. The same applies to thei dynamic stiffness, when assemblage is performed. This assemblage enforcin compatibility and equilibrium formulated in the frequency domain links the dy namic stiffness matrix of the unbounded soil at the structure-soil interface to tha at the fictitious interface (Figure 4). This relationship involves the dynamic stiff ness matrix of the cell, a bounded domain, which can be expressed by its stati stiffness and mass matrices.

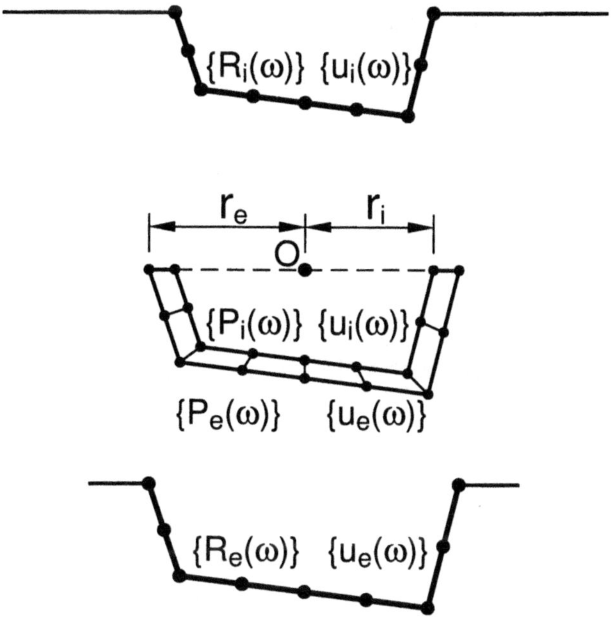

Figure 5. Finite element cell between interior and exterior boundaries and corresponding interfaces of unbounded soil

The force-displacement relationship of the finite element cell located between the interior and exterior boundaries (Figure 5) is written after partitioning as

$$\begin{bmatrix} [S_{ii}(\omega)] & [S_{ie}(\omega)] \\ [S_{ei}(\omega)] & [S_{ee}(\omega)] \end{bmatrix} \left\{ \begin{matrix} \{u_i(\omega)\} \\ \{u_e(\omega)\} \end{matrix} \right\} = \left\{ \begin{matrix} \{P_i(\omega)\} \\ \{P_e(\omega)\} \end{matrix} \right\} \tag{49}$$

with the nodal force amplitudes $\{P(\omega)\}$. The dynamic stiffness matrix equals

$$[S(\omega)] = [K] - \omega^2[M] \tag{50}$$

with the static stiffness matrix $[K]$ and the mass matrix $[M]$ of the finite-element cell. The interaction force-displacement relationship of the unbounded soil at the interfaces corresponding to the interior and exterior boundaries is formulated as (equation (1))

$$\{R_i(\omega)\} = [S_i^\infty(\omega)]\{u_i(\omega)\} \tag{51a}$$

$$\{R_e(\omega)\} = [S_e^\infty(\omega)]\{u_e(\omega)\} \tag{51b}$$

Note that by using the same displacement amplitudes at the interfaces (equations (49) and (51)), compatibility is enforced. Formulating equilibrium at the interior and exterior boundaries relates the interaction force amplitudes of the unbounded soil to the force amplitudes of the cell

$$\{P_i(\omega)\} = \{R_i(\omega)\} \tag{52a}$$

$$\{P_e(\omega)\} = -\{R_e(\omega)\} \tag{52b}$$

Eliminating $\{P(\omega)\}$, $\{R(\omega)\}$ and $\{u_e(\omega)\}$ from equations (49), (51) and (52) leads to

$$([S_i^\infty(\omega)] - [S_{ii}(\omega)] + [S_{ie}(\omega)]([S_e^\infty(\omega)] + [S_{ee}(\omega)])^{-1}[S_{ei}(\omega)])\{u_i(\omega)\} = 0 \tag{53}$$

Since equation (53) is satisfied for an arbitrary $\{u_i(\omega)\}$, the coefficient matrix must vanish, yielding

$$[S_i^\infty(\omega)] = [S_{ii}(\omega)] - [S_{ie}(\omega)]([S_e^\infty(\omega)] + [S_{ee}(\omega)])^{-1}[S_{ei}(\omega)] \tag{54}$$

This relationship derived by assemblage links the dynamic stiffness matrices at the two interfaces i and e. After taking the limit of the infinitesimal cell width $w \to 0$ and using the equation based on similarity (equation (48)) permits the dynamic stiffness matrix at the structure-soil interface to be expressed as a function of the property matrices (static stiffness and mass matrices) of the finite-element cell. As the limit $w \to 0$ is performed, the property matrices are expressed as a function of the coefficient matrices $[E^0]$, $[E^1]$, $[E^2]$ and $[M^0]$ (equations (14) and (15)). This leads to the scaled boundary finite-element equation in dynamic stiffness (equation (23)). Historically, this equation was called the *consistent infinitesimal finite-element cell equation*, reflecting this mechanically based derivation based on the infinitesimal cell.

Note that the limit is performed analytically, which enforces the equations exactly in the radial direction. This also means that boundary conditions in the radial direction such as free surfaces (Figure 4) and fixed surfaces are satisfied exactly without any additional discretisation.

The mechanically based derivation starts with a discrete formulation (equation (49)). This appealing feature for engineers skips the derivation from the continuous formulation in the governing differential equations. Taking the analytical limit, a continuous formulation (equation (23)) is again established. Thus, this derivation first discretises and then performs an analytical limit in the radial direction, which represents a detour!

6. Analytical Solution in Frequency Domain [6]

The linear ordinary differential equations such as the scaled boundary finite-element equation in displacement can be solved to a large extent analytically. After determining the integration constants in the general solution by enforcing boundary conditions, this analytical procedure allows results to be calculated selectively, e.g. the dynamic stiffness matrix for a specific frequency or the displacement in a specific point. This is in contrast to a numerical procedure, where, for example the dynamic stiffness matrices for all frequencies from a very large value down to the specific frequency or the displacements in all points located from the structure-soil interface to the specific point must be determined.

In statics, analytical solutions are readily available. This includes body loads [7]. As mentioned in the Introduction, only dynamics is addressed in this paper.

The boundary conditions for the unbounded soil which have to be enforced in the solution of the scaled boundary finite-element equations in displacement (equation (16)) are discussed. The independent variable is equal to $\bar{\xi} = \omega\xi$. At infinity the *radiation condition* must be enforced [21], which states that no *energy* may be radiated from infinity into the soil towards the structure-soil interface. This means that for $\bar{\xi} \to \infty$ the rate of energy transmission must be positive. The latter is proportional to the quadratic form using the imaginary part of the dynamic stiffness matrix. For $\bar{\xi} \to \infty$, from the high frequency asymptotic expansion of the dynamic stiffness (equation (25)), it follows that the *dashpot matrix* $[C_\infty]$ *must be positive definite*. This is the radiation condition for the unbounded soil with many degrees of freedom on the structure-soil interface. A free surface can be present. At the other boundary, the structure-soil interface ($\xi = 1$), either the displacements or the interaction forces (nodal forces) are prescribed.

For dynamic unbounded soil-structure-interaction analysis, the dynamic stiffness matrix $[S^\infty(\omega)]$ at the structure-soil interface ($\xi = 1$) must be calculated (Section 2).

The analytical solution of the scaled boundary finite-element equation in displacement (equation (16)) proceeds as follows. A transformation to first order ordinary differential equations with twice the number of unknowns is performed. With the independent variable $\bar{\xi}$ (which is related to the dimensionless frequency a (equation (17))

$$\bar{\xi}[X(\bar{\xi})],_{\bar{\xi}} = -[Z][X(\bar{\xi})] - \bar{\xi}^2 \begin{bmatrix} 0 & 0 \\ [M^0] & 0 \end{bmatrix} [X(\bar{\xi})] \tag{55}$$

results, with the unknown

$$\{X(\bar{\xi})\} = \begin{Bmatrix} \xi^{0.5(s-2)}\{u(\xi)\} \\ \xi^{-0.5(s-2)}\{Q(\xi)\} \end{Bmatrix} \tag{56}$$

and the internal nodal forces $\{Q(\xi)\}$ specified in equation (18). The Hamiltonian matrix [Z] equals

$$[Z] = \begin{bmatrix} [E^0]^{-1}[E^1]^T - \frac{s-2}{2}[I] & -[E^0]^{-1} \\ -[E^2] + [E^1][E^0]^{-1}[E^1]^T & -[E^1][E^0]^{-1} + \frac{s-2}{2}[I] \end{bmatrix} \tag{57}$$

The following eigenvalue problem is solved (actually the same as for the static case)

$$[Z]\begin{bmatrix}[\Phi_{11}] & [\Phi_{12}]\\ [\Phi_{21}] & [\Phi_{22}]\end{bmatrix} = -\begin{bmatrix}[\Phi_{11}] & [\Phi_{12}]\\ [\Phi_{21}] & [\Phi_{22}]\end{bmatrix}\begin{bmatrix}-\lceil\lambda_i\rfloor & \\ & \lceil\lambda_i\rfloor\end{bmatrix} \tag{58}$$

The real parts of $\lceil\lambda_i\rfloor$ are smaller than zero. The series solution of equation (55) equals [22]

$$[X(\bar{\xi})] = [\Phi]\left(\sum_{k=0}[A_k]\bar{\xi}^{2k}\right)\bar{\xi}^{\lceil\Lambda\rfloor}\bar{\xi}^{[U]}\begin{Bmatrix}\{c_1\}\\ \{c_2\}\end{Bmatrix} \tag{59}$$

with $\lceil\Lambda\rfloor$ denoting the negative values of the eigenvalues of $[Z]$ in equation (57) arranged in descending order of their real parts and $\{c_1\}$, $\{c_2\}$ being integration constants. The coefficient matrices $[A_k]$ and $[U]$ are determined as specified in references [22] and [6].

Equation (59) represents two independent sets of solutions. The two sets of integration constants of the general solution follow from the boundary conditions at two values of $\bar{\xi}$ denoted as $\bar{\xi}_i$ (i for interior) and $\bar{\xi}_e$ (e for exterior). For the unbounded soil $\xi_i = 1$ and $\bar{\xi}_e \to \infty$ should be chosen. However, problems occur when enforcing the radiation condition for $\bar{\xi}_e \to \infty$. No obvious choice of $\{c_1\}$ and $\{c_2\}$ satisfying this condition exists. To calculate $[X(\bar{\xi})]$ for $\bar{\xi} \to \infty$ would require an infinite number of terms if the series solution (equation (59)) were used. This is obviously not feasible. As an alternative, the high frequency asymptotic expansion of the dynamic stiffness matrix $\xi^{2-s}[S^{\infty}(\omega,\xi)]$ (equation (25)) is applied, which satisfies the radiation condition at infinity. The unbounded soil of Figure 2a is replaced by the bounded domain of Figure 6. On its exterior boundary ξ_e, which is similar to the structure-soil interface ($\xi_i = 1$), the nodal force-displacement relationship determined by $[S_e^{\infty}(\omega,\xi_e)]$ (calculated for a large but finite $\bar{\xi}_e = \omega\xi_e$) serves as the boundary condition. This procedure leads to the integration constants $\{c_1\}$, $\{c_2\}$. The series solution follows from equation (59). As both the amplitudes of the displacements and of the internal nodal forces (interaction forces) are known, the dynamic stiffness matrix is determined straightforwardly.

The dynamic stiffness matrix of the unbounded soil on the boundary can be calculated as (equation (54))

$$[S^{\infty}(\omega)] = [S_{ii}(\omega)] - [S_{ie}(\omega)]([S_e^{\infty}(\omega)] + [S_{ee}(\omega)])^{-1}[S_{ei}(\omega)] \tag{60}$$

with the submatrices $[S_{ii}(\omega)]$, $[S_{ie}(\omega)]$, $[S_{ei}(\omega)]$ and $[S_{ee}(\omega)]$ of the bounded

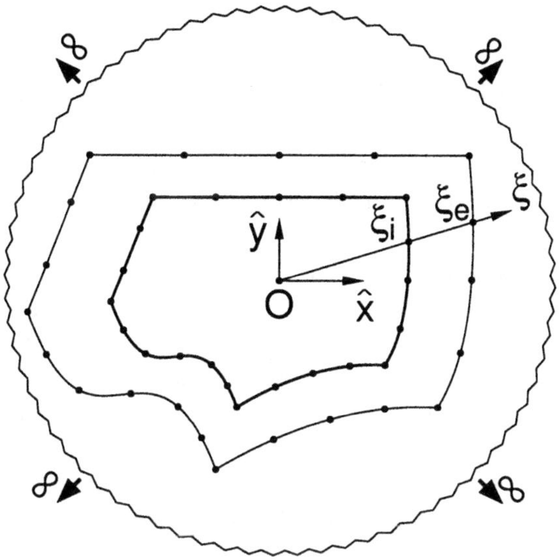

Figure 6. Unbounded soil on exterior of boundary at $\xi_i = 1$ replaced by adjacent bounded soil with similar exterior boundary at ξ_e where high frequency asymptotic expansion of dynamic stiffness matrix is enforced

medium $\xi_i = 1 \leq \xi \leq \xi_e$, determined analytically, equal to

$$\begin{bmatrix} [S_{ii}(\omega)] & [S_{ie}(\omega)] \\ [S_{ei}(\omega)] & [S_{ee}(\omega)] \end{bmatrix} = \begin{bmatrix} -[X_{21}(\omega)] & -[X_{22}(\omega)] \\ \xi_e^{\frac{s-2}{2}}[X_{21}(\bar{\xi}_e)] & \xi_e^{\frac{s-2}{2}}[X_{22}(\bar{\xi}_e)] \end{bmatrix} \times \begin{bmatrix} [X_{11}(\omega)] & [X_{12}(\omega)] \\ \xi_e^{-\frac{s-2}{2}}[X_{11}(\bar{\xi}_e)] & \xi_e^{-\frac{s-2}{2}}[X_{12}(\bar{\xi}_e)] \end{bmatrix}^{-1} \quad (61)$$

An analytical solution is also possible for equation (35), yielding $\{u(\bar{\xi})\}$ for $\bar{\xi} > \bar{\xi}_e$. The far field displacements can thus be determined. In these first order differential equations, the boundary condition at $\bar{\xi}_e$ follows from the series expansion in equation (59). The solution procedure for equation (35) in the form of a series is analogous to that for equation (55).

Thus, it is possible to calculate analytically the response throughout the unbounded soil avoiding numerical discretisation in the $\bar{\xi}$ direction. The high frequency asymptotic expansion of the dynamic stiffness matrix in the far field has to be introduced as an approximation. Room for improvement by eliminating this approximation thus exists!

7. Numerical Solution in Frequency and Time Domains [2], [14]

The governing ordinary differential equations can also be solved numerically i the $\bar{\xi}$ direction. Thus, a numerical procedure is applied in all directions. The trans formation to the scaled boundary coordinates ξ, η, ζ does, however, permit th radiation condition to be introduced rigorously for $\xi \to \infty$, which is a significan advantage.

To be able to enforce the boundary condition at infinity and at the structure-soi interface, the scaled boundary finite-element equation in displacement (equatio (16)) is not solved directly. It is replaced by the two first order differential equa tions, the scaled boundary finite-element equation in dynamic stiffness (equatio (22)) and the interaction force-displacement relationship (equation (21)).

In the first step, equation (22) is solved, incorporating the radiation condition The boundary condition (starting value) is calculated for a very large but finit $\bar{\xi}_h$ from the high frequency asymptotic expansion of the dynamic stiffness matri (equation (25))

$$\xi_h^{2-s}[S^{\infty}(\omega,\xi_h)] \approx \mathrm{i}\bar{\xi}_h[C_{\infty}] + [K_{\infty}] + \sum_{j=1}^{m} \frac{1}{(\mathrm{i}\bar{\xi}_h)^j}[A_j] \qquad (62$$

The radiation condition is satisfied (either by constructing a positive definite $[C_{\infty}$ from an eigenvalue problem (equation (26)) or using the impedances (equatio (34)). The non-linear first order ordinary differential equation (22) is then solve starting from $\bar{\xi}_h$ for decreasing $\bar{\xi}$ down to the structure-soil interface ($\xi_i = 1$ – $\bar{\xi}_i = \omega$). A fourth order Runge-Kutta scheme [23] is applied. At the beginnin for large $\bar{\xi}$ where the variables vary smoothly, an adaptive integration step siz is determined. Later on, for smaller $\bar{\xi}$ a fixed integration step size is selected This yields the dependent variable $\xi^{2-s}[S^{\infty}(\omega,\xi)]$ as a function of $\bar{\xi}$ (or of th dimensionless frequency a or a_0). The error introduced through the boundary a $\bar{\xi}_h$ diminishes for decreasing $\bar{\xi}$.

In the second step, equation (21) is solved with the known $\xi^{2-s}[S^{\infty}(\omega,\xi)$ from the first step. The boundary condition at the structure-soil interface can be ex pressed in displacement amplitudes. If interaction force amplitudes are prescribed displacement amplitudes follow from solving equation (1) using the dynamic s tiffness matrix. The linear first order ordinary differential equation (equation (21) is then solved from $\bar{\xi} = \omega$ for increasing $\bar{\xi}$. A forward Euler scheme is applied The same integration step size as in the first step is used, as $\xi^{2-s}[S^{\infty}(\omega,\xi)]$ mus be known. This yields the dependent variable $\{u(\bar{\xi})\}$.

A numerical scheme can also be used to solve equation (35) for increasin $\bar{\xi}$ yielding the far field displacement amplitudes. The boundary condition at $\bar{\xi}$ follows either from the second step or from the analytical solution (Section 6).

To determine the response matrix to a unit impulse of accelerations $[M^{\infty}(t)$ at distinct time stations the integral equation (46) can be discretised for increasin time. A (linear) Lyapunov equation results with the coefficient matrix independent of the time step. At each time station a back substitution is performed.

8. Extensions [2]

8.1. INCOMPRESSIBLE ELASTICITY

The saturated soil when undrained behaves as a nearly incompressible material in a one-phase formulation. Incompressible elasticity (Poisson's ratio equal to 0.5) requires a special approach in the displacement based finite element method. The elasticity matrix is decomposed into the shear and volumetric parts. Selective reduced integration is used with a Poisson's ratio very close to 0.5. The same procedure can also be applied to the scaled boundary finite-element method. As a further step, the limit of Poisson's ratio equal to 0.5 can be enforced *analytically*. The formulation can thus be streamlined.

The response of an incompressible unbounded soil is *instantaneous* in the entire domain owing to the infinite dilatational wave velocity. This manifests itself by a mass which does not appear in the compressible case. The high frequency response for the incompressible soil is dominated by concentrated masses located in the nodes on the structure-soil interface whereby dashpots are also present.

8.2. VARIATION OF MATERIAL PROPERTIES IN RADIAL DIRECTION

In all derivations up to now the unbounded soil is assumed to be homogeneous in the radial direction towards infinity, that is the elasticity matrix $[D]$ and the mass density ρ are constant in the radial direction. As an extension, $[D]$ and ρ are assumed to be power functions of r. The shear modulus and mass density vary as

$$G(r) = G_0 \left(\frac{r}{r_0}\right)^g \tag{63a}$$

$$\rho(r) = \rho_0 \left(\frac{r}{r_0}\right)^m \tag{63b}$$

G_0 and ρ_0 correspond to the structure-soil interface with the characteristic length (radial coordinate) r_0. The powers g and m are real numbers.

For this case, the dimensionless frequency is defined as (equation (17))

$$a_0 = \frac{\omega}{c_{s0}} \frac{r^{1-\frac{g}{2}+\frac{m}{2}}}{r_0^{-\frac{g}{2}+\frac{m}{2}}} \tag{64}$$

with $c_{s0} = \sqrt{G_0/\rho_0}$.

As an example of the results, the scaled boundary finite-element equation in dynamic stiffness on the structure-soil interface (corresponding to equation (23) for the homogeneous case) is given below

$$([S^{\infty}(\omega)] + [E^1])[E^0]^{-1}([S^{\infty}(\omega)] + [E^1]^T) - (g + s - 2)[S^{\infty}(\omega)] \\ - \left(1 - \frac{g}{2} + \frac{m}{2}\right) \omega [S^{\infty}(\omega)]_{,\omega} - [E^2] + \omega^2 [M^0] = 0 \quad (65)$$

The coefficient matrices $[E^0]$, $[E^1]$, $[E^2]$, $[M^0]$ are calculated with the material properties at the structure-soil interface.

The solution procedures for the homogeneous case still apply.

8.3. REDUCED SET OF BASE FUNCTIONS [13]

To increase the computational efficiency of the scaled boundary finite-element method, the displacement amplitudes $\{u(\bar{\xi})\}$ in equation (16) are represented by a reduced set of base functions $[\Phi]$ and corresponding amplitudes $\{v(\bar{\xi})\}$

$$\{u(\bar{\xi})\} = [\Phi]\{v(\bar{\xi})\} \quad (66)$$

Equation (16) is transformed to

$$[e^0]\bar{\xi}^2\{v(\bar{\xi})\}_{,\bar{\xi}\bar{\xi}} + ((s-1)[e^0] - [e^1] + [e^1]^T)\bar{\xi}\{v(\bar{\xi})\}_{,\bar{\xi}} \\ + ((s-2)[e^1]^T - [e^2])\{v(\bar{\xi})\} + [m^0]\bar{\xi}^2\{v(\bar{\xi})\} = 0 \quad (67)$$

with the coefficient matrices

$$[e^0] = [\Phi]^T[E^0][\Phi \quad (68a)$$

$$[e^1] = [\Phi]^T[E^1][\Phi \quad (68b)$$

$$[e^2] = [\Phi]^T[E^2][\Phi \quad (68c)$$

$$[m^0] = [\Phi]^T[M^0][\Phi] \quad (68d)$$

As the dynamic stiffness at low frequency dominates in many cases the dynamic response in soil-structure interaction, the reduced set of base functions $[\Phi]$ is selected from the solution of equation (16) for statics. The first few eigenfunctions corresponding to the lowest rate of decay of displacements in the radial direction determine $[\Phi]$ with the eigenvalues $\lceil\Lambda\rfloor$, which are specified in equations (57) and (58)

$$\begin{bmatrix} [E^0]^{-1}[E^1]^T - \frac{s-2}{2}[I] & -[E^0]^{-1} \\ -[E^2] + [E^1][E^0]^{-1}[E^1]^T & -[E^1][E^0]^{-1} + \frac{s-2}{2}[I] \end{bmatrix} [\Phi] = -[\Phi]\lceil\Lambda\rfloor \quad (69)$$

This reduction in size from $\{u(\bar{\xi})\}$ to $\{v(\bar{\xi})\}$ enables realistic dynamic soil-structure interaction problems with many degrees of freedom to be analysed.

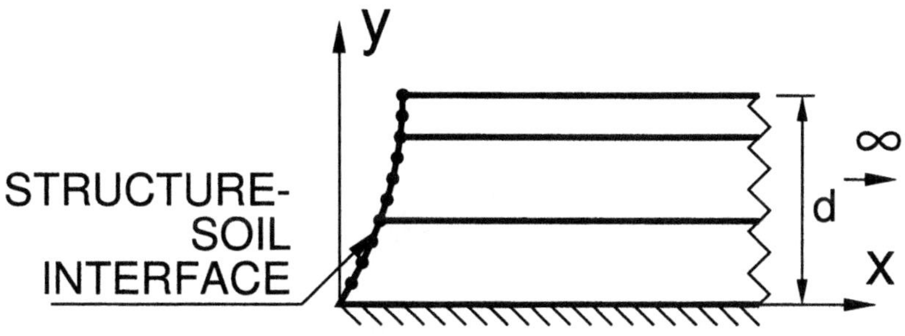

Figure 7. Two dimensional layered unbounded soil with structure-soil interface discretised with line finite elements

8.4. TWO DIMENSIONAL LAYERED UNBOUNDED SOIL

A layered unbounded soil is often encountered in practice (Figure 7). The unbounded soil is enclosed by two parallel boundaries extending to infinity (e.g. free and fixed) and the structure-soil interface which can be curved. Often the material properties vary in the direction perpendicular to the boundaries extending to infinity.

For the two dimensional case the scaling centre is located at infinity. In this special case, the so-called consistent boundary method, also called the thin layer method [24], leads to the same relations. The dimensionless frequency is defined as

$$a_0 = \frac{\omega d}{c_s} \tag{70}$$

with the (constant) depth d of the soil layers. a_0 is independent of the location of the structure-soil interface. This permits a streamlined formulation.

The scaled boundary finite-element method in displacement equals

$$[E^0]\{u(\xi)\},_{\xi\xi} + ([E^0] - [E^1] + [E^1]^T)\{u(\xi)\},_{\xi} \\ - [E^2]\{u(\xi)\} + \omega^2[M^0]\{u(\xi)\} = 0 \tag{71}$$

which is a linear second order ordinary differential equation with constant coefficients (compared to equation (13) with variable coefficients).

The scaled boundary finite-element method in dynamic stiffness is formulated as

$$([S^\infty(\omega)] + [E^1])[E^0]^{-1}([S^\infty(\omega)] + [E^1]^T) - [E^2] + \omega^2[M^0] = 0 \tag{72}$$

Compared to equation (23) with $s = 2$, the term with the derivative $\omega[S^\infty(\omega)],_\omega$ is missing. This allows $[S^\infty(\omega)]$ at distinct frequencies to be determined directly.

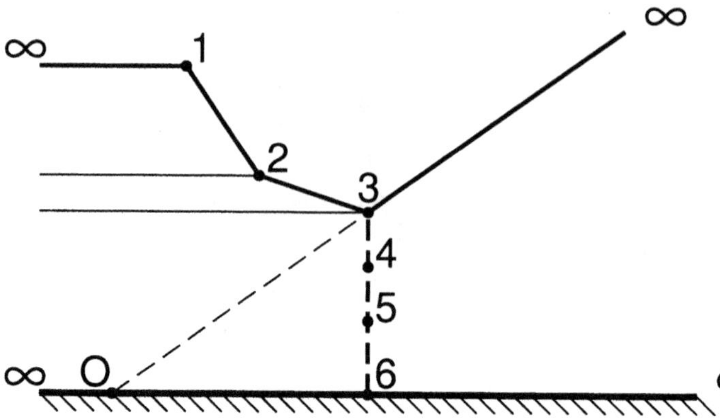

Figure 8. Substructuring with additional discretisation on common boundary

8.5. SUBSTRUCTURING

For the unbounded soil, the scaling centre is chosen in a zone from which the discretised structure-soil interface is visible (see Figure 2). To avoid discretisation of the side faces such as a free surface, the scaling centre lies at the intersection of the extensions of the side faces. Thus limitations exist.

A further class of problems can be analysed using substructuring, which for unbounded domains is, however, limited. Each substructure has its own scaling centre, and spatial discretisation is also required on the common boundaries.

A two dimensional example consisting of a site fixed at its base, with parallel and inclined free surfaces is shown in Figure 8. Nodes 1, 2, 3 define the structure-soil interface, where the dynamic stiffness matrix is to be calculated. Two substructures are selected, the first structure with parallel layers on the left of line 1, 2, 3, 4, 5, 6 with the scaling centre located on the right at infinity, and the second substructure on the right of the line 3, 4, 5, 6 with the inclined free surface and fixed base, determining its scaling centre O. Additional degrees of freedom are introduced on the common boundary in nodes 4, 5, 6. An analysis of the scaled boundary finite-element equation in dynamic stiffness for the first substructure (equation (72)) yields the dynamic stiffness with degrees of freedom in nodes 1 to 6, an analogous computation for the second substructure using equation (23) leads to the dynamic stiffness matrix with degrees of freedom in nodes 3 to 6. Assembling the two dynamic stiffness matrices and eliminating the degrees of freedom in nodes 4, 5, 6 yields the final result, the dynamic stiffness matrix of the site at the structure-soil interface with nodes 1, 2, 3.

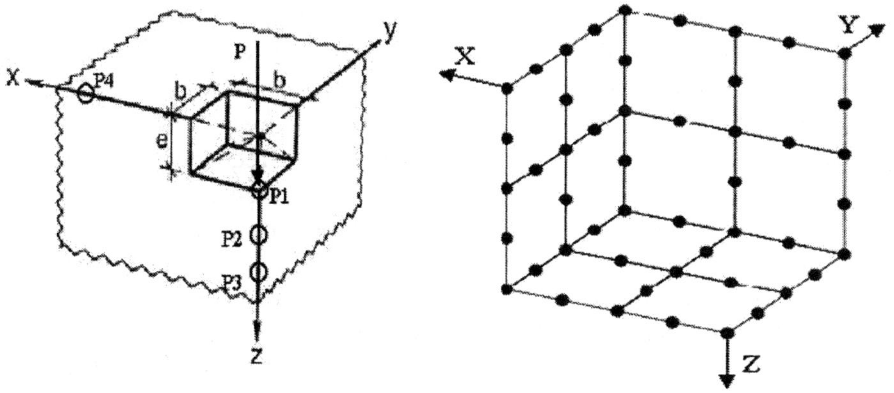

Figure 9. One quarter of square prism embedded in half space: (a) geometry with scaling centre; (b) finite element discretisation of structure-soil interface

9. Numerical Examples

9.1. PRISM FOUNDATION EMBEDDED IN HALF SPACE [14]

The first example is mainly solved numerically. As a truly three dimensional problem, a square prism of length $2b$ embedded with depth e in a homogeneous half space governed by the vector wave equation of elastodynamics is addressed. Due to symmetry, only a quarter of the embedded prism shown in Figure 9 is addressed. A rigid structure-soil interface (base mat and side walls) is introduced. A vertical harmonic load $P(\omega)$ of frequency ω acts in the centre of the base mat. The vertical displacements of the base mat (Point P_1), below the foundation (Points P_2 and P_3) and on the free surface (Point P_4) are to be calculated.

The dimensions of the foundation follow from $b = 0.5$m, $e = 0.5$m. The material properties of the half space equal shear wave velocity $c_s = 250$m/s, Poisson's ratio $\nu = 0.4$ and mass density $\rho = 2000\text{kg/m}^3$, resulting in a dilatational wave velocity $c_p = c_s\sqrt{2(1-\nu)/(1-2\nu)} \approx 2.45\, c_s \approx 612$m/s and a Rayleigh wave velocity $c_R = 0.94\, c_s = 235$m/s. The frequency f of the harmonic load varies from 0 to 200Hz in increments of 10Hz ($\omega = 2\,\pi\, f$). The result points are specified by the coordinates $P_1(0,0,0.5)$, $P_2(0,0,1.5)$, $P_3(0,0,3)$, $P_4(3,0,0)$.

For comparison, the results of a boundary element method analysis in the frequency domain [25] are used. Constant boundary elements with the fundamental solution of the full space are applied. Besides the structure-soil interface, the free surface must also be discretised up to a distance of 6m with boundary elements of lengths varying from 0.25m to 0.4m. For a quarter of the embedded foundation almost 300 boundary elements are introduced.

In the scaled boundary finite-element method, the scaling centre is chosen at the origin of the coordinate system. This permits the free surface of the half spac to be identified as a side face (Figure 2b) without any discretisation. The spatia discretisation is thus limited to the interface of the foundation. Due to symmetry only a quarter is analysed. The finite element mesh of this structure-soil interfac using 12 8-node elements is shown in Figure 9b. This leads to 49 nodes wit 129 degrees of freedom after enforcing the symmetry boundary conditions. Th shortest wave length calculated for the highest frequency $f = 200$Hz and wit c_s equals $c_s/f = 1.25$m. With the distance between two adjacent nodes equal t 0.125m, 10 nodes per wave length are thus present. (An analysis performed usin a finer mesh with 27 8-node elements yielding 274 degrees of freedom essentiall confirms the results).

The numerical solution procedure of Section 7 is applied.

The analysis for this three dimensional case ($s = 3$) proceeds as follows. Th order of the vector $\{u(\bar{\xi})\}$ is equal to the number of degrees of freedom, th is 129, and the coefficient matrices $[E^0], [E^1], [E^2], [M^0]$ and dynamic stiffnes matrix $[S^{\infty}]$ are of order 129×129. Solving the eigenvalue problem (equatio (27)) permits $[C_{\infty}]$ (equation (26)), $[K_{\infty}]$, $[A_1]$ ($j = 1$) to be calculated. The hig frequency asymptotic expansion of the dynamic stiffness matrix evaluated at $\bar{\xi}$ = 30 $c_s/\ b$ yields the boundary condition $(1/\xi)[S^{\infty}]$ (equation (25)). This allow equation (22) to be solved numerically for decreasing $\bar{\xi}$ down to $\bar{\xi} = 0$ leading t $(1/\xi)[S^{\infty}]$ as a function of $\bar{\xi}$ for all $\bar{\xi}$. In particular, at the structure-soil interfac $\xi = 1$ $[S^{\infty}]$ of order 129×129 is a function of ω corresponding to the dynami stiffness matrix of a flexible interface. Enforcing the rigid interface constrains th degrees of freedom. This condition of all vertical displacement amplitudes bein equal to the scalar $u_0(\omega)$ and all horizontal displacement amplitudes vanishin defines the vector $\{\phi\}$, leading to

$$\{u(\omega)\} = \{\phi\}\ u_0(\omega) \tag{73}$$

The corresponding dynamic stiffness coefficient of the rigid interface follows as

$$S^{\infty}(\omega) = \{\phi\}^T[S^{\infty}]\{\phi\} \tag{74}$$

It is customary to introduce the dimensionless frequency at the structure-soil in terface $\xi = 1$, $a_0 = \omega\ b/\ c_s$, to non-dimensionalise S^{∞} with the static-stiffnes coefficient K^{∞} ($= S^{\infty}(\omega = 0)$) and to apply the following decomposition

$$S^{\infty}(a_0) = K^{\infty}(k(a_0) + ia_0c(a_0)) \tag{75}$$

with the dimensionless spring coefficient $k(a_0)$ and damping coefficient $c(a_0)$.

With $K^{\infty} \approx 6.86\ 10^8$ N/m, $k(a_0)$ and $c(a_0)$ are plotted as a function a_0 in Figure 10. In addition the results avoiding solving the eigenvalue probler (equation (27)) and with $[K_{\infty}] = [A_1] = 0$ in the asymptotic high frequenc expansion (equation (25)) are presented. In this case $[C_{\infty}]$ is directly construct ed from the impedances. Distributed dashpots with the coefficients ρc_p for th

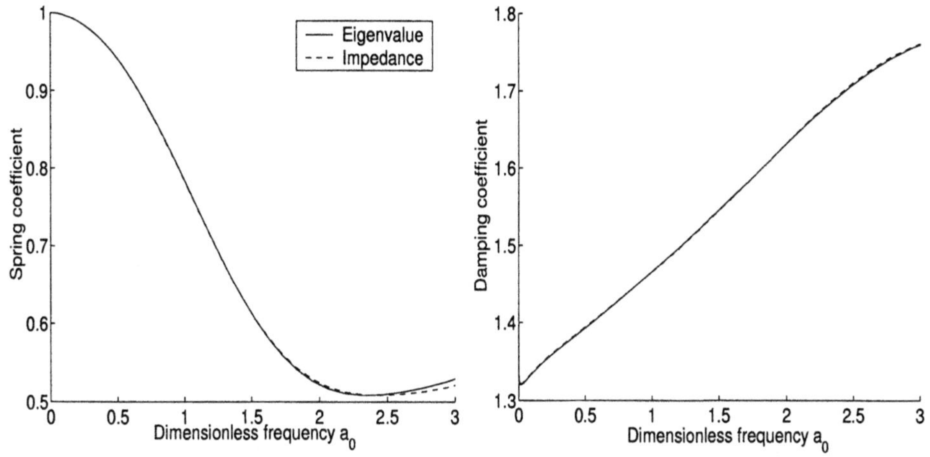

Figure 10. Vertical dynamic stiffness coefficient of embedded prism foundation

degrees of freedom perpendicular to the finite element and ρc_s for those in the plane of the finite element act on the structure-soil interface. The $[C_\infty]$ matrix of each finite element involves integrations of the products of shape functions. It is thus equal to the standard two dimensional mass matrix but multiplied by the corresponding wave velocity. Equation (22) is again solved numerically. Good agreement of the dynamic stiffness coefficient in Figure 10 is achieved.

It is worth noting that equation (22) with the boundary condition (equation (25)) involves an independent variable $\bar{\xi}$ which is the product of ξ with ω. For the various ω, this operation has to be performed only once. Each coefficient of $(1/\xi)[S^\infty]$ could be decomposed and plotted as in Figure 10 as a function of $\bar{\xi}$ or of the dimensionless frequency at ξ, $\omega b\xi/c_s = b\bar{\xi}/c_s$.

Applying the load with the amplitude $P(\omega)$ to the foundation with a rigid interface results in the vertical displacement amplitude $u_0(\omega)$ of the foundation

$$u_0(\omega) = \frac{P(\omega)}{S^\infty(\omega)} \tag{76}$$

Equation (73) then yields the boundary condition at the interface $\bar{\xi} = \omega$ enabling the integration of equation (21) for increasing $\bar{\xi}$ up to the value corresponding to the result point. $(1/\xi)[S^\infty]$ for each $\bar{\xi}$ has been determined earlier by solving equation (22). The corresponding displacement amplitudes $u(\bar{\xi})$ are complex.

The vertical displacement magnitudes $|u_z(\omega)| = \sqrt{(\text{Re}u_z(\omega))^2 + (\text{Im}u_z(\omega))^2}$, non-dimensionalised with $K^\infty/P(\omega)$, are plotted as a function of the dimensionless frequency a_0 in the four result points in Figure 11. The result in Point P_1 (base mat) corresponds to $u_0(\omega)$ and thus to S^∞ (equation (76)). Good agreement with the boundary element solution in all points and for all frequencies is achieved.

The variation of the vertical displacement for a specific frequency along the vertical z-axis and the horizontal x-axis is represented in Figures 12 and 13. The

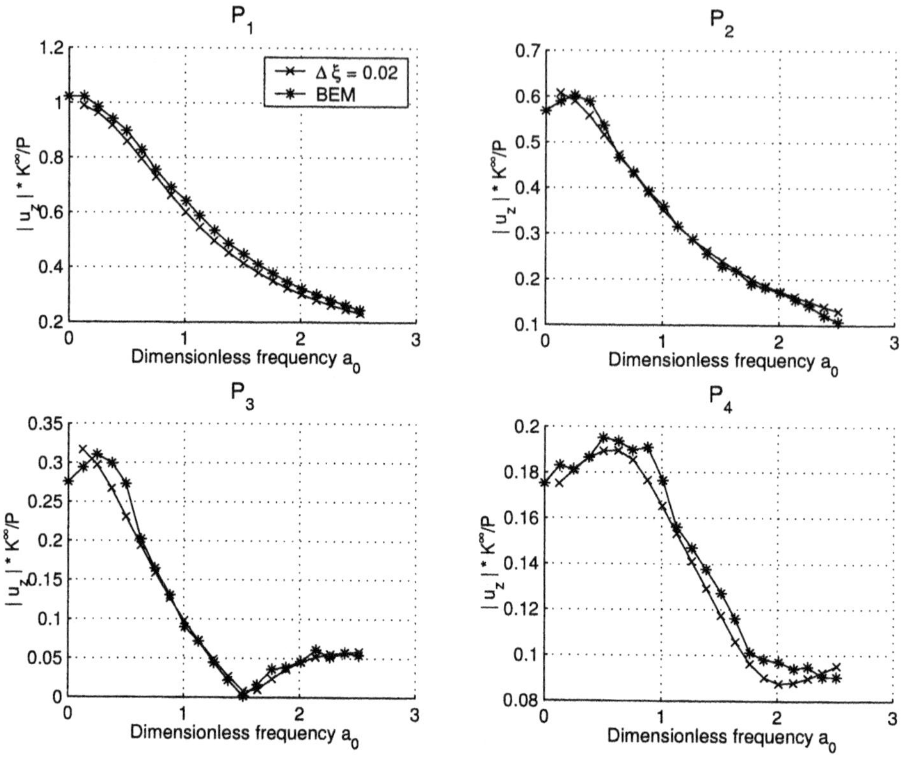

Figure 11. Dimensionless magnitude of vertical displacement

real and imaginary parts are non-dimensionalised by multiplying by the factor $K^{\infty}/P(\omega)$. The vertical and horizontal distances are also non-dimensional. The frequencies $f = 100$Hz and $f = 200$Hz are processed. Note that the variation both in the horizontal and vertical directions is governed by the Rayleigh-wave velocity c_R. The corresponding wave length equals c_R/f which for example for $f = 200$Hz results in 1.17m, or in non-dimensional form by dividing by e or b in 2.35. This is clearly visible in Figures 12 and 13. Thus, for this intermediate to high frequency range, the motion is caused by surface waves.

9.2. SPHERICAL CAVITY IN FULL SPACE WITH SPHERICAL SYMMETRY [6]

This second example is solved analytically.

A spherical cavity of radius r embedded in a full space with shear modulus G, Poisson's ratio $\nu = 0.25$ and mass density ρ is examined (Figure 14). The wall of the spherical cavity corresponds to the structure-soil interface and the full space to the unbounded soil. The constant normal displacement on the structure-soil interface is denoted by u_0. Spherical symmetry exists. The dynamic stiffness coefficient at the structure-soil interface is to be calculated.

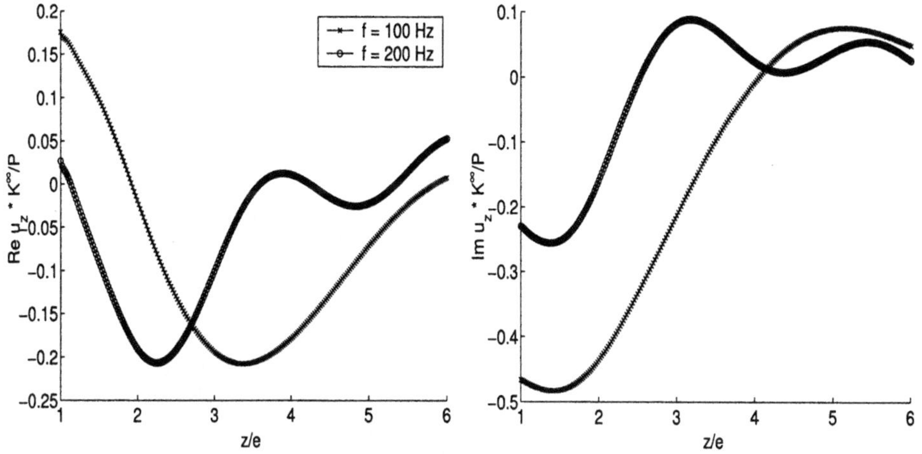

Figure 12. Amplitude of vertical displacement along vertical z-axis (below base-mat)

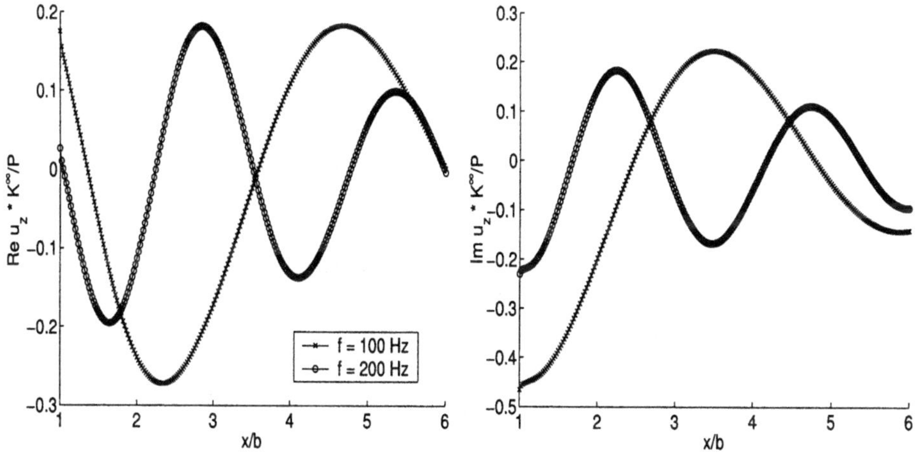

Figure 13. Amplitude of vertical displacement along horizontal x-axis (free surface)

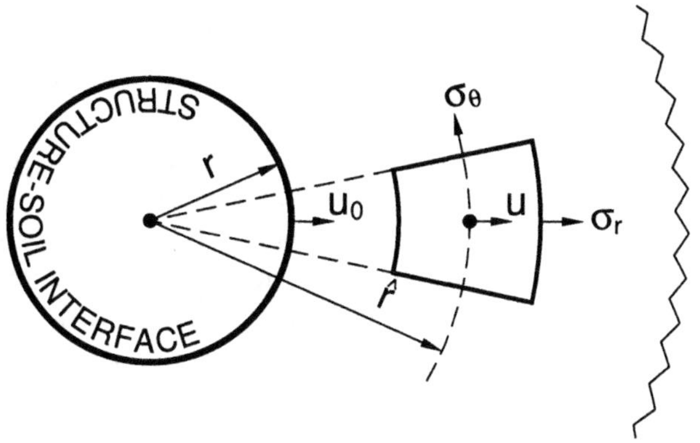

Figure 14. Spherical cavity embedded in full space (section) with symmetric waves

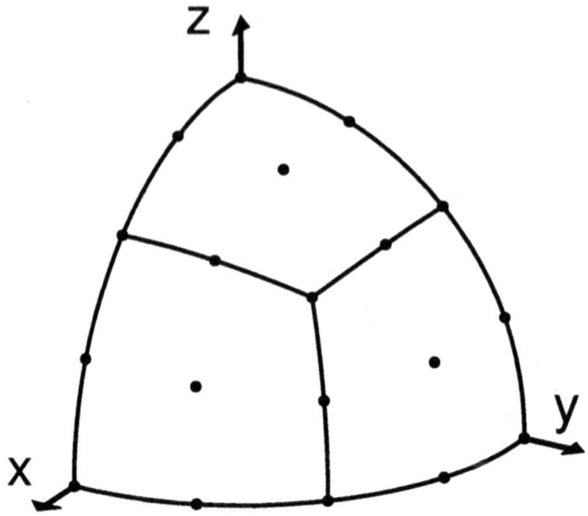

Figure 15. Finite element mesh of one octant of structure-soil interface (wall of cavity)

The analytical solution of the dynamic stiffness coefficient equals

$$S^{\infty}(a_0) = 16\pi G r (k(a_0) + \mathrm{i} a_0 c(a_0)) \tag{77}$$

with the dimensionless spring and damping coefficients

$$k(a_0) = 1 - \frac{1-\nu}{2(1-2\nu)} \frac{a_0^2}{1+a_0^2} \tag{78a}$$

$$c(a_0) = \frac{1-\nu}{2(1-2\nu)} \frac{a_0^2}{1+a_0^2} \tag{78b}$$

and the dimensionless frequency

$$a_0 = \frac{\omega r}{c_p} \tag{79}$$

The spherical cavity is solved with the scaled boundary finite-element method as a three dimensional problem with nine node surface finite elements. The mesh of one octant consists of three finite elements as shown in Figure 15. On the boundary 294 degrees of freedom are introduced. The analytical solution procedure of Section 6 is applied. The bounded domain between the interior boundary with radius r ($\xi_i = 1$) and the exterior boundary with radius $2r$ ($\xi_e = 2$), a hollow sphere, is considered. In the series solution of equation (59) 18 terms ($k = 17$) are selected. The submatries $[S_{ii}(\omega)]$, $[S_{ie}(\omega)]$, $[S_{ei}(\omega)]$, $[S_{ee}(\omega)]$ are of order 294×294. After enforcing the uniform radial displacement, the four dynamic stiffness coefficients (equation (61)) follow. An analytical solution exists for comparison. Excellent agreement, e.g. for $S_{ii}(a_0)$ is achieved up to $a_0 = 6$, as shown in Figure 16, although a strong dependency on a_0 exists.

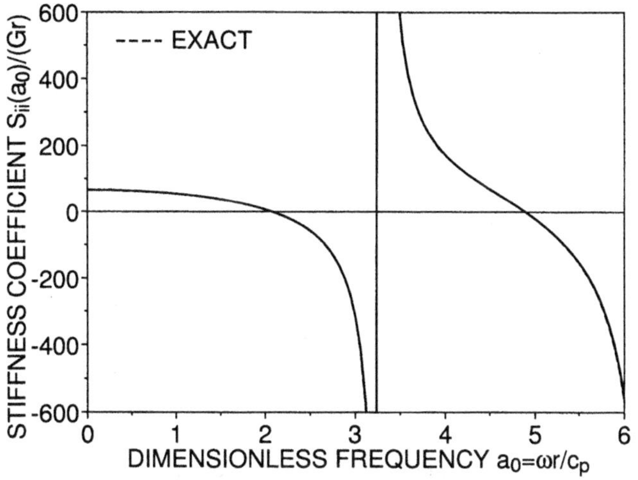

Figure 16. Dynamic stiffness coefficient of hollow sphere (diagonal coefficient for interior boundary)

The high frequency asymptotic expansion for $(1/\xi_e)[S_e^{\infty}(\omega, \xi_e)]$ of order 294×294 (equation (25)) is evaluated at $\bar{\xi}_e = 3c_p/r$ with $m = 1$. The dynamic stiffness matrix $[S^{\infty}(\omega)]$ of order 294×294 follows from equation (60)) with the dynamic stiffness matrix $[S_{ii}(\omega)]$, $[S_{ie}(\omega)]$, $[S_{ei}(\omega)]$, $[S_{ee}(\omega)]$ of the corresponding hollow sphere (with $\xi_i = 1$) calculated with 7 terms ($k = 6$) in the power series of equation (59). The corresponding dynamic stiffness coefficient of the spherical cavity $S^{\infty}(a_0)$ with enforced uniform radial symmetry agrees well with the analytical solution (equations (77) and (78)) as is seen in Figure 17. It can be concluded that $S^{\infty}(a_0)$ can be directly calculated accurately for any a_0 based on the high frequency asymptotic expansion. Note that the error decreases for diminishing a_0.

9.3. IN-PLANE MOTION OF SEMI-INFINITE WEDGE [13]

This third example demonstrates the efficient use of a reduced set of base functions (Section 8.3).

As an example, the in-plane motion of a semi-infinite wedge (Figure 18) with shear modulus G, Poisson's ratio $\nu = 0.25$ and mass density ρ is addressed. One of the boundaries extending to infinity is fixed, and the other is a free surface. The structure-soil interface is an arc of radius r_0 with an opening angle $\alpha = 30°$. On the structure-soil interface a linear function in the circumferential direction of the horizontal motion $u_0(\theta)$ and zero vertical motion are prescribed. The structure-soil interface is discretized with 8 3-node line elements, which leads to 32 degrees of freedom. The equivalent spring and damping coefficients defined as in equation (75) with $K = 0.752G$ and $a_0 = \omega r_0/c_s$ are plotted in Figure 19 as solid lines.

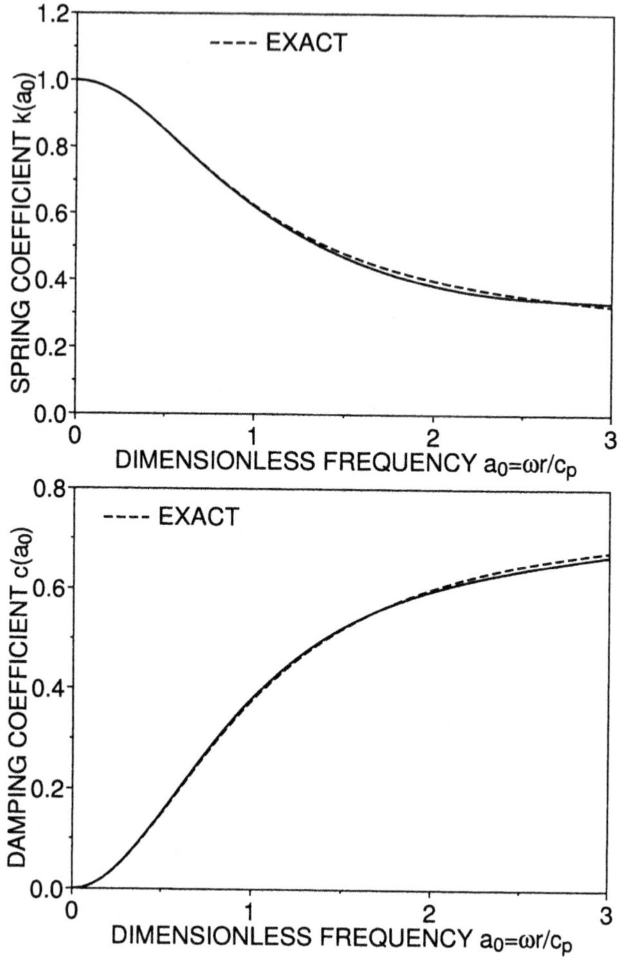

Figure 17. Dynamic stiffness coefficient of spherical cavity

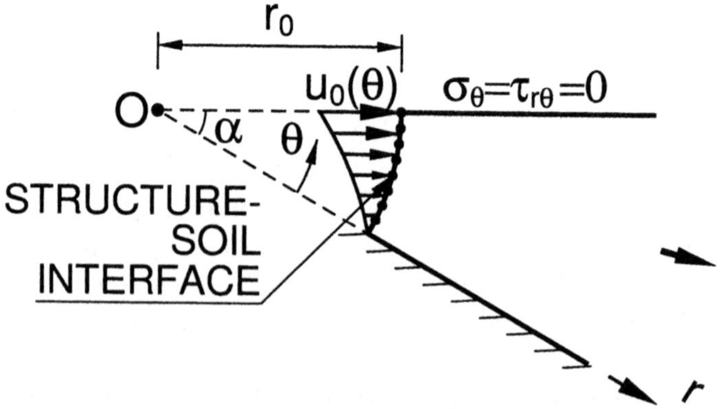

Figure 18. In-plane motion of semi-infinite wedge with prescribed horizontal displacement

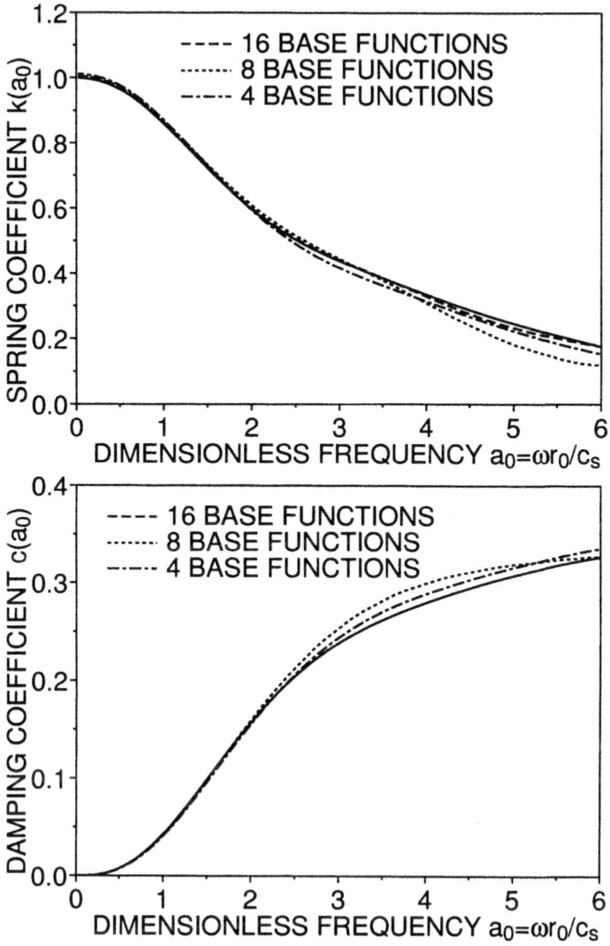

Figure 19. Dynamic stiffness coefficient of in-plane motion of semi-infinte wedge

The coefficients obtained with reduced numbers of base functions (equation (67)) are also shown. The result with only 4 base functions is close to that with the full set of base functions for $a_0 < 2$.

9.4. IN-PLANE MOTION OF CIRCULAR CAVITY IN FULL PLANE [2]

This fourth example addresses the incompressible unbounded soil (Section 8.1).

As a two dimensional problem with an analytical solution, a circular cavity embedded in a full plane representing the unbounded soil is considered (Figure 20). The shear modulus is denoted by G, Poisson's ratio by ν, and the mass density by ρ, leading to the shear wave velocity $c_s = \sqrt{G/\rho}$ and the dilatational wave velocity $c_p = c_s\sqrt{2(1-\nu)/(1-2\nu)}$. On its rigid wall, the structure-soil

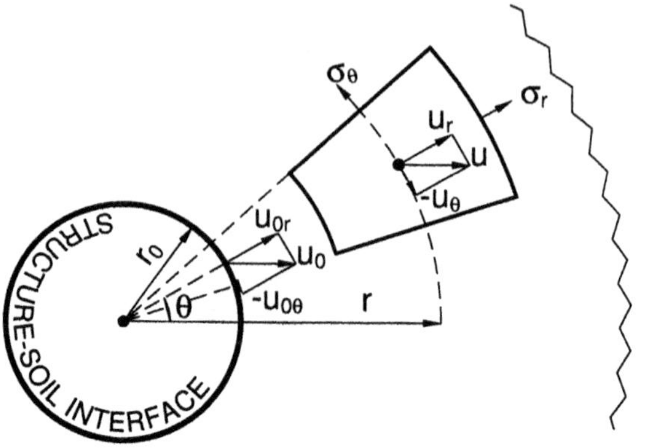

Figure 20. Circular cavity embedded in full plane with prescribed translational motion

interface, a constant horizontal displacement u_0 is enforced.

The analytical solution for the compressible case of the dynamic stiffness coefficient is

$$S^{\infty}(a_0) = \pi G a_0^2 \frac{\alpha + \beta - 4}{\alpha\beta - \alpha - \beta} \tag{80}$$

where

$$\alpha = a_0 \frac{H_0^{(2)}(a_0)}{H_1^{(2)}(a_0)} \tag{81a}$$

$$\beta = \frac{c_s}{c_p} a_0 \frac{H_0^{(2)}(\frac{c_s}{c_p} a_0)}{H_1^{(2)}(\frac{c_s}{c_p} a_0)} \tag{81b}$$

with the dimensionless frequency $a_0 = \omega r_0 / c_s$. $H_0^{(2)}$ and $H_1^{(2)}$ are the second kind Hankel functions of the zeroth and first order, respectively. For the incompressible case ($\nu = 0.5$)

$$S^{\infty}(\omega) = -\pi r_0^2 \rho \omega^2 + 4\pi G a_0 \frac{H_1^{(2)}(a_0)}{H_0^{(2)}(a_0)} \tag{82}$$

applies.

In the scaled boundary finite-element analysis only a quarter of the structure-soil interface is discretised with 4 3-node line elements of equal length. For $\nu = 1/3$, $S^{\infty}(\omega)$, non-dimensionalised with the shear modulus, is decomposed into $k(a_0)$ and $c(a_0)$. Good agreement (Figure 21) with the analytical solution (equation (80)) results. For the incompressible case the same also applies for the scaled boundary finite-element result (Figure 22) when compared with the analytical solution (equation (82)). The effect of the mass is clearly visible.

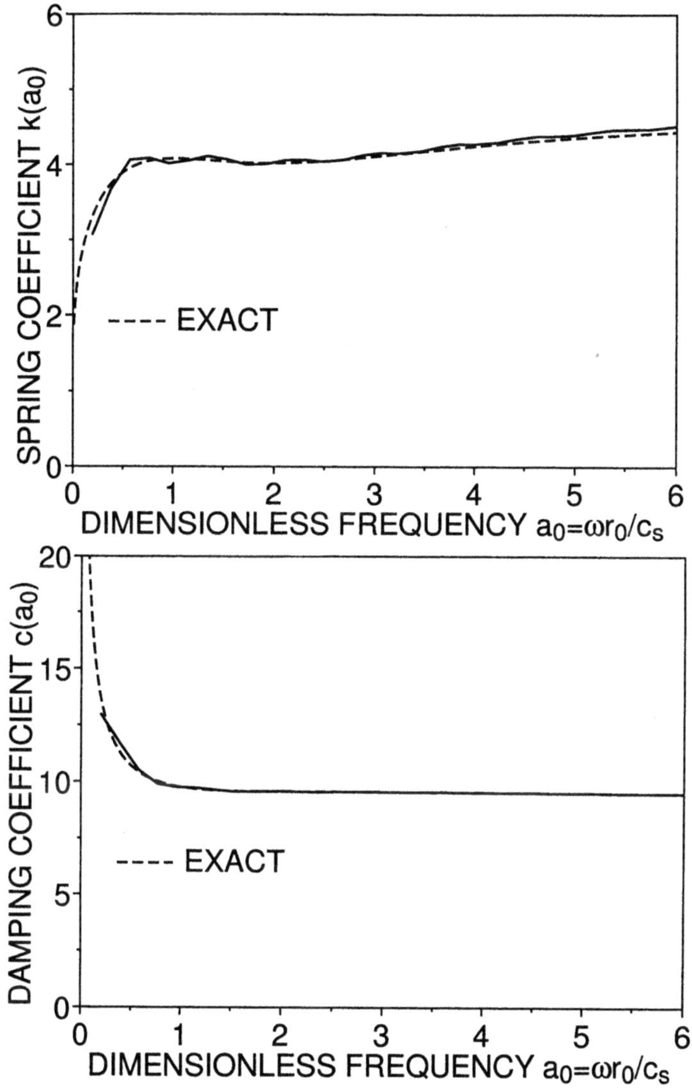

Figure 21. Dynamic stiffness coefficients of circular cavity embedded in compressible full plane

9.5. OUT-OF-PLANE MOTION OF CIRCULAR CAVITY IN FULL PLANE WITH HYSTERETIC DAMPING [15]

This fifth example addresses the unbounded soil with hysteretic damping (Section 4.5).

The out-of-plane motion of a circular cavity of radius r_0 embedded in a full plane with shear wave velocity c_s and hysteretic damping ratio ζ is examined. On the structure-soil interface ($\xi = 1$) the displacement is described as $\cos n\theta$ with the circumferential angle θ and n an integer. The analytical solution for the

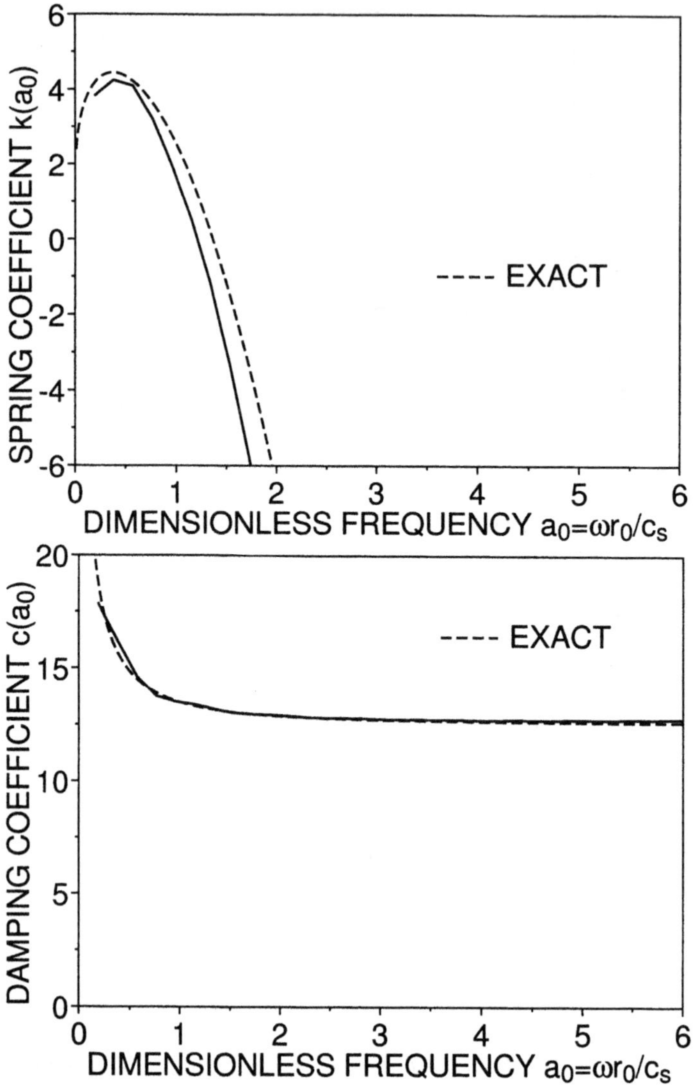

Figure 22. Dynamic stiffness coefficients of circular cavity embedded in incompressible full plane

displacement amplitude equals

$$u_n(\xi, \theta) = c_n H_n^{(2)} \left(\frac{\omega \xi r_0}{c_s^*} \right) \cos n\theta \tag{83}$$

with $c_s^* = \sqrt{1 + 2\mathrm{i}\zeta} c_s$ and c_n following from the boundary condition at $\xi = 1$.

In the scaled boundary finite-element method the total boundary is discretised with 20 2-node finite elements. The high frequency asymptotic expansion for $[S^\infty(\bar{\xi})]$ (equation (38)) is evaluated at $\bar{\xi} = 8c_s/r_0$. Equation (37) is integrated with an increment $\Delta\bar{\xi} = 0.1c_s/r_0$ down to $\xi = 1$ leading to only 70 steps.

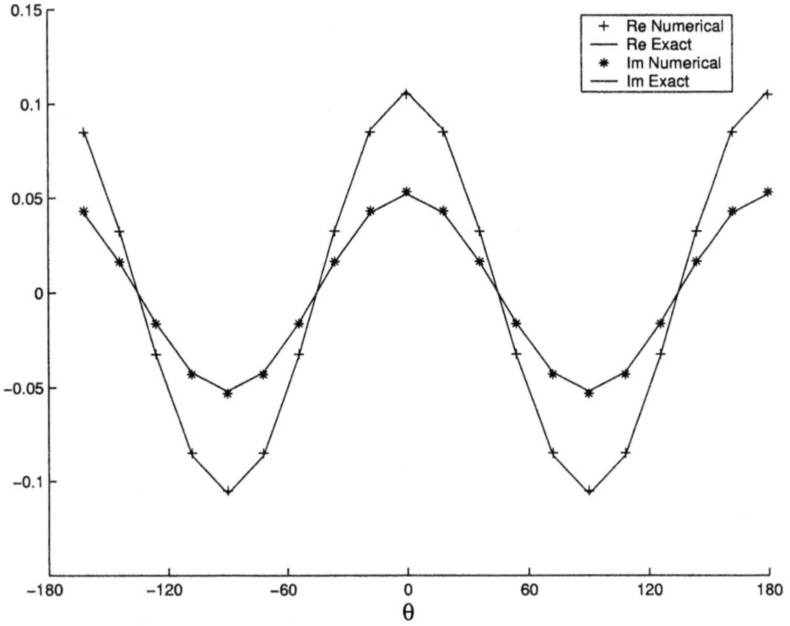

Figure 23. Displacement amplitudes on circumference at $8r_0$ for circular cavity with hysteretic damping

After enforcing the displacement boundary condition at $\xi = 1$, equation (40) is integrated with the same increment. For $n = 2$, $\omega r_0/c_s = 1$ and $\zeta = 0.05$, the displacements on the circumference as a function of θ are plotted in Figure 23. Good agreement with the analytical solution (equation (83)) results.

10. Bounded Medium [16]

The scaled boundary finite-element method can also be applied to model a bounded medium. Its striking advantages are preserved. Some features are mentioned in the following.

In general, the scaling centre O is chosen in the interior of a bounded medium (Figure 24a). The (non-dimensional) radial coordinate ξ points from O towards the boundary. The domain is specified by $0 \le \xi \le 1$. The discretisation is again restricted to the boundary S. As a special case, the scaling centre can be selected on the boundary (Figure 24b), which avoids discretisation on the adjacent straight side faces A. This concept is very attractive for calculating stress singularities in fracture mechanics. Placing the scaling centre at the crack tip, no discretisation of the adjacent (straight) crack faces is required. Since an analytical solution in the ξ direction exists, the stress intensity factors can be calculated directly from their definition, the power of the singularity being calculated in the analysis.

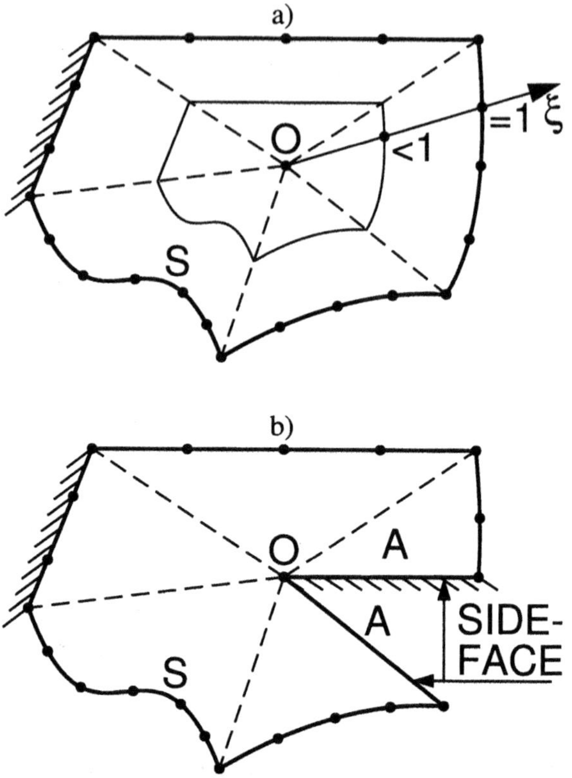

Figure 24. a) Scaling centre within bounded medium with spatial discretisation on boundary; b
Scaling centre on boundary without spatial discretisation on adjacent side faces

For a bounded medium, the salient concept of Section 3 still applies. The tw derivations of Section 4 and 5, appropriately modified, are valid. In particular the scaled boundary finite-element method in displacement (equation (13)) hold without any modification. As the boundary for a bounded medium is a positiv face, a sign change occurs in equation (19), which also reverses the signs o the right hand side of equation (21). This means that in the scaled boundar finite-element equation in dynamic stiffness (equations (22) and (23)) the sig of the dynamic stiffness matrix is reversed ($[S^\infty(\omega)]$ replaced by $-[S^b(\omega)]$ wit superscript b for bounded).

The familiar static stiffness matrix $[K^b]$ and mass matrix $[M^b]$ follow fro substituting the low frequency expansion

$$[S^b(\omega)] = [K^b] - \omega^2[M^b] \tag{84}$$

in the scaled boundary finite-element equation in dynamic stiffness as

$$([K^b] - [E^1])[E^0]^{-1}([K^b] - [E^1]^T) + (s-2)[K^b] - [E^2] = 0 \tag{85}$$

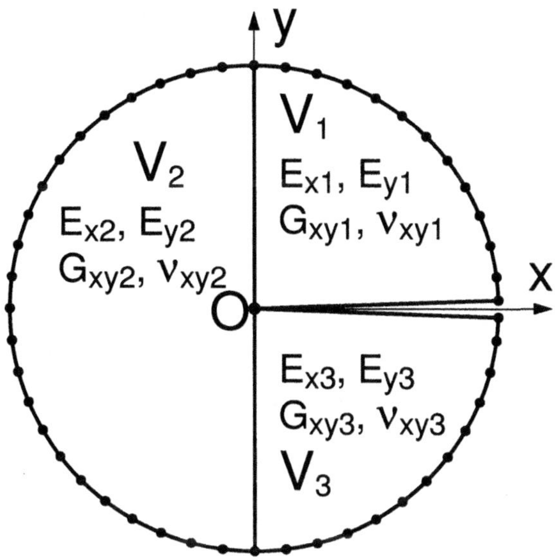

Figure 25. Orthotropic bimaterial plate with crack normal to and terminating at interface

$$((-[K^b] + [E^1])[E^0]^{-1} - 0.5s[I])[M^b] + [M^b]([E^0]^{-1}(-[K^b] + [E^1]^T) - 0.5s[I]) + [M^0] = 0 \quad (86)$$

A unique semi-positive definite solution for $[K^b]$ of the quadratic matrix equation (equation (85)) exists. Equation (86) is linear in the unknown $[M^b]$.

Alternatively, the scaled boundary finite-element equation in displacement can be solved for the static case ($\omega = 0$). The static-stiffness matrix follows as (see equation (58))

$$[K^b] = [\Phi_{21}][\Phi_{11}]^{-1} \quad (87)$$

and the mass matrix is calculated based on its definition using the static displacements as the shape functions.

To demonstrate the features and accuracy of the scaled boundary finite-element method, an orthotropic bimaterial plate in plane stress with a crack normal to and terminating at the material interface (Figure 25) is considered. In the scaled boundary finite-element method the scaling centre O coincides with the crack tip. Note that no discretization is required not only on the crack faces but also on the material interface. The analytical solution of the displacement around the crack tip is expressed as a series of power functions [26] as in the scaled boundary finite-element method. For various combinations of materials specified by 4 constants E_x, E_y, G_{xy}, ν_{xy} with the same properties for V_1 and V_3 the two powers of singularity calculated with the scaled boundary finite-element method are compared in Table I with the analytical solution. Excellent agreement is achieved.

TABLE I. Powers of singularity at crack tip

No. of materials		Calculated		Analytical	
V_1, V_3	V_2	λ_I	λ_{II}	λ_I	λ_{II}
1	2	0.44942	0.62362	0.44948	0.62368
3	2	0.34301	0.46762	0.34301	0.46760
7	8	0.25474	0.46589	0.25477	0.46589
2	1	0.57752	0.35329	0.57744	0.35327
2	3	0.63088	0.54417	0.63088	0.54417
9	10	0.38447	0.48739	0.38447	0.48739
10	9	0.56860	0.50487	0.56860	0.50487

11. Concluding Remarks

Both the finite element and boundary element methods exhibit disadvantages. Addressing these disadvantages leads to the scaled boundary finite-element method, much as an oyster responds to a grain of sand. Or in other words, the scaled boundary finite-element method *combines the advantages of the finite element and boundary element methods*. Most attractive features of both methods are kept. For the finite element method they are that no fundamental solution is required and thus expanding the scope of application, for instance to anisotropic materials without any increase in complexity, that singular integrals are avoided, that symmetry of the results is automatically satisfied and that no fictitious eigenfrequencies occur for an unbounded soil. For the boundary element method they are that the spatial dimension is reduced by one since only the boundary is discretised with surface elements, reducing the data preparation and computational effort, that the boundary conditions at infinity (radiation condition) are satisfied exactly and that no approximation other than that of the surface elements on the boundary is introduced. In addition, the scaled boundary finite-element method presents *appealing features of its own*: an analytical solution in the radial direction inside the domain is achieved, allowing for instance accurate stress intensity factors to be determined directly, and no spatial discretisation of certain free and fixed boundaries and interfaces between different materials is required.

Having selected the scaling centre as the origin, the distance to a point is represented by the radial coordinate ξ. Discretizing the boundary (structure-soil interface) with surface elements determines the other two circumferential coordinates η, ζ. Applying the scaled boundary transformation to the geometry the partial differential equations in Cartesian coordinates are formulated in the local coordinates ξ, η, ζ. Using a *numerical* approach, the weighted residual technique of finite elements, in the two circumferential directions η, ζ parallel to the boundary, results in linear second order ordinary differential equations in displacements with the radial coordinate ξ as the independent variable. For this scaled boundary finite-element equation in displacement an *analytical* solution exists. Thus, the core of the scaled boundary finite-element method consists of transforming the *partial* differential equations to *ordinary* differential equations which can be solved *analytically*. Only the boundary is discretized with curved surface elements. The scaled boundary finite-element method is thus a *semi-analytical* procedure for solving partial differential equations.

In a nutshell, the scaled boundary finite-element method is a *semi-analytical fundamental-solution-less boundary element method based on finite elements*. Paraphrasing the title of the famous paper by Professor Zienkiewicz, the best of both worlds (mariage à la mode) is achieved in two ways: with respect to the analytical and numerical methods and with respect to the finite element and boundary element methods within the numerical procedures.

The analytical solution based on a series expansion of the scaled boundary finite-element equation in displacement has two sets of integration constants. One set is determined by the conditions on the structure-soil interface and the other by the radiation condition at infinity. To enforce the latter rigorously, a high frequency asymptotic expansion for the dynamic stiffness matrix is applied.

As an alternative, the linear second order differential equation, the scaled boundary finite-element equation in displacement, is not solved directly but replaced by two first order differential equations, a nonlinear one, the scaled boundary finite-element equation in dynamic stiffness, more precisely dynamic stiffness divided by the radial coordinate in three dimensions, and a linear one, the interaction force-displacement relationship. A numerical procedure consisting of two steps can be chosen. In the first step, the scaled boundary finite-element equation in dynamic stiffness is solved for decreasing product of radial coordinate and frequency down to the structure-soil interface. The high frequency expansion of the dynamic stiffness satisfying the radiation condition serves as the starting value (boundary condition). This boundary condition follows either from solving an eigenvalue problem or directly from the radiation condition using impedances. This leads to the dynamic stiffness. In the second step, the interaction force-displacement relationship is integrated numerically for increasing product of radial coordinate and frequency. The boundary condition is formulated at the structure-soil interface involving for example applied loads. This yields the displacements.

The scaled boundary finite-element method, of course, also exhibits certain disadvantages. Where there is light, there is shadow! These restrictive properties

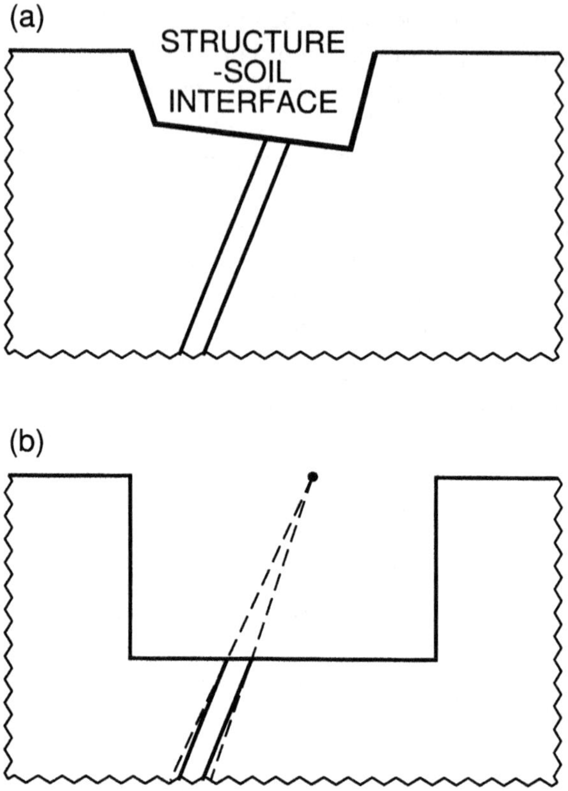

Figure 26. (a) Original dynamic soil-structure-interaction problem with inclined interfaces: (b) Approximate representation enforcing scaling of unbounded soil with structure-soil interface moved outwards

are listed as follows:

Geometry. The novel method is based on scaling (similarity). For a bounded medium several substructures, each with its own scaling centre, can be introduced which yields additional boundaries between two adjacent substructures. For each substructure, its boundary must be visible from its scaling centre. This is a very powerful extension, permitting for instance multiple cracks with stress singularities at their tips to be analysed rigorously.

For an unbounded medium, substructuring is limited (Section 8.5). An approximate representation of the unbounded soil satisfying scaling (similarity) is illustrated in Figure 26. Strongly inclined parallel interfaces between parts of a half plane are present. By moving the structure-soil interface outwards, a model approximately satisfying scaling is constructed. The further away the structure-soil interface is chosen, the better the approximation becomes. The procedure calculates rigorously the dynamic system with the dashed interfaces up to infinity.

This approach should be compared with that of truncating the discretisation of the interfaces extending to infinity as in the boundary element method, where it is not clear what dynamic system is actually analysed from the trucation point up to infinity.

Eigenvalue problem. In the solution an eigenvalue problem must be solved, in contrast to finite element and boundary element methods. This means that the scaled boundary finite-element method is not competitive for standard bounded media with smooth stress variations. However, in the presence of stress singularities, the solution of the eigenvalue problem determines the power of the stress singularities, leading to high accuracy without any special measures. This is not the case for finite element and boundary element methods which use predetermined polynomials to interpolate the displacements. A very large number of elements or special techniques which, for example, incorporate the power of the singularities, which must thus be known a priori, are necessary. For the unbounded soil, solution of the eigenvalue problem allows the radiation condition to be satisfied exactly also in complex situations (anisotropy, incompressible soil, presence of free surfaces and material interfaces). The eigenvalue problem can be avoided by constructing the high frequency limit of the dynamic stiffness of compressible soil using impedances.

Unit-impulse response. In a boundary element method working in the time domain, a transient excitation can be processed directly. This is not possible in the scaled-boundary finite-element method, where unit impulse response matrices are calculated first. A transient analysis then involves convolution integrals with a computational effort which is proportional to the square of the number of time steps. These can be avoided by using a rational approximation and corresponding implementation. For instance, the dynamic behaviour of the unbounded soil is then represented by a recursive evaluation of the interaction forces or by a spring-dashpot-mass model leading to a solution of linear differential equations, both with a computational effort which is proportional to the number of time steps.

Spatial discretisation in circumferential directions. The structure-soil interface (boundary) is discretised spatially with surface elements, leading to a sufficient number of nodes to represent the shortest wave length in the circumferential directions (In the radial direction, an analytical formulation is present without any discretisation). For a surface parallel to the structure-soil interface in the unbounded soil with a constant radial coordinate the same number of nodes will be present due to scaling. Although the distance between the nodes has increased according to the scaling law, the accuracy does not deteriorate. This is remarkable!

12. References

1. Song Ch. and Wolf J. P., Consistent infinitesimal finite-element cell method: three-dimensional vector wave equation, *International Journal for Numerical Methods in Engineering*, **39** (1996), 2189–2208.
2. Wolf J. P. and Song Ch., *Finite-Element Modelling of Unbounded Media*, John Wiley & Sons Chichester, 1996, reprinted 1997, 1999, 2000, 331 pages.
3. Song Ch. and Wolf J. P., The scaled boundary finite-element method – alias consistent infinitesimal finite-element cell method – for elastodynamics, *Computer Methods in Applied Mechanics and Engineering*, **147** (1997), 329–355.
4. Wolf J. P. and Song Ch., Unit-impulse response of unbounded medium by scaled boundary finite-element method, *Computer Methods in Applied Mechanics and Engineering*, **15** (1998), 355–367.
5. Song Ch. and Wolf J. P., The scaled boundary finite-element method – alias consistent infinitesimal finite-element cell method – for diffusion, *International Journal for Numerical Methods in Engineering*, **45** (1999), 1403–1431.
6. Song Ch. and Wolf J. P., The scaled boundary finite-element method: analytical solution in frequency domain, *Computer Methods in Applied Mechanics and Engineering*, **16** (1998), 249–264.
7. Song Ch. and Wolf J. P., Body loads in the scaled boundary finite-element method, *Computer Methods in Applied Mechanics and Engineering*, **180** (1999), 117–135.
8. Deeks A. J. and Wolf J. P., Semi-analytical elastostatic analysis of unbounded two-dimensional domains, *International Journal for Numerical and Analytical Methods in Geomechanics*, submitted.
9. Wolf J. P. and Song Ch., The scaled boundary finite-element method – a semi-analytical fundamental-solution-less boundary element method, *Computer Methods in Applied Mechanics and Engineering*, 190(2001) 5551-5568.
10. Wolf J. P. and Song Ch., The scaled boundary finite-element method – a primer: derivations *Computers & Structures*, **78** (2000), 191–210.
11. Song Ch. and Wolf J. P., The scaled boundary finite-element method – a primer: solution procedures, *Computers & Structures*, **78** (2000), 211–225.
12. Wolf J. P. and Song Ch., The semi-analytical scaled boundary finite-element method to model unbounded soil, *Proceedings 11th European Conference on Earthquake Engineering*, Paris A.A. Balkema, 1998.
13. Wolf J. P. and Song Ch., Some cornerstrones of dynamic soil-structure interaction *Engineering Structures*, 24(2002), 13-28.
14. Wolf J. P., Far-field displacements in 3-d soil in scaled boundary finite-element method, *Wav 2000*, Bochum, A.A. Balkema, 2000.
15. Wolf J. P. and Moussaoui F., Far-field displacements of soil in scaled boundary finite-element method, *Proceedings 10th International Conference of the International Association fo Computer Methods and Advances in Geomechanics*, Tucson, AZ, A.A. Balkema, 2001.
16. Song Ch. and Wolf J. P., Semi-analytical representation of stress singularity as occurring in cracks in anisotropic multi-materials with the scaled boundary finite-element method *Computers & Structures*, (in press).
17. Song Ch. and Wolf J. P., Semi-analytical evaluation of dynamic stress-intensity factors *Computational Mechanics, New Frontiers for the New Millemum, Proceedings of the Fir Asian-Pacific Congress on Computational Mechanics*, Valliappan, S. and Khalili N. (editors Vol.2, 1041-1046.
18. Wolf J. P. and Huot, F.G., On modeling unbounded saturated poroelastic soil with the s caled boundary finite-element method, *Computational Mechanics, New Frontiers for the Ne*

Millemum, Proceedings of the First Asian-Pacific Congress on Computational Mechanics, Valliappan, S. and Khalili N. (editors), Vol.2, 1047-1056.

19. Deeks A. J. and Wolf J. P., Stress recovery and error estimation for the scaled boundary finite-element method, *International Journal for Numerical Methods in Engineering*, (in press).
20. Deeks A. J. and Wolf J. P., An *h*-hierarchical adaptive procedure for the scaled boundary finite-element method, *International Journal for Numerical Methods in Engineering*, (in press).
21. Sommerfeld A., *Partial Differential Equations in Physics*, Chapter 28, Academic Press, New York, (1949).
22. Gantmacher F. R., *The Theory of Matrices, Vol. 2*, Chelsea, New York, 1977.
23. Press W. H., Flannery B. P., A. Teukolsky S., and Vetterling W. T., *Numerical Recipes*, Chapter 15, Cambridge University Press, Cambridge, (1988).
24. Waas G., *Linear Two-Dimensional Analysis of Soil Dynamics Problems in Semi-Infinite Layered Media*, PhD dissertation, University of California, Berkeley, CA, 1972.
25. Friedrich K. and Schmid G., *Personal Communication*, 2000.
26. Chen D. H. and Harada K., Stress singularities for crack normal to and terminating at bimaterial interface on orthotropic half-plates, *International Journal of Fracture*, **81** (1996), 147–162.

CHAPTER 4
BEM ANALYSIS OF SSI PROBLEMS IN RANDOM MEDIA

G.D. MANOLIS[1] and C.Z. KARAKOSTAS[2]
[1]Department of Civil Engineering, Aristotle University
P.O. Box 502, GR 540 06, Thessaloniki, Greece
(Tel: + 30 31 995663, fax: + 30 31 995769, email: gdm@civil.auth.gr)
[2]Institute of Engineering Seismology and Earthquake Engineering, P.O. Box 53
GR 551 02 Finikas, Thessaloniki, Greece
(Tel: + 30 31 476081, fax: + 30 31 476085, email: christos@itsak.gr)

1. Introduction

The concept of a random medium is not a mathematical abstraction. Seismological studies, for instance, show the existence of coda (Latin cauda, meaning trail) waves in recorded accelerograms as the most compelling evidence supporting a random heterogeneous structure of the earth's lithosphere [1]. More specifically, S-coda waves are continuous wave trains trailing the passage of shear (S) waves in an accelerogram, whose amplitude envelop gradually decreases with increasing time. They are the product of superposition of incoherent waves scattered by randomly distributed heterogeneities in the earth. Viewed from another angle, two basic mechanisms of seismic wave attenuation exist : First, a scattering mechanism due to randomly occurring changes in an otherwise uniform (or gradually varying) geological stratum which basically distributes energy; and second, an intrinsic mechanism which converts vibration energy into heat and is the cumulative result of a number of phenomena (for example, presence of voids, two-phase materials, a

W.S. Hall and G. Oliveto (eds.), Boundary Element Methods For Soil-Structure Interaction, 175–233.

crystalline structure of the rock formations etc.). Thus, the ability to manipulate random variables is paramount in formulating rational wave propagation models.

Problems involving random media are governed by stochastic differential equations. The key assumption [2] in the solution of such equations is the decomposition of the differential operator into deterministic plus random parts. Formal inversion of the deterministic part is accomplished through the use of Green's functions, and the stochastic differential equation with its boundary conditions can be recast as a random integral equation. Solution of the random integral equation can then be accomplished iteratively through use of the resolvent kernel, which in turn is defined through a Neumann series expansion. Alternatively, the dependent variable can be expanded in series, which under certain conditions is equivalent to the aforementioned Neumann series or to a Born approximation [3]. Finally, approximate solutions can be generated by applying the expectation operator to the random integral equation and then using various closure approximations, by perturbations with the usual restriction of small fluctuations about a mean value, or by other techniques [2].

In the case of wave motions, which form the theoretical background for all soil-structure interaction (SSI) problems, two basic representations of material stochasticity are possible, that is the medium can be viewed as a random collection of scatterers or as a random continuum [4]. More specifically, in the former case the scatterer is a random distribution of many well defined particles such as spheres, while in the latter case the medium has properties which vary randomly and continuously in space and possibly in time. Propagation of elastic waves in a random, continuous medium that differs only slightly from the homogeneous case was originally considered by Karal and Keller [5] using the random integral equation formulation previously discussed. They derived an effective wave propagation constant which indicates that an originally coherent wave is now continuously scattered by the randomness and converted into an incoherent wave with diminishing propagation velocity. This technique was subsequently extended to multi-layered systems through the use of transfer matrices by Chu et al. [6] and by Hryniewicz [7]. Randomly layered media have also been considered by Kotulski [8], who employed a complex transfer matrix approach in conjunction with a homogenization process to derive an equation for the effective amplitude and wave number of elastic waves in a stratified slab and by Kohler et al. [9], who used asymptotic methods for stochastic differential equations to compute power spectra for receivers in a randomly layered half-space overlying a homogeneous half-space, with the source placed in the latter medium. Techniques by which the probability density function, the space correlation function or other statistical measures governing wave propagation of acoustic or elastic waves in media modelled by random configurations of a large number of densely packed, identical scatterers of finite size have also been developed [10-12]. Finally, a detailed review of various methods of analysis for waves in continuous as well in discrete stochastic media and for wave scattering by stochastic surfaces can be found in Sobczyk‘ [13], where techniques such as perturbations, the Born approximation, methods based on geometric optics, the parabolic equation approximation, homogenization methods and the functional approach for large parameter fluctuations are presented.

Two broad classes of numerical approaches for evaluating the dynamic response of continuous media can be distinguished, namely simulation techniques [14] and perturbation methods [15]. The former techniques are considered to yield exact solutions at the expense of high computational effort, but provide limited insight into the sensitivity of the system to different parameter uncertainties. The latter methods are more versatile and can be easily integrated with existing deterministic solution techniques such as finite elements [15,16] and boundary elements [17,18]. They provide satisfactory second-moment statistics of the system's response subject to the assumption of small randomness. Associated with perturbation methods, however, are questions regarding the accuracy and convergence of higher order statistical moments, especially within the context of transient problems, where it is known that secular terms arise [19]. When using perturbations, numerical implementation is through a Taylor series expansion of key system parameters. A more versatile way, but for single degree-of-freedom (dof) systems, is through a Fourier series expansion with the coefficients of the expansion evaluated numerically via the fast Fourier transformation [20].

In an effort to overcome the limitation of small parameter uncertainty, series expansions of the random response of a system in polynomials that are orthogonal with respect to the expectation have appeared [21-24]. In particular, Sun [21] used this type of expansion to study a particular class of ordinary differential equations with one random parameter, while

Jensen and Iwan [22] extended this approach to study the dynamic response of single and multiple dof structural systems with uncertain parameters to both deterministic and modulated Gaussian white-noise excitations. Similar parameter representations, but within the context of finite element solutions of static problems, have been used by Lawrence [23] and by Spanos and Ghanem [24]. In all cases, medium stochasticity is accounted for by means of a random process, thus departing from older work in which medium stochasticity was viewed as a random variable. Spanos and Ghanem [24], for instance, represented a unidimensional random process through its covariance function, which admits a spectral decomposition in terms of the eigensolution of an integral equation that uses this covariance as its kernel. If the random process is written as the algebraic sum of a mean term plus a random term with zero mean and non-zero covariance, then the latter part can be expressed as a series with the eigenvectors as base vectors. For the case of a Gaussian process, this series can be shown to converge. Lawrence [23] also used a similar expansion in terms of orthogonal random variables with vanishing higher order moments, plus deterministic base functions in terms of Legendre polynomials, but the convergence properties of this expansion are not clear. Other approaches are also possible. For instance, an averaged Green's function is derived for the scalar wave equation in a unidimensional stochastic continuum which is independent of the magnitude of the random fluctuations by Belyaev and Ziegler [25]. Here, the scalar wave equation with a random coefficient is solved by means of the Dyson integral equation, and through successive approximations, a closed form solution for the averaged Green's function is obtained for various correlation functions of the random material field. In closing, the major problem with both stochastic boundary and finite element methods is of computational nature, since the numerical effort increases rapidly past the second order approximation of the relevant random variables [26]. In general, boundary element method (BEM) stochastic analyses are more appropriate for problems involving a continuum, while their finite element counterparts are better used in problems addressing the structural component level. Finally, an extensive review on modelling of the ground as a random continuum can be found in Manolis [27], while the various BEMs used in geomechanics are listed in Manolis et al. [28].

We consider here two analyses. In the first for 'small' amounts of soil randomness, we formulate a direct BEM for a class of SSI problems which encompasses the dynamic response of underground openings under plane-strain conditions subjected to general transient loads such as those induced by earthquakes. The resulting solution comprises a mean vector plus a covariance matrix for the displacements and tractions that develop at the cavity interface. The perturbation method is used for expanding all the dependent variables, including the fundamental solutions, about their mean values. Substitution of these expansions in the appropriate boundary integral equations, along with conventional numerical integration schemes, yields a compact BEM solution scheme that is valid for a wide range of problems. The entire methodology is defined in the Laplace transform domain and an efficient inverse transformation algorithm is employed for reconstructing the transient response. A couple of numerical examples for underground openings in a stochastic geological medium under wave induced motions and surface loads, serve to illustrate the proposed methodology [29]. In the second analyses for 'large' amounts of randomness, stochastic methods based on Taylor series expansions of the dependent variables about a mean value, which retain first or at most second order terms, is inadequate. Thus, an improvement comes through the introduction of polynomial chaos transformations, where all stochastic variables which may have arbitrary distribution functions, are expanded in terms of an orthogonal polynomial basis which is a function of a random variable. The particular choice of basis is dictated by the standardised distribution function desired for representing the key dependent variable of the problem. This path is followed here, and appropriate fundamental solutions have already been constructed for the case of horizontally polarised shear waves [30]. Although the necessary BEM formulation for the polynomial chaos method is not available at present, a numerical example utilising the aforementioned fundamental solutions serves to illustrate the power of this particular approach in representing large medium randomness.

2. Review of the Literature

In this section, we review the literature on soil dynamics and dynamic soil-structure-interaction for the case where the ground is assumed to be a random medium.

2.1. RANDOM LOADING

The case where the only part of the problem that is random is the loading is rather straightforward. For a linear system excited by a forcing function that has a Gaussian distribution, the response will also have a Gaussian distribution and is thus completely described by a mean value and a standard deviation. Many of the methodologies developed for the corresponding deterministic problem can be easily recycled to handle random loads, especially if stationary conditions can be assumed and if the problem is solved in the frequency domain. Vanmarke [31] has suggested several possible applications of random vibration theory for solving problems in soil dynamics, including determination of non-linear soil response and assessment of liquefaction potential. As far as geotechnical applications are concerned, we mention the work of Gazetas et al. [32] on the non-linear hysteretic response of earth dams to non-stationary stochastic excitation described by the Kanai-Tajimi spectrum, of Luco and Wong [33] on the dynamic response of a rigid square foundation to random seismic excitations by postulating a spatial coherence function for the ground motion, of Pais and Kausel [34] on the stochastic response of embedded foundations through superposition of wave trains each characterized by a spectral density function, and of Hao [35] on the response of a multiply supported rigid plate to ground motions whose spatial variation is described by the coherency model of Harichandran and Vanmarke [36].

2.2. MONTE CARLO SIMULATIONS

Monte-Carlo simulations are extensively used in many scientific fields and are regarded as the "exact" solution to problems involving stochastic media. They are, however, computationally expensive. They seem best suited for problems with many random variables that are correlated, where other analytical or numerical techniques turn out to be impractical. In civil engineering they find widespread use in the analysis of non-linear structures subjected to seismic motions, ocean waves and wind turbulence. Typical applications in geomechanics where Monte Carlo simulations are efficient are settlement of foundation systems consisting of many individual footings such as pile groups [37]. This is so because the spatial proximity of the individual footings results in correlated relative settlements, even if the total foundation rests on homogeneous soil.

2.3. RANDOM BOUNDARIES

In geotechnical design, the presence of random boundaries is of minor importance compared to soil and/or loading randomness. As a result, work on this topic is virtually non-existent. There is, however, interest in random boundary conditions when it comes to acoustic and electromagnetic wave propagation, especially in conjunction with rough seabed topography, mountainous topography and the presence of scatterers with very irregular surfaces. The methods for treating a random surface are either of the perturbation type, where the surface is described in terms of a mean value plus a small fluctuation about the mean [38] or of the non-perturbation type, where the random wave field, regarded as a functional of the randomly rough surface is generated by means of Gaussian random measure and requires the use stochastic functional calculus [39]. In all cases, the objective is to compute the various statistical measures of the scattered waves from the stochastic wave field. Other applications of random boundaries are in the field of structural mechanics, as is the case of determining the variation in stresses and strains caused by fluctuations between the idealized and the actual geometry of a structural component [40].

2.4. SOIL MODELLING

Several methods for incorporating uncertainty in an overall reliability evaluation of geotechnical performance are proposed in Tang et al [41]. Two basic groups of uncertainty are identified, namely uncertainty in soil properties [42] and uncertainty in geotechnical design. In the former case, procedures based on the Bayesian methodology have been developed [43-45] so as to synthesize site exploration data in assessing geological anomalies such as occurrence probability and size distribution of unexpected material within an otherwise homogeneous soil deposit. In the latter case, a full-scale Monte Carlo simulation procedure can be performed to incorporate the material uncertainty in performance reliability calculations. Due to the excessive amount of computation often required by these simulations, approximate methods such as use of correction factors stemming from first-order analyses for modifying the deterministic performance function are used [46]. Examples studied using approximate methodologies include raft foundation settlement in soil with random soft pockets and slope stability analysis with progressive failure due to the presence of randomly located soft zones along a potential slip surface. More advanced methods [47] employ random fields to model the soil material state. Probabilistic analyses are then used to generate soil profile statistics, which serve as the starting point for reliability calculations such as evaluation of exceedance probability of a given threshold value by the geotechnical design's response. Applications include the frictional capacity of individual axial piles and pile groups in soil layers with random properties and tunnel construction through regions that contain a random distribution of boulders. Markov theory has been employed by Benaroya [48] as a framework for understanding the role of uncertainties in the dynamic behaviour of soils. In particular, the Markov state transition matrix, which is the probabilistic counterpart of the transfer matrix concept employed in deterministic mechanics, is used for establishing the evolution of a soil state given its present state. This type of approach finds application in the dynamic constitutive modelling of soil, for example stress-strain behaviour of soil specimens of random structure subjected to loading and unloading cycles.

2.5. FOUNDATIONS

A probabilistic model for stability analysis of deepwater gravity-based platform foundations appears in Ronold [49]. In particular, a model that accounts for uncertainty in the platform foundation design and in the soil properties is developed. The ocean storm loading is given in terms of spectral densities for the horizontal force and overturning moment on the platform and is derived from the spectral density of the wave energy. The input to a first order reliability analysis of the platform is a limit state function specified in terms of basic stochastic variables and deterministic parameters. Finally, platform foundation stability is expressed in terms of a failure probability, which is updated from available measurements of pore pressure and foundation stiffness. The statistics of individual pile and pile group settlement are determined in Quek et al. [50] by combining a hybrid approach with first order perturbations. The soil is assumed to be linear elastic and isotropic, while its shear modulus is a random field characterized by a mean value, a variation and a scale of fluctuation. The piles are represented by rod finite elements and are coupled with soil flexibility coefficients stemming from a separate finite element analysis. In order to obtain a measure of assurance of the serviceability of the foundation, a reliability analysis based on a first order, second moment method is used to produce a reliability index plus the probability of unserviceable behaviour. In Drumm et al [51], a one dimensional finite element code is used to analyse the lateral response of drilled shaft foundations (caissons). The response of the pier is linear elastic, while the soil response is non-linear (the generalized Ramberg-Osgood model is used) and exhibits natural variability of the shear modulus. The approximate procedure used is that of Rosenblueth [52], whereby the continuous probability density function (pdf) of the response is modelled by three discrete points so that the mean and standard deviation can be calculated from a discrete point estimate of the input random variables. The final results obtained are the mean and variation of the pier deflections. Other applications include the work of Baecher and Ingra [53] who evaluated the uncertain displacements of an infinite strip footing using a first-order technique and of Nakagiri and Hisada [54] regarding the behaviour of a pipeline resting on an uncertain Winkler foundation.

2.6. SLOPE STABILITY

A methodology for constructing seismic slope failure stability matrices is derived in Lin [55]. This task requires the synthesis of the following three concepts: site seismic hazard, static stability of existing slopes and landslide potential of various slopes under different ground shaking intensities. The methodology focuses on the last concept and employs a probabilistic sliding block which allows for a systematic incorporation of uncertainties associated with both ground excitation and strength of soil. The ground excitation is described by its peak acceleration and earthquake magnitude, while the scatter exhibited by the ground motion is represented via an equivalent stationary motion model. The extent of damage to a slope is finally defined in terms of earthquake induced permanent displacements. Finally, the effect of uncertainty in key soil parameters, such as cohesion and angle of internal friction, on predicting the shear strength of soil and subsequently the factor of safety of slopes is examined in a number of publications [56-58] using both two- and three-dimensional earth slope models as well as the finite element method [59,60].

2.7. CONSOLIDATION

The development and dissipation of excess pore pressure in a soil under external loads is examined in Koppula [61] for a random variable representation of the consolidation coefficient. Closed form expressions for the mean and variance of the pressure as functions of the mean time factor are derived under uni-dimensional conditions by approximating the consolidation coefficient through a generalized gamma probability density function. A probabilistic analysis based on nonstationary random vibration theory is developed in Kavazanjian and Wang [62] for determining the seismic site response and liquefaction potential of layered horizontal soil deposits subject to vertically propagating shear waves. The nonstationarity of the ground motions is separated into a deterministic amplitude modulating function and a stationary random process. Subsequently, the beam analogy is used to develop transfer functions for the soil deposit, while pore pressure development is monitored through a weighted averaging method based upon the expected incremental pore pressure for each cycle of loading. Seismic fragility curves are finally constructed that express the liquefaction potential as a function of the root-mean-square (RMS) accelerations developed at a particular site.

2.8. SOIL-STRUCTURE INTERACTION

The kinematic interaction manifested between the ground and its supporting structure is investigated in Hoshiya and Ishii [63,64] by recourse to earthquake records at the ground and structure levels. The methodology developed first separates the kinematic interaction (assuming a massless function) from the total dynamic interaction using deconvolution. Subsequently, the unknown foundation input motion is expressed by a moving average (MA) in terms of the ground (free-field) motion, thus representing the filtering effect of kinematic interaction. The structure itself is modelled by a multiple dof system so as to account for the total dynamic interaction. The parameters of the multiple dof system are treated as random variables within the context of Monte-Carlo simulations, while the coefficients of the MA model are identified through a Kalman filter for each realization of the above parameters. Finally, a minimization of the RMS error between the observed response and the calculated response yields the best estimate of the structural system's parameters and of the MA model coefficients. The probabilistic analysis of deeply buried structures subjected to a stress wave loading is reported in Harren and Fossum [65], who couple the finite element method with the fast probability integration technique. The soil is modelled under plane strain conditions and randomness is considered in both loading and in the composition of the soil stratum. The above procedure requires much less computational effort than traditional Monte-Carlo simulations. Finally, another finite element application is for evaluating the uncertain displacement and stress fields in a random soil continuum [66].

2.9. EARTHQUAKE SOURCE MECHANISM

The accurate determination of ground motions due to earthquakes is of paramount importance in predicting the dynamic response of geotechnical designs. Seismological studies of source mechanisms and wave propagation models enable the generation of realistic strong motions based on deterministic source-to-site characteristics. These models can be converted to stochastic ones by regarding the various filter parameters (such as the apparent duration of fault rupture, the dislocation rise time, the distance to source, the rate of wave amplitude decay, etc.) to be random variables. Examples include construction of empirical Green's functions that approximate the impulse response of the ground between earthquake source and observation site that can be used for generation of synthetic accelerograms [67], derivation of a stochastic model for strong ground motions based on the normal mode theory [68], and the stochastic time-predictable model with Weibull [69] or lognormal [70] distributions of event inter-arrival for representing characteristic earthquake events that occur repeatedly over relatively regular intervals of time with small variations in magnitude. More recent work [71] assumes that the slip rate along major tectonic faults is a random function described by its mean and coefficient of variation, which in turn are determined from available geological data. A semi-Markovian stochastic model, whose parameters are estimated using Bayesian statistical methods, is used to characterize the resulting sequence of earthquakes. This methodology was applied to earthquakes occurring in the Mexican subduction zone.

By processing the strong motion data recorded by the large-scale digital array SMART1, Harichandran and Vanmarke [36] were able to determine the frequency-dependent spatial correlation of earthquake-induced ground motions. The key idea is to visualize accelerograms from the same seismic event as samples from space-time random fields. The cross-spectral density function model for space-time random fields proposed above can be used within the context of representing propagating earthquake motions as correlated stationary autoregressive random processes with zero mean [72]. A generalization of these ideas finds application in the random vibration analysis of non-linear structures subjected to loads associated with natural phenomena such as wind turbulence, ocean waves and earthquake ground motions. Efficient algorithms have been proposed [73-75] for the generation of records of a multivariate stationary process with a specified (target) spectral matrix. These algorithms are based on an approximation of the particular process as the output to white noise input of autoregressive (AR) systems. The AR model can subsequently become the basis for efficient and reliable autoregressive moving average (ARMA) or purely moving average (MA) approximations. Models of earthquake excitation as nonstationary stochastic processes have been developed by numerous authors in both time and frequency domain, as pointed out in the survey by Kozin [76]. Selection of the particular domain is arbitrary since both cases involve modulation of the amplitude of a stationary noise process and both cases may encounter problems in certain ranges of the response spectrum. Examples of individual earthquake records treated as samples from an underlying population characterized by an ARMA model can be found in the work of Cakmak and his co-workers [77,78]. Finally, earthquake hazard estimation can be performed either within the framework of ARMA models [79] or within the framework of refined Bayesian models that incorporate newly developed information from geophysical and geological studies along with historical data [80,81].

2.10. PROBABILISTIC RESPONSE SPECTRA

The earthquake source mechanism models can be extended a step further through the generation of response (or structural design) spectra that are compatible with statistical descriptions of the ground motions [82]. The essence of this problem is to relate the power spectrum of the stochastic seismic motion to the design spectrum. As shown in Spanos and Vargas-Loli [83], a particular earthquake record can be viewed as a realization of a nonstationary stochastic process with an evolutionary power spectrum. An approximate solution is subsequently used to calculate the probability density function of the response of a lightly damped, single dof oscillator to the stochastic earthquake input. The last step in this procedure is to match the target spectrum of the input with the maximum values of the response statistics that define the design spectrum. Such design spectra can be used in the stochastic dynamic analysis of any structure (including geotechnical ones) under ground motions, provided its natural frequencies are known. Finally, the description of structural response in terms of probabilistic spectra is quite vast and more information can be found elsewhere [84].

3. Integral Equation Formulation

3.1. THEORETICAL BACKGROUND

The equation governing soil stochasticity is the following general stochastic differential equation:

$$L(\gamma)[u(x,\gamma)] = f(x,\gamma) \tag{1}$$

In the above, L is a differential operator of order n with random coefficients. Furthermore, u is the dependent variable, f is the forcing function and argument γ denotes a random quantity. In most cases, u can be identified with a displacement component and x with a spatial variable. Equation (1) is, of course, accompanied by the appropriate boundary (and initial for time-dependent problems) conditions.

The key assumption [85] is that the operator L can be decomposed into a deterministic part D and a zero-mean random part $R(\gamma)$ so that equation (1) becomes

$$(D + R(\gamma))[u(x,\gamma)] = f(x,\gamma) \tag{2}$$

with

$$\left.\begin{aligned} D[u] &= \sum_{i=0}^{n} \alpha_i(x) d^i u / dx^i \\ R(\gamma)[u] &= \sum_{i=0}^{n} \beta_i(x,\gamma) d^i u / dx^i \end{aligned}\right\} \tag{3}$$

where coefficients α_i, β_i depend on x with $\alpha_n = 1$ and $\beta_n = 0$. If a Green's function $G(x, x_1)$ exists for the deterministic operator D, then the stochastic differential equation is equivalent to the random integral equation

$$u(x,\gamma) + \int_{-\infty}^{\infty} N(x, x_1, \gamma) u(x_1, \gamma) dx_1 = H(x,\gamma) \tag{4}$$

where

$$\left.\begin{aligned} H(x,\gamma) &= \int_{-\infty}^{\infty} f(x_1,\gamma) G(x, x_1) dx_1 \\ N(x, x_1, \gamma) &= G(x, x_1) \sum_{i=0}^{n} \beta_i(x_1,\gamma) d^i / dx_1^i \end{aligned}\right\} \tag{5}$$

Equation (4) is a Volterra integral equation of the second kind with random kernel N and a random generalized forcing function H.

3.2. FORMAL SOLUTION

At this stage, there are a number of options available regarding the solution of equation (4). Before we proceed with approximate techniques, we will first discuss the closed-form solution. Following the deterministic case, the resolvent kernel Γ of N is defined through the following Neumann series [85]:

$$\Gamma(x, x_1, \gamma) = -\sum_{k=1}^{\infty} N_k(x, x_1, \gamma) \tag{6}$$

In the above, the iterated kernels N_k are given by the recurrence relation

$$N_k(x, x_1, \gamma) = \int_{-\infty}^{\infty} N(x, x_2, \gamma) N_{k-1}(x_2, x_1, \gamma) dx_2 \qquad k = 2,3,... \tag{7}$$

with $N_1 = N$. Thus, the formal solution of equation (4) can be written as

$$u(x, \gamma) = H(x, \gamma) + \int_{-\infty}^{\infty} \Gamma(x, x_1, \gamma) H(x, \gamma) dx_1 \tag{8}$$

where it has been assumed that the Neumann series of equation (6) converges uniformly. Although the above solution methodology is quite general and applicable to the case where equation (1) is a vector equation, the convergence of the Neumann series plus the construction of the resolvent kernel are difficult to establish. For the particular case where only the coefficient β_0 in the operator R is nonzero, convergence of the formal solution given by equation (8) has been established [85].

An alternative solution of the random integral of equation (4) is through a series expansion of the dependent variable u in the form

$$u(x, \gamma) = -\sum_{k=1}^{\infty} u_k(x, \gamma) \tag{9}$$

Since the above decomposition is not unique, a recursive relation is chosen so that u_k is an explicit function of u_{k-1} only. For the linear differential operator of equation (2), the decomposition can be written as follows:

$$\left.\begin{aligned} & u_0(x, \gamma) = f(x, \gamma) \\ & u_k(x, \gamma) = \int_{-\infty}^{\infty} N_k(x, x_1, \gamma)\, u_{k-1}(x_1, \gamma)\, dx_1 \qquad k = 1,2,... \end{aligned}\right\} \tag{10}$$

If the decomposition in equation (10) is written explicitly and the result is compared with equation (8), it becomes obvious that this series solution is equivalent to the formal solution involving the resolvent kernel.

Consider now the first three terms in equation (10), that is,

$$u(x,\gamma)=H(x,\gamma)-\int_{-\infty}^{\infty}N(x,x_1,\gamma)H(x_1,\gamma)dx_1 + \int_{-\infty}^{\infty}\int_{-\infty}^{\infty}N(x,x_1,\gamma)N(x_1,x_2,\gamma)H(x_2,\gamma)dx_2dx_1 \tag{11}$$

Since we are dealing with a stochastic problem, the above solution must be recast in terms of the expectation $E\{g(x,\gamma)\}=<g(x)>$ of all the random variables $g(x,\gamma)$ involved. This operation denotes statistical averaging and its application to both sides of equation (11) results in

$$<u(x)>=<H(x)>-\int_{-\infty}^{\infty}<N(x,x_1)H(x_1)>dx_1 + \int_{-\infty}^{\infty}\int_{-\infty}^{\infty}<N(x,x_1)N(x,x_2)H(x_2)>dx_1dx_2 \tag{12}$$

If the random operator R and the forcing function f are statistically independent and if the former is a zero-mean process, then equation (12) reads as

$$<u(x)>=<H(x)>-\int_{-\infty}^{\infty}\int_{-\infty}^{\infty}<N(x,x_1)N(x_1,x_2)><H(x_2)>dx_1dx_2 \tag{13}$$

where $<NN>$ is the correlation function for the random process N and needs to be specified.

3.3. CLOSURE APPROXIMATION

It should be noted that the solution methodology described so far fails to work for the case of a zero forcing function. To overcome this difficulty it is necessary to go back to the original random integral in equation (4) and apply the expectation operator. The result is

$$<u(x)>=<H(x)>-\int_{-\infty}^{\infty}G(x,x_1)<R(x_1)u(x_1)>dx_1 \tag{14}$$

Unfortunately, $<Ru>$ is unknown and will not separate into $<R><u>$. The approximation that has been used for $<Ru>$ is to invert the differential operator in equation (2), apply the random operator to both sides, and finally take the expectation of both sides of the equation. This results in

$$\begin{aligned} <R(x)u(x)> &= <R(x)D^{-1}(x)f(x)> - <R(x)D^{-1}(x)R(x)u(x)> \\ &= - <R(x)D^{-1}(x)R(x)><u(x)> \end{aligned} \tag{15}$$

The only new approximation involved in the above equation is the closure approximation $<RD^{-1}Ru> = <RD^{-1}R><u>$, which is of a higher order than an approximation of the type $<Ru> = <R><u>$. As a result of the above, equation (14) now reads as

$$\begin{aligned} <u(x)> = <H(x)> + \int_{-\infty}^{\infty} G(x, x_1) <R(x_1) \\ \int_{-\infty}^{\infty} G(x_1, x_2) R(x_2) dx_2 > <u(x_1)> dx_1 \end{aligned} \tag{16}$$

where the correlation function $<RR>$ of the random operator R needs to be specified. The above closed-form solution for the average displacement $<u>$ invokes the additional closure approximation when compared to the formal solution given by equation (13), but has the advantage that it is applicable to the case of a zero forcing function.

4. Vibrations in Random Soil Media

The success of approximate solution methodologies for the differential equations governing soil stochasticity depends on the physical properties of the particular problem at hand. Thus, we shall focus our attention from now on to wave motions in a three-dimensional random medium. This problem has widespread applications in earthquake engineering, soil-structure interaction and other related subjects.

4.1. PROBLEM STATEMENT

The governing equation for elastic wave propagation in a three-dimensional medium under time harmonic conditions is Helmholtz's equation

$$\nabla^2 \varphi(\mathbf{x}, \omega) + k^2 \varphi(\mathbf{x}, \omega) = f(\mathbf{x}, \omega) \tag{17}$$

In the above, φ is the displacement potential, f is the forcing function, $\mathbf{x} = (x, y, z)$ is the position vector and ω is the frequency of vibration. Laplace's operator ∇^2 is equal to $\nabla \cdot \nabla$ where ∇ is the gradient. In addition, the wave number is $k = \omega / c$, where c is the wave propagation speed. We distinguish two cases of elastic waves: (i) Longitudinal (or pressure) waves, where the irrotational component of the displacement vector $u(\mathbf{x}, \omega)$ is equal to $\nabla \varphi$ and (ii) Transverse (or shear) waves, in which case equation (17) is a vector equation and the equivoluminal component of $u(\mathbf{x}, \omega)$ is equal to $\nabla \times \varphi$. In the former case, $c_p^2 = (\lambda + 2\mu)/\rho$, while in the latter case $c_s^2 = \mu / \rho$, where λ and μ are the Lamé elastic constants and ρ is the density of the medium. In addition, the case of anti-plane strain, where the only non-zero displacement component is $u_z(x, y, \omega)$, is governed by equation (17) written directly for u_z and with $c = c_s$. The usual type of boundary conditions respectively are

$$\varphi = \overline{\varphi} \text{ on } S_1 \quad \text{and} \quad \partial \varphi / \partial n = \overline{q} \text{ on } S_2 \tag{18}$$

for the displacement potential and for its flux, where $S = S_1 + S_2$ is the total surface and $\mathbf{n}$ is the outward pointing, unit normal vector on S. The homogeneous form of the boundary conditions (18) respectively denotes a rigid surface and a traction-free surface. Furthermore, in the presence of unbounded media the scattered waves must obey the radiation condition.

4.2. GROUND RANDOMNESS

In order to solve realistic problems in geomechanics, the elastic constants and the density must be position-dependent. As far as wave propagation problems are concerned, this requires a detailed description of very large soil and/or rock volumes, which is a nearly impossible task for all practical purposes. Furthermore, solutions of equation (17) are known only for very specific forms of $k(x)$. Thus, the assumption of stochasticity offers an attractive alternative, in view of the fact that there exist models for describing medium randomness from other engineering disciplines [4].

A typical such model is to assume that the wave number has a small fluctuation about its mean homogeneous value k_0 given by

$$k(x,\gamma) = k_0(x) + \varepsilon k_1(x,\gamma) + \varepsilon^2 k_2(x,\gamma) + \ldots \tag{19}$$

where k_1, k_2,... are zero-mean random processes and $\varepsilon << 1$. There are also three more possible sources of randomness stemming from the initial conditions, the boundary conditions and the forcing function. The first case is irrelevant for a steady-state problem, while the other two cases are discussed in the next section. Finally, it should be mentioned that in order to solve for the random response statistics, it is necessary to at least prescribe the autocorrelation function of k_1. For that purpose, the following models are commonly used:

(i) Exponential function, where

$$\langle k_1^2(\gamma)\rangle = \sigma_k^2 \exp(-|\xi|/\lambda_k) \tag{20a}$$

with σ_k^2 the variance of k_1, λ_k the correlation length and $\xi = R_j - R_i$;

(ii) delta function, where

$$\langle k_1^2(\gamma)\rangle = \sigma_k^2 \delta(\xi) \tag{20b}$$

and

(iii) piecewise linear function, where

$$\langle k_1^2(\gamma)\rangle = \sigma_k^2 (1 - |\xi|/\lambda_k) \quad \text{if} \quad |\xi| < \lambda_k \tag{20c}$$

and zero otherwise. Note that R_i and R_j are the positions of two arbitrary points in the medium and ξ is their relative distance.

4.3. ANALYTICAL SOLUTION

The case of harmonic wave propagation through randomly structured ground, where the origin of disturbance is a point source, was solved analytically by Karal and Keller [5] for the case of a constant shear modulus and a random density of the type

$$\rho(R,\gamma) = \rho_0(1+\rho_1(R,\gamma)) \tag{21}$$

where ρ_1 is a zero-mean fluctuation about mean ρ_0 and R is the radial co-ordinate.

As a result, the wave number is

$$k^2 = k_0^2(1+\rho_1(R,\gamma)) \tag{22}$$

with $k_0 = \omega / c_s$ and $c_s = (\mu / \rho_0)^{1/2}$. The governing equation for this problem is equation (17) in spherical co-ordinates, which results in a uni-dimensional differential equation in terms of the radial co-ordinate. By comparing the aforementioned governing equation with the differential operator decomposition of equations (2) and (3) we see that the only non-zero coefficients are

$$\alpha_0 = k_0^2, \quad \alpha_1 = 2/R, \quad \alpha_2 = 1 \quad \text{and} \quad \beta_0 = k_0^2 \rho_1(R,\gamma) \tag{23}$$

with variable **x** replaced by variable R. Since there is no forcing function in this problem, the closed-form solution given by equation (16) must be used. Furthermore, a common choice for the correlation function $< RR >$ is a simple decaying exponential.

For a plane wave solution with amplitude A and wave number k, the solution $< u(R) >= A\exp(-kR)$ is substituted into the governing equation and integration finally yields

$$k^2 = k_0^2 - i\sigma_k^2 k_0^4 k^{-1}\left(\left(\xi^{-1} - ik_0 - ik\right)^{-1} + \left(\xi^{-1} - ik_0 + ik\right)^{-1}\right)/2 \tag{24}$$

The above result indicates that the true random wave number k is a complex quantity. As a consequence of the presence of the imaginary component, there is attenuation in the mean field $<u>$ as if a damping mechanism were at work. For large values of the quantity $k_0\xi$ compared to $1/\sigma_k$, equation (24) simplifies to

$$k = k_0\left(1 + i/2k_0\xi\right) \tag{25}$$

The attenuation coefficient is thus inversely proportional to the separation distance ξ, while the real part is constant and as such is independent of the correlation function chosen for $<RR>$.

4.4. APPROXIMATE SOLUTION TECHNIQUE

The most common approximate solution technique is the perturbation method [86], which takes full advantage of the structure of equation (19). There are, of course, other approximate techniques available such as the Taylor series expansion that still requires a small random fluctuation of the medium properties about their deterministic values and is used in conjunction with the stochastic finite element method [15]. Other techniques for discrete parameter systems include Fourier transforms coupled with normal mode analysis [87], exact solutions of the Fokker-Plank equation that governs the probability density function of the problem [88], and scattering expansions where the randomness is expressed in terms of a deterministic distribution of inhomogeneities in the otherwise homogeneous medium [11]. In what follows, we will apply the perturbation method to both the differential operator and the boundary integral equation governing SH wave induced ground vibrations.

4.4.1. *BEM Approach with Volume Integrals*

A consequence of medium randomness as described by equation (19) is that the response is also random and can be expanded using perturbation about the mean (deterministic) value φ_0 as

$$\varphi(\mathbf{x},\omega,\gamma)=\varphi_0(\mathbf{x},\omega)+\varepsilon\varphi_1(\mathbf{x},\omega,\gamma)+\varepsilon^2\varphi_2(\mathbf{x},\omega,\gamma)+\ldots \tag{26}$$

Substitution of equations (19) and (26) in the wave equation (17) and sorting of powers of ε yields the following system of equations:

$$\left.\begin{aligned}
&\nabla^2\varphi_0+k_0^2\varphi_0=f_0\\
&\nabla^2\varphi_1+k_0^2\varphi_1=f_1-2k_0k_1\varphi_0=\bar{f}_1\\
&\nabla^2\varphi_2+k_0^2\varphi_2=f_2-\left(k_1^2+2k_0k_2\right)\varphi_0-k_0^2\varphi_1=\bar{f}_2
\end{aligned}\right\} \tag{27}$$

The above system pre-supposes that the forcing function is also random and can be decomposed in a manner analogous to that of equation (26). If not, then the components f_1, f_2, etc, are absent. Furthermore, the zeroth order equation is accompanied by the mean boundary conditions on φ and $\partial\varphi/\partial n$. If the boundary conditions are deterministic, then all higher order equations are subjected to homogeneous boundary conditions. Otherwise, those equations are subjected to random boundary conditions of the corresponding order. Finally, the important thing to note is that the deterministic wave operator appears in the left-hand side of all equations. This allows for a unified treatment using boundary integral equation formalism.

We will now consider the case encompassing both zeroth and first order terms. The solution scheme that will be outlined can easily be modified to include higher order terms such as φ_2. We start by recasting the differential operator for the wave equation as a boundary integral equation [89] of the form

$$\alpha(r_S)\,\varphi_0(r_S)=\int\{G_0(R)\,q_0(r)-F_0(R)\,\varphi_0(r)\}\,dS(r)+\int G_0(R)\,f_0(r)\,dV(r) \tag{28}$$

In the above, r_S and r respectively are the field point and the integration point, while $R=|r-r_S|$ is the relative distance between them. Also, α is the jump term equal to 0.5 for a smooth boundary and V is the volume of the body in question. The fundamental solution G_0 for a deterministic homogeneous medium is

$$G_0(R) = \exp(ik_0R)/4\pi R \tag{29}$$

in three dimensions and

$$G_0(R) = iH_0^{(1)}(k_0R)/4c^2 \tag{30}$$

in two dimensions, where $i = \sqrt{-1}$ and $H_0^{(1)}$ is Hankel's function of the first kind and zero order. Fundamental solutions for the homogeneous half-space or half-plane can be respectively constructed from those of equations (29) and (30) by using the method of images. Finally, $F_0(R) = \partial G_0(R)/\partial n$ and the surface integral involving F_0 is understood in a Cauchy principal-value sense.

Next, an equation involving the first order terms is written by replacing φ_0, q_0 and f_0 by φ_1, q_1, and $\bar{f}_1$, respectively, in equation (28). At this stage, there are two routes available. The first one is to apply the expectation operator at the boundary integral equation level and reconstitute response statistics later, while the second one is to solve the boundary integral equations first, reconstitute a total solution from a sum of the zeroth and first order terms, and then apply the expectation operator. The latter route is simpler and will be preferred. Following routine numerical processing of equation (28) using boundary element concepts [89], the following system of algebraic equations is obtained:

$$A_{ij}z_j = B_{ij}y_j + C_{ik}f_k \tag{31}$$

In the above, z_j contain the unknown nodal potentials φ and fluxes q; y_j are the prescribed nodal boundary conditions of equations (18); f_k are nodal values of the forcing function; and A_{ij}, B_{ij}, C_{ik} are coefficient matrices. Note that indices i, j range from 1 to the total number of surface nodes and k ranges from 1 to the total number of volume nodes. Solving equation (31) for z_j and introducing matrix notation gives

$$\mathbf{z} = \mathbf{A}^{-1}\mathbf{B}\mathbf{y} + \mathbf{A}^{-1}\mathbf{C}\mathbf{f} = \mathbf{D}\mathbf{y} + \mathbf{H}\mathbf{f} \tag{32}$$

where upper case letters denote matrices and lower case letters denote vectors.

The expectation operator E is now applied to the total solution $\mathbf{z}$ which is reconstituted from the zeroth $\mathbf{z}^0$ and the first $\mathbf{z}^1(\gamma)$ order solutions, both of which have the general form shown in equation (32). In particular,

$$E\{\mathbf{z}\} = E\{\mathbf{z}^0 + \varepsilon\mathbf{z}^1\} = \mathbf{D}\left(\mathbf{y}^0 + \varepsilon\langle\mathbf{y}^1\rangle\right) + \mathbf{H}\left(\mathbf{f}^0 + \varepsilon\langle\tilde{\mathbf{f}}^1\rangle\right) \tag{33}$$

The above expression simplifies for a problem with no exterior forcing function (that is, $\mathbf{f}^0 = 0$) and for a first order solution about a zero mean, which will be the case considered for the rest of this section. It then reads as

$$E\{\mathbf{z}\} = m_i^z = \mathbf{D}\mathbf{y}^0 \tag{34}$$

which is nothing more than the deterministic solution of the problem for mean values of φ and q. Next, the covariance matrix K_{ij}^z of the response $\mathbf{z}$ is obtained as

$$K_{ij}^z = E\{z_i z_j\} = \Big\langle \left[D_{ik}\left(y_k^0 + \varepsilon y_k^1\right) + H_{il}\left(f_l^0 + \varepsilon f_l^1\right) \right] \cdot \left[D_{jm}\left(y_m^0 + \varepsilon y_m^1\right) + H_{jn}\left(f_n^0 + \varepsilon f_n^1\right) \right] \Big\rangle \tag{35}$$

Although the above expression in general gives rise to 16 terms, there is considerable simplification for the conditions stated previously, plus the fact that the zeroth order is deterministic, so that the final answer is

$$K_{ij}^z = \varepsilon^2 \Big[D_{ik} D_{jm} \langle y_k^1 y_m^1 \rangle + H_{il} H_{jn} \langle f_l^1 f_n^1 \rangle + D_{ik} H_{jn} \langle y_k^1 f_n^1 \rangle + H_{il} D_{jm} \langle f_l^1 y_m^1 \rangle \Big] \tag{36}$$

The above equations give second order statistics for the general, mixed boundary value problem of vibration in a random medium. These response statistics are constructed from solution of the usual deterministic problem plus a problem that requires volume integration of the mean values previously obtained and under homogeneous boundary conditions if the original boundary conditions are deterministic. With that in mind, $\mathbf{y}^1=\mathbf{0}$ and equation (36) further simplifies to

$$K_{ij}^z = \varepsilon^2 H_{il} H_{jn} < f_l^1 f_n^1 > \tag{37}$$

4.4.2. *BEM Approach without Volume Integrals*

If the governing equation (17) does not have a forcing function and if volume integrals are to be avoided, then a boundary integral equation representation can be written as follows:

$$a(\mathbf{r}_s)\,\varphi(\mathbf{r}_s,\gamma) = \int \{G(R,\gamma)\,q(\mathbf{r},\gamma) - F(R,\gamma)\,\varphi(\mathbf{r},\gamma)\}\,dS(\mathbf{r}) \tag{38}$$

It is clear that kernels G and F will be functions of a random parameter γ if the original wave operator in equation (17) contains a random coefficient k. These two kernels are now expanded as

$$\left.\begin{aligned} G(R,\gamma) &= G_0(R) + \varepsilon G_1(R,\gamma) + \varepsilon^2 G_2(R,\gamma) + \ldots \\ \text{and} \qquad & \\ F(R,\gamma) &= F_0(R) + \varepsilon F_1(R,\gamma) + \varepsilon^2 F_2(R,\gamma) + \ldots \end{aligned}\right\} \tag{39}$$

By substituting the above expansions along with equation (26) and an identical expression for $q(x,\gamma)$ in equation (38) and equating powers of ε, the following zeroth and first order solutions are obtained:

$$\left.\begin{aligned} a\varphi_0(\mathbf{r}_s) &= \int \{\, G_0(R)\, q_0(\mathbf{r}) - F_0(R)\;\; \varphi_0(\mathbf{r})\}\;\; dS(\mathbf{r}) \\ a\varphi_1(\mathbf{r}_s,\gamma) &= \int \{\, G_0(R)\, q_1(\mathbf{r},\gamma) - F_0(R)\;\; \varphi_1(\mathbf{r},\gamma)\}\;\; dS(\mathbf{r}) \\ &\quad + \int \{\, G_1(R,\gamma)\, q_0(\mathbf{r}) - F_1(R,\gamma)\;\; \varphi_0(\mathbf{r})\}\;\; dS(\mathbf{r}) \end{aligned}\right\} \tag{40}$$

As before, the first of equations (40) is the deterministic solution for mean values of φ and q. Boundary integral equations for higher order terms follow the convolution-like pattern of equations (40) with the highest order term multiplying the zeroth order fundamental solution and so forth. Equations (40) are now discretized following routine procedures to become

$$\left.\begin{aligned} A_{ij}^{0} z_{j}^{0} &= B_{ij}^{0} y_{j}^{0} \\ A_{ij}^{0} z_{j}^{1}(\gamma) + A_{ij}^{1}(\gamma)\ z_{j}^{0} &= B_{ij}^{0} y_{j}^{1}(\gamma) B_{ij}^{1}(\gamma)\, y_{j}^{0} \end{aligned}\right\} \tag{41}$$

The above equations can now be solved for the unknowns $\mathbf{z}$ in terms of the prescribed boundary conditions $\mathbf{y}$ to yield

$$\left.\begin{aligned} \mathbf{z}^{0} &= (\mathbf{A}_{0})^{-1}\,\mathbf{B}^{0}\mathbf{y}^{0} = \mathbf{D}\mathbf{y}^{0} \\ \mathbf{z}^{1}(\gamma) &= (\mathbf{A}^{0})^{-1}\{\mathbf{B}^{0}\mathbf{y}^{1}(\gamma) + \mathbf{B}^{1}(\gamma)\,\mathbf{y}^{0} - \mathbf{A}^{1}(\gamma)\,\mathbf{z}^{0}\} = \mathbf{D}\mathbf{y}^{1}(\gamma) + \mathbf{M}(\gamma)\,\mathbf{y}^{0} - \mathbf{N}(\gamma)\,\mathbf{z}^{0} \end{aligned}\right\} \tag{42}$$

Although random boundary conditions can be included, for example $\mathbf{y}^{1}(\gamma)$, we consider again the deterministic boundary condition case. Then the first order solution satisfies homogeneous boundary conditions and the total solution can now be reconstituted as

$$\mathbf{z} = \mathbf{z}^{0} + \varepsilon\,\mathbf{z}^{1} = (\mathbf{D} + \varepsilon\mathbf{M}(\gamma) - \varepsilon\,\mathbf{N}(\gamma)\,\mathbf{D})\,\mathbf{y}^{0} = (\mathbf{D} + \varepsilon\,\mathbf{Q}(\gamma))\,\mathbf{y}^{0} \tag{43}$$

Application of the expectation operator to the above equation gives equation (34) again for the mean value of $\mathbf{z}$, since matrix $\mathbf{Q}$ depends on $k_1(\gamma)$ that has a zero mean. Finally, the covariance matrix of the response $\mathbf{z}$ is

$$\begin{aligned} K_{ij}^{z} &= E\{z_1 z_j\} = \langle (D_{im} + \varepsilon Q_{im}(\gamma))\, y_m^0 (D_{jn} + \varepsilon Q_{jn}(\gamma))\, y_n^0 \rangle \\ &= D_{im} D_{jn} + \varepsilon^2 \langle Q_{im}(\gamma) Q_{jn}(\gamma) \rangle\, y_m^0\, y_n^0 \end{aligned} \tag{44}$$

It should be emphasized that only surface integrations are required in the construction of the system matrices $\mathbf{D}$ and $\mathbf{Q}(\gamma)$. Furthermore, any assumption on the correlation function for $k_1(\gamma)$ filters into the first and higher order kernels and makes analytical integrations all but impossible.

4.4.3. *General Comments*

Care must be exercised, however, when using perturbation methods because occasionally secular terms are generated which lead to divergent solutions. One such case is described in Askar and Cakmak [19] and involves plane waves in the infinite space. For this example, only the displacement potential φ is perturbed. With reference to the system of equations (27) for the one-dimensional case ($\mathbf{x} = x$), the homogeneous solution is $\varphi_0 = \exp(ik_0 x)$ which in turn generates a first order solution φ_1 involving the term $x\exp(ik_0 x)$. This latter term is secular and diverges as $x \to \infty$ thus restricting the validity of the solution to small values of x. A further analysis of this problem reveals that this particular perturbation solution is the limit of the actual solution of an integro-differential equation which results from substituting a plane wave solution in equation (16) and solving for the true wave number $k(x,\gamma)$. Thus, the perturbation solution is obtained from $k(x,\gamma)$ in the limit as the product $k\lambda_k$ goes to zero. For soft soils with wave velocities around 100 m/sec and frequencies of vibration of up to 5 Hz, which results in rather large values for the wave number, the perturbation method fails for any reasonable value of the correlation length λ_k. The only way to remedy the difficulty is to remove the secular terms. This is achieved by perturbing an eigenvalue of the problem (in this case it would be the wave number) about its mean value and determining the correction terms by imposing orthogonality of the right-hand sides of the system of equations (27), for all terms past the zeroth one, with respect to φ_0.

4.5. STOCHASTIC FIELD SIMULATIONS

Once the mean and the covariance of the response have been determined, it is possible to obtain a realization of the random field $\varphi(\mathbf{x},\omega,\gamma)$ by assuming that it is a uni-dimensional, uni-variate stochastic process [14]. This procedure will be illustrated, for the sake of convenience, for the case of wave propagation in a full-space where the response φ can be represented directly by the random three-dimensional Green's function $G(R,\gamma)$ of equation (39). We begin by defining a zero-mean random field as

$$G'(R,\gamma) = G(R,\gamma) - G_0(R) \tag{45}$$

so that $m_{G'} = 0$ and the covariance is, for the exponential correlation function of equation (20a)

$$K_{G'}(\xi) = -\ \varepsilon^2 \sigma_k^2 \exp(i\, k_0\, (2R+\xi)) \exp(-|\xi|/\lambda_k) \,/\, (4\pi)^2 \qquad (46)$$

The above expression is a function of the relative distance ξ between two receivers and, as such, is the autocovariance of $G'(R,\gamma)$.

It is well known that the autocovariance function $K_{G'}(\xi)$ and the power spectral density function (psdf) $S_{G'}(s)$ form a Fourier transform pair, that is,

$$S_{G'}(s) = \frac{1}{2\pi} \int_{-\infty}^{\infty} K_{G'}(\xi) \exp(-\mathrm{i}s\xi)\, \mathrm{d}\xi \qquad (47a)$$

and

$$K_{G'}(\xi) = \int_{-\infty}^{\infty} S_{G'}(s) \exp(+\mathrm{i}s\xi)\, \mathrm{d}s \qquad (47b)$$

where s is the Fourier transform parameter corresponding to ξ. Once the direct transform of equation (47a) has been performed, usually numerically, then realization of the random motion $G'(R, \gamma)$ is given as follows:

$$\tilde{G}_c'(R) = \sqrt{2} \sum_{j=1}^{N} c_j \cos(s_j\, R + \Phi_j) \qquad (48)$$

where

$$c_j = \sqrt{2 S_{G'}(s_j)\Delta s} \quad \text{and} \quad s_j = j\Delta s \qquad (49)$$

In the above, Δs is the increment in s equal to $2\pi/(N\Delta\xi)$ with $\Delta\xi$ the increment in ξ and Φ_j is a random phase angle which is uniformly distributed in the interval $(0,2\pi)$. As the number of samples N increases the realization $\tilde{G}_c'$ becomes a more realistic representation of the original random variable G' as the whole process becomes equivalent to a Monte Carlo simulation. Furthermore, the envelope $\tilde{G}_e'$ of the random vibration can be found as

$$\tilde{G}_e^{'}(R) = ((\tilde{G}_c^{'}(R))^2 + (\tilde{G}_s^{'}(R))^2)^{1/2} \tag{50}$$

where realization $\tilde{G}_s^{'}$ is given by equations (48) and (49) with the cosine replaced by a sine. Finally, all that remains to be done is to add the mean to the above realizations.

The psdf $S_{G'}(s)$ can be found in closed form for the case of an exponential correlation for $k_1(\gamma)$ (see equation (46)) as

$$S_{G'}(s) = \frac{Z}{\pi} \frac{1/\lambda_k}{1/\lambda_k^2 + (k_0 - s)^2} \tag{51}$$

where

$$Z = -\varepsilon^2 \sigma_k^2 \exp(i2k_0 R)/(4\pi)^2 \tag{52}$$

For the case of a delta correlation (see equation (20b)), we have a constant psdf in s, that is

$$S_{G'}(s) = Z/(2\pi) \tag{53}$$

and for other types of correlation functions, such as the piecewise linear function of equation (20c), standard discrete fast Fourier transform (FFT) algorithms can be used.

4.6. NUMERICAL EXAMPLE

The governing equation for wave propagation in the half-plane is equation (17) with $\mathbf{x} = (x,y)$. This example appears in Manolis [27] and focuses on the anti-plane strain case so that the displacement potential φ is identified with the displacement component $w=u_z(x,y,\omega)$. Thus, the boundary condition resulting from the presence of the horizontal surface of the ground S is that the stress component

$$\sigma_{yz} = \mu\ \partial u_z / \partial y = 0 \quad \text{at} \quad y=0 \tag{54}$$

and as such is identical to the homogeneous version of equations (18). Furthermore, it is assumed that randomness is manifested against a homogeneous background, that is

$$k(\gamma) = k_o + \varepsilon k_1(\gamma) + \varepsilon^2 k_2(\gamma) + \ldots \tag{55}$$

This example was solved by perturbing the boundary integral equation statement, in conjunction with stochastic field simulations.

As shown in Figure 1a, the source of seismicity is a point source placed at $\mathbf{r}_s$ and at a depth d from the surface. Then, the ground motions w can be identified with the Green's function $G(R,\gamma)$ expanded as

$$G(R,\gamma) = G_0(R) + \varepsilon\, G_1(R,\gamma) + \varepsilon^2\, G_2(R,\gamma) + \ldots \tag{56}$$

where $R = |\,\mathbf{r}-\mathbf{r}_s\,|$ and $\mathbf{r}$ is the receiver. The above expression is obtained by perturbing the random wave number $k(\gamma)$ and involves a Taylor series expansion of $H_0^{(1)}$ about k_0.

The first term $G_0(R)$ is obviously that of equation (30), while the second order term is obtained by using the recurrence relations for the derivatives of $H_0^{(1)}$ given in Abramowitz and Stegun [90] as

$$G_1(R,\gamma) = -ik_1(\gamma)\, R\, H_1^{(1)}(k_0R) \,/\, 4c_s^2 \tag{57}$$

where $H_1^{(1)}$ is the Hankel function of first kind and first order. Next, by applying the expectation operator to equation (56) we have that the mean solution m_G is $G_0(R)$ and the covariance of the response is

$$K_{G'}(\xi) = \varepsilon^2 \sigma_k^2 R(R+\xi)\; H_1^{(1)}(k_0R+k_0\xi)\; H_1^{(1)}(k_0R)\; \exp(-|\xi|/\lambda_k) \,/\, (4c_s^2)^2 \tag{58}$$

for an exponential autocorrelation function for $k_1(\gamma)$. The above covariance can now be used within the context of a stochastic realization process.

Consider a point source at a depth of 10 km below the surface and a receiver whose original position changes along a line parallel to the surface S in increments of 0.1 km for a total distance of 10 km also. The reference frequency of vibration ω considered is a low one and equal to

2.31 rad/sec (0.37 Hz), which results in a mean wave number k_0 value of 0.5 km^{-1} for a typical soil shear wave velocity c_s=4.62 km/sec. Reference values for the statistical description of the half-plane are σ_k^2=0.25 km^{-2} and λ_k=0.1 km. These values denote weak to intermediate statistical correlation. As far as the realization process is concerned, we employed N=256 sample values and an increment $\Delta s=2\pi/(N\Delta\xi)$=0.63 km^{-1}, which was dictated by the choice $\Delta\xi$=0.04 km. Figures 1b and 1c investigate the sensitivity of the envelope of the wave motion w to different values of correlation length λ_k and of sample size N at k_0=0.5 km^{-1}. Concurrently plotted is the mean amplitude of vibration $|w|$ versus radial distance R from the source. The envelope was determined from the stochastic realization process of equation (48) based on the autocovariance function of equation (58), but without the scaling factor ε. It is observed that the envelope is rather insensitive to changes in the correlation length in the neighbourhood of $R/50$ to $R/200$. Furthermore, an increase in the sample size for the discrete FFT used for obtaining the psdf gives a somewhat smoother envelope profile at large distances from the source.

Finally, Figure 1d plots the amplitude of the random motion w_r for two different values of the wavenumber, namely 0.1 and 1.0 km^{-1}. In order to obtain w_r, the envelope must be superimposed on the mean motion w by choosing an appropriate value for the scaling factor ε. This, however, requires physical measurements or some form of quantitative analysis at the site of interest [45-47]. In the absence of such information, the envelope was directly added to the mean value. We observe here that the presence of randomness influences the mean motion along the entire path of propagation R and at all frequencies of vibration. In general, this influence is more pronounced the further away one moves from the source (and the mean motion starts to decrease) and at higher frequencies of vibration. Finally, the presence of the horizontal surface is not felt at a depth of 10 km.

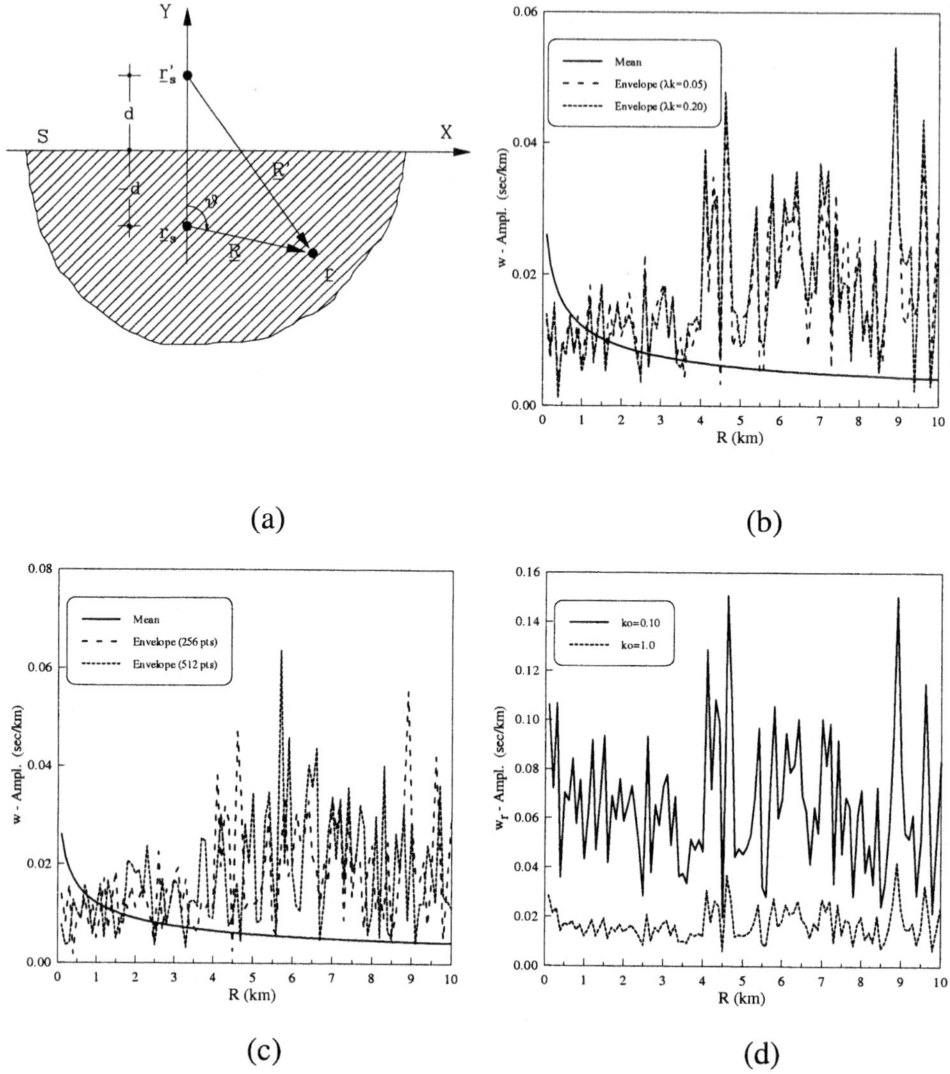

Figure 1. SH waves in a random medium : (a) Source-to- receiver geometry, (b), (c) two cases for mean amplitude and envelope function at k_0=0.5 km^{-1} and (d) total displacement wave signal at k_0=0.1 and 1.0 km^{-1} for ε=1.0.

5. BEM Formulation Based on Perturbations

In this section, we will develop the BEM for stochastic problems using perturbations.

5.1. BACKGROUND

The BEM for a 2D medium is developed here. Firstly, the fundamental laws of linear elastodynamics combine to give the well-known Navier-Cauchy equations

$$(\lambda+\mu)\, u_{i,ij} + \mu\, u_{j,ii} + \rho\, b_j = \rho\, \ddot{u}_j \tag{59}$$

which hold for an elastic, isotropic and homogeneous medium of volume V and surface S. In equations (59), λ and μ are the Lamé coefficients, u the displacement, b the body force per unit volume and ρ the material density. The above equations can be re-written in terms of the wave propagation velocities $c_1^2 = (\lambda + 2\mu) / \rho$ (for P-waves) and $c_2^2 = \mu / \rho$ (for S-waves). It is assumed that the summation convention holds for repeated indices (i,j=1,2 in 2D). Furthermore, we prescribe boundary conditions

$$u_i(\mathbf{x}, t) = U_i \quad \text{for } \mathbf{x} \in S_1 \quad \text{and} \quad t_i(\mathbf{x}, t) = \sigma_{ij} n_j = T_i \quad \text{for } \mathbf{x} \in S_2 \tag{60}$$

on surface $S=S_1 \cup S_2$, where t_i are the tractions and n_j is the outward normal vector to the surface. Also, the use of bold symbols indicates vectors. Finally, zero body forces and homogeneous initial conditions are assumed in what follows.

Numerical solution is achieved through recourse to the fundamental integral equation [89]

$$c_{ij}(\xi)\, u_i(\xi,t) = \int_S \{G_{ij}(\mathbf{x},\xi,t) * t_i(\mathbf{x},t) - F_{ij}(\mathbf{x},\xi,t) * u_i(\mathbf{x},t)\}\, dS \tag{61}$$

where ξ is the position of the receiver, $\mathbf{x}$ that of the source and t denotes time. G_{ij} and F_{ij} are the Green's functions, with the former corresponding to displacements and the latter to tractions. Finally, c_{ij} is the jump term which depends on the smoothness of the surface, while symbol $*$ denotes time convolution.

The solution will be formulated in the Laplace transform domain. A Laplace transform pair is defined as

$$\bar{f}\,(\mathrm{x},s) = \int_0^\infty f(\mathrm{x},t)\,\mathrm{e}^{-st}\,dt \quad \text{and} \quad f\,(\mathrm{x},t) = \frac{1}{2\pi i}\int_{\beta-i\omega}^{\beta+i\omega} \bar{f}\,(\mathrm{x},s)\,\mathrm{e}^{st}\,ds \tag{62}$$

where s is the transform parameter, $\mathrm{i}=\sqrt{-1}$ and $\beta > 0$. Using thi transformation, equations (61) are written as follows:

$$c_{ij}(\xi)\;\bar{u}_i\;(\xi,s) = \int_S \{\;\bar{G}_{ij}(\mathbf{x},\xi,s)\;\bar{t}_i\,(\mathbf{x},s) - \bar{F}_{ij}(\mathbf{x},\xi,s)\;\bar{u}_i(\mathbf{x},s)\;\}\;dS \tag{63}$$

The evolution of the problem in time is thus taken into account b discretizing the transformed parameter s as $n\Delta s$, n=1,2,...N, where N i the total number of the transformed steps. Then, equations (63) ar numerically solved for a range of values of s, and a spectrum is obtaine for the transformed displacements $\bar{u}_i$ and tractions $\bar{t}_i$. Finally, in order t invert the results back to the time domain, numerical invers transformation techniques are used.

5.2. FORMULATION

The extension for the case of a random medium is achieved through us of the method of perturbations. The basic assumption is that the wav propagation velocities are functions of random parameter γ, and can b expanded as

$$c_1(\gamma) = c_1^{\,m} + c_1^{\,f}(\gamma) \qquad c_2(\gamma) = c_2^{\,m} + c_2^{\,f}(\gamma) \tag{64}$$

where c_i^m is a mean reference value and $c_i^f(\gamma)$ is the fluctuation about thi mean. If the fluctuation has a zero mean and a known, nonzero varianc a simple representation for the wave propagation velocities is

$$c_1(\gamma) = c_{01}\,(1+\gamma) \qquad c_2\,(\gamma) = c_{02}\,(1+\gamma) \tag{65}$$

where c_{01}, c_{02} are mean velocities. For γ we have that $< \gamma > = 0$ and $< \gamma^2$ $= \sigma_\gamma^2$, with $< >$ denoting the expectation operator.

As a next step, all problem variables are expanded in Taylor series a follows:

$$\overline{G}_{ij}(r,s,\gamma) = \overline{G}_{ij}{}^{m} + \gamma\, \overline{G}_{ij}{}^{f} = \overline{G}_{ij}(r,s,0) + \gamma\, \frac{\partial \overline{G}_{ij}(r,s,\gamma)}{\partial \gamma}\Big|_{\gamma=0} \tag{66}$$

$$\overline{u}_{i}(r,s,\gamma) = \overline{u}_{i}{}^{m} + \gamma\, \overline{u}_{i}{}^{f} = \overline{u}_{i}(r,s,0) + \gamma\, \frac{\partial \overline{u}_{i}(r,s,\gamma)}{\partial \gamma}\Big|_{\gamma=0} \tag{67}$$

The expansions for $\overline{F}_{ij}$ and $\overline{t}_i$ are exactly analogous. Furthermore, superscripts m and f denote mean and fluctuation parts, respectively, while $r = |\mathbf{x}\text{-}\xi|$ is the relative distance between source and receiver. Application of the expectation operator yields a mean value for the first fundamental solution

$$< \overline{G}_{ij}(r,s,\gamma) > = \overline{G}_{ij}{}^{m} = \overline{G}_{ij}(r,s,0) \tag{68}$$

which coincides with the deterministic value of $\overline{G}_{ij}$. This is a consequence of the simple, first order perturbation expansion used. If higher order terms were also included, then the mean value of the stochastic problem would not coincide with the solution of the deterministic one. Of course, the same comment holds true for the remaining variables. Furthermore, the covariance between two receivers at r_k and r_l is

$$\begin{aligned} &< [\overline{G}_{ij}(r_k,s,\gamma) - \overline{G}_{ij}{}^{m}(r_k,s)]\, [\overline{G}_{ij}(r_l,s,\gamma) - \overline{G}_{ij}{}^{m}(r_l,s)] > \\ &\qquad = \sigma_\gamma^2\, [\overline{G}_{ij}{}^{f}(r_k,s)]\, [\overline{G}_{ij}{}^{f}(r_l,s)] \end{aligned} \tag{69}$$

where σ_γ^2 is the variance of the random variable γ. The variance can finally be obtained from the covariance by setting $r_k = r_l = r$.

The integral equation (63), defined for the stochastic problem, assumes the following form:

$$c_{ij}(\xi)\, \overline{u}_i(\xi,s,\gamma) = \int_S \{ \overline{G}_{ij}(r,s,\gamma)\, \overline{t}_i(\mathbf{x},s,\gamma) - \overline{F}_{ij}(r,s,\gamma)\, \overline{u}_i(\mathbf{x},s,\gamma) \}\, dS \tag{70}$$

Substitution of the expansions for the system variables in the above equation leads to two systems of equations, a zero order (deterministic) and a first order (stochastic) one, given below as

$$c_{ij}\,\bar{u}_i^{\,m}(\xi,s) = \int_S \{\,\bar{G}_{ij}^{\,m}(r,s)\,\bar{t}_i^{\,m}(\mathbf{x},s) - \bar{F}_{ij}^{\,m}(r,s)\,\bar{u}_i^{\,m}(\mathbf{x},s)\}\,dS \qquad (71)$$

and

$$c_{ij}\,\bar{u}_i^{\,f}(\xi,s,\gamma) = \int_S \{[\,\bar{G}_{ij}^{\,m}(r,s)\,\bar{t}_i^{\,f}(\mathbf{x},s,\gamma) + \gamma\,\bar{G}_{ij}^{\,f}(r,s)\,\bar{t}_i^{\,m}(\mathbf{x},s)] - [\,\bar{F}_{ij}^{\,m}(r,s)\,\bar{u}_i^{\,f}(\mathbf{x},s,\gamma) + \gamma\,\bar{F}_{ij}^{\,f}(r,s)\,\bar{u}_i^{\,m}(\mathbf{x},s)\,]\}\,dS \qquad (72)$$

The solution of these systems of equations is achieved numerically, essentially in the same manner as described for the case of the deterministic problem, and is explained in detail below. For simplicity, we use the symbols $\mathbf{G_o}$ for $\bar{G}_{ij}^{\,m}$ and $\mathbf{G_1}$ for $\gamma\bar{G}_{ij}^{\,f}$, and similar ones for the remaining problem variables. Using the BEM, the following matrix equations are obtained, where [ ] denotes a matrix and { } a vectorial quantity :

$$[\mathbf{G_o}]\,\{\mathbf{t_o}\} = [\mathbf{F_o}]\,\{\mathbf{u_o}\} \qquad (73)$$

and

$$[\mathbf{G_o}]\,\{\mathbf{t_1}\} + [\mathbf{G_1}]\,\{\mathbf{t_o}\} = [\mathbf{F_o}]\,\{\mathbf{u_1}\} + [\mathbf{F_1}]\,\{\mathbf{u_o}\} \qquad (74)$$

In a well-defined problem, half the variables $\{\mathbf{t_o}\}$, $\{\mathbf{u_o}\}$, $\{\mathbf{t_1}\}$, $\{\mathbf{u_1}\}$ are known from the boundary conditions.

Let us here assume that displacements u are known on part S_1 of the boundary, while tractions t are known on the remaining part S_2. Assuming an appropriate partitioning of equations (73) and (74), we have

$$[\,\mathbf{S_o}\,]\{\,\mathbf{b_o}\,\} = [\,\mathbf{T_o}\,]\{\,\mathbf{x_o}\,\} \qquad (75)$$

and

$$[\,\mathbf{S_o}\,]\,\{\,\mathbf{b_1}\,\} + [\,\mathbf{S_1}\,]\,\{\mathbf{b_o}\} = [\,\mathbf{T_o}\,]\,\{\mathbf{x_1}\} + [\,\mathbf{T_1}\,]\,\{\mathbf{x_o}\} \qquad (76)$$

where $\{\mathbf{x_o}\}$ and $\{\mathbf{x_1}\}$ are the unknown boundary variables and $\{\mathbf{b_o}\}$ and $\{\mathbf{b_1}\}$ are the prescribed boundary conditions. We note that in the present work we deal with deterministic boundary conditions, and hence the fluctuation terms $\{\mathbf{u_1}\}$ on S_1 and $\{\mathbf{t_1}\}$ on S_2 are zero, that is, $\{\mathbf{b_1}\}= \{\mathbf{0}\}$. Thus

$$\{\mathbf{x_o}\}= [\ \mathbf{T_o}\ ]^{-1}\ [\ \mathbf{S_o}\ ]\ \{\mathbf{b_o}\}= [\ \mathbf{A}\ ]\ \{\mathbf{b_o}\} \tag{77}$$

and

$$\{\mathbf{x_1}\}=(\ [\ \mathbf{T_o}\ ]^{-1}\ [\ \mathbf{S_1}\ ] - [\ \mathbf{T_o}\ ]^{-1}\ [\ \mathbf{T_1}\ ]\ [\ \mathbf{T_o}\ ]^{-1}\ [\ \mathbf{S_o}\ ]\ )\ \{\mathbf{b_o}\}= [\ \mathbf{W}\ ]\ \{\mathbf{b_o}\} \tag{78}$$

The mean value vector and the covariance matrix of the free variables of the problem are now evaluated as follows :

$$\{\ \mathbf{m_x}\ \} = <\{\mathbf{x}\}> = <\{\mathbf{x_0}\} + \{\mathbf{x_1}\}> = <\{\mathbf{x_0}\}> = \{\mathbf{x_0}\}= [\mathbf{A}]\ \{\mathbf{b_o}\} \tag{79}$$

and

$$[\ \mathbf{K_x}\ ] = \ <\{\mathbf{x_1}\}\{\mathbf{x_1}\}^{\mathrm{T}}> = \sigma_\gamma^{\ 2}\,(\ [\ \mathbf{Y}\ ]\ \{\mathbf{b_o}\}\ \{\mathbf{b_o}\}^{\mathrm{T}}[\ \mathbf{Y}\ ]^{\mathrm{T}}\ ) \tag{80}$$

where we note that the random parameter γ can be isolated as a common factor (that is, $\{\mathbf{x_1}\}= \sigma_\gamma\ [\mathbf{Y}]\ \{\mathbf{b_o}\}$). Any assumption can be made for the variance of γ, for example $\sigma_\gamma^{\ 2} = \sigma_\gamma^{\ 2}\ \exp(-|r|/l_{cor})$, where r is the relative distance between two receivers and l_{cor} is the correlation length [4].

5.3. FUNDAMENTAL SOLUTIONS

The Green's functions for 2D wave propagation in the Laplace transformed domain have been derived elsewhere [89]. These functions are given here for the purpose of completeness. We have that

$$\overline{G}_{ij}\,(r,s,\gamma) = \frac{1}{2\pi\rho c_2(\gamma)^2}\ \{\ \psi(r,s,\gamma)\ \delta_{ij} - \varphi(r,s,\gamma)\ r_{,i}\ r_{,j}\ \} = \frac{1}{2\pi\rho c_2(\gamma)^2}\ V(r,s,\gamma) \tag{81}$$

and

$$\overline{F}_{ij}(r,s,\gamma) = \frac{1}{2\pi} \{ [\frac{\partial\psi(r,s,\gamma)}{\partial r} - \frac{\varphi(r,s,\gamma)}{r}] [\delta_{ij} \frac{\partial r}{\partial n} + r_{,j}\, n_i]$$
$$- (\frac{2\varphi(r,s,\gamma)}{r}) [n_j\, r_{,i} - 2\, r_{,i}\, r_{,j} \frac{\partial r}{\partial n}] - (2 \frac{\partial\varphi(r,s,\gamma)}{\partial r}) [r_{,i}\, r_{,j} \frac{\partial r}{\partial n}]$$
$$+ [\frac{\partial\psi(r,s,\gamma))}{\partial r} - \frac{\partial\varphi(r,s,\gamma)}{\partial r} - \frac{\varphi(r,s,\gamma)}{r}] [\frac{c_1(\gamma)^2}{c_2(\gamma)^2} - 2] (r_{,i}\, n_j) \}$$
$$= \frac{1}{2\pi} \{ [W(r,s,\gamma)] [w] - [X(r,s,\gamma)[x]$$
$$- [Y(r,s,\gamma)[y] + [Z(r,s,\gamma)] [\frac{c_1(\gamma)^2}{c_2(\gamma)^2} - 2] [z] \} \tag{82}$$

where

$$\psi(r,s,\gamma) = K_0(\frac{sr}{c_2(\gamma)}) + \frac{c_2(\gamma)}{sr} \{ K_1(\frac{sr}{c_2(\gamma)}) - \frac{c_2(\gamma)}{c_1(\gamma)} K_1(\frac{sr}{c_1(\gamma)}) \} \tag{83}$$

and

$$\varphi(r,s,\gamma) = K_2(\frac{sr}{c_2(\gamma)}) - \frac{c_2(\gamma)^2}{c_1(\gamma)^2} K_2(\frac{sr}{c_1(\gamma)}) \tag{84}$$

In the above expressions, K_n are modified Bessel functions of order n [90].

The mean value for the first fundamental solution is simply

$$< \overline{G}_{ij}(r,s,\gamma) > = \overline{G}_{ij}{}^{m} = \overline{G}_{ij}(r,s,0) = \frac{1}{2\pi\rho c_{02}^2} \{ \psi(r,s,0)\, \delta_{ij} - \varphi(r,s,0)\, r_{,i}\, r_{,j}\} \tag{85}$$

while the covariance is given according to equation (69). Finally, the variance is obtained from the covariance by setting $r_k = r_l = r$.

The fluctuation term required by equation (69) is computed as follows:

$$\overline{G}_{ij}{}^{f}(r,s,0) = \frac{\partial \overline{G}_{ij}(r,s,\gamma)}{\partial\gamma}|_{\gamma=0}$$
$$= \frac{1}{2\pi\rho} \{ [\frac{\partial}{\partial\gamma}(\frac{1}{c_2^2})] V(r,s,\gamma) + \frac{1}{c_2^2} \frac{\partial V(r,s,\gamma)}{\partial\gamma} \} |_{\gamma=0}$$

$$= \frac{1}{2\pi\rho} \{ D(\gamma)\, V(r,s,\gamma) + \frac{1}{c_2^2}\, E(r,s,\gamma) \} |_{\gamma=0} \tag{86}$$

The various terms appearing above are evaluated in [29].

The mean for the second fundamental solution follows along the same lines as above, that is $\overline{F}_{ij}{}^m(r,s,\gamma) = \overline{F}_{ij}(r,s,\gamma=0)$. For computation of the covariance, the fluctuation term must also be evaluated. Specifically, we have that

$$\overline{F}_{ij}{}^f(r,s,\gamma) = \frac{\partial \overline{F}_{ij}(r,s,\gamma)}{\partial \gamma} |_{\gamma=0}$$

$$= \frac{1}{2\pi r} \{ [\frac{\partial}{\partial \gamma}(W\, r)]|_{\gamma=0}\ [w] - [\frac{\partial}{\partial \gamma}(X\, r)]|_{\gamma=0}\ [x]$$

$$- [\frac{\partial}{\partial \gamma}(Y\, r)]|_{\gamma=0}\ [y] + [\frac{c_{01}^2}{c_{02}^2} - 2]\, [\frac{\partial}{\partial \gamma}(Z\, r)]|_{\gamma=0}\ [z] \tag{87}$$

The various terms appearing above are given in [29]. All these solutions have been tested against results given by Monte-Carlo simulations and have been found to be quite accurate, within the limits set by the perturbation approach (that is for values of $c^f(\gamma)$ up to about 10 – 20% of c_0).

5.4. NUMERICAL EXAMPLES

5.4.1. *Circular Unlined Tunnel Enveloped by a Pressure Wave*

As shown in Figure 2a, a circular unlined tunnel of radius R=5 m at some depth in the ground is enveloped by a compressive pressure wave propagating along the negative X-direction. Due to the wave, stresses along the propagation path (σ_{xx}= -10 MPa) as well as in the lateral direction ($\sigma_{yy}= \nu\sigma_{xx}/(1-\nu)$) develop. The mean material properties of the

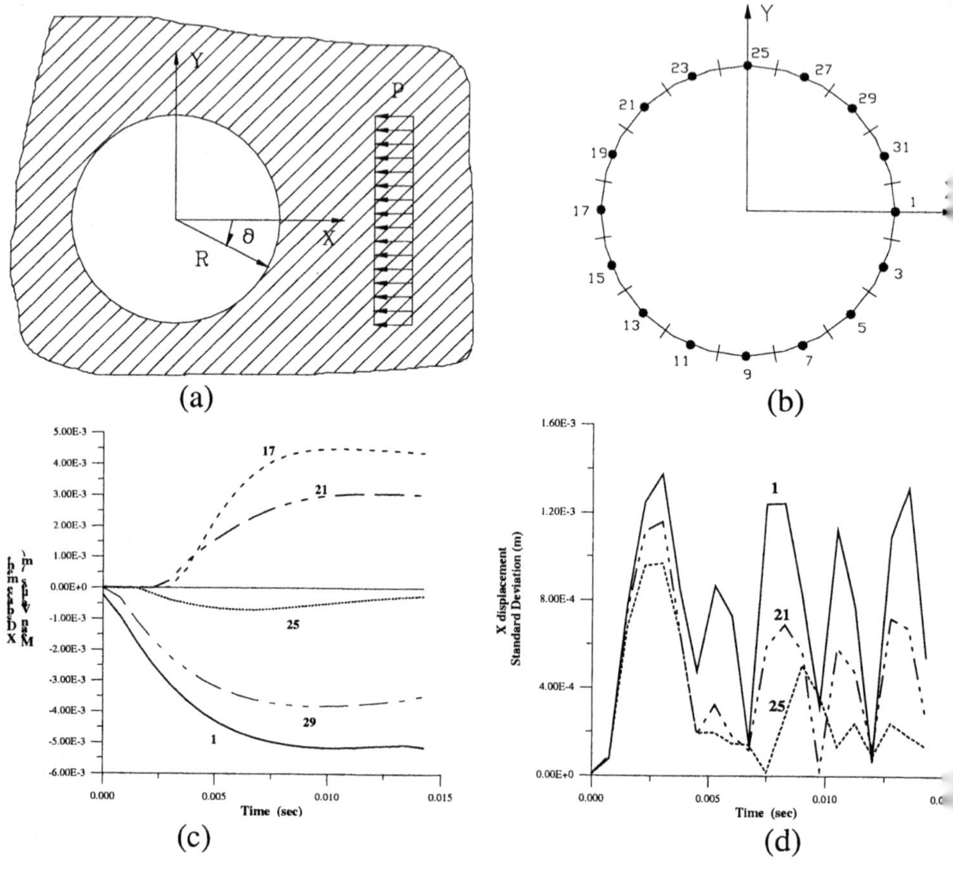

Figure 2. Circular tunnel under a P-wave : (a) geometry, (b) discretization, (c) mean values and (d) standard deviations of horizontal displacements u_x versus time.

surrounding stiff soil are $\lambda=\mu=8000$ MPa and $\rho=2550$ kg/m^3, whic correspond to mean wave propagation velocities $c_{01}=3068$ m/sec an $c_{02}=1770$ m/sec. Time $t=0$ is the instant of arrival of the wave at station 1 The statistical measures of the tunnel response (mean and standar deviation of the displacements) are evaluated for a given variance of th wave velocity. Specifically, an assumption of constant value $\sigma_\gamma^2=0.10$ i made for the variance of the random parameter γ, which implies that th standard deviations of the two wave velocities c_1 and c_2 are 970 m/se and 560 m/sec, respectively. The total time span examined is 0.015 sec that is triple the time necessary for the wave to fully envelop the tunnel and a time discretization into 20 equal time steps is used. For the discretization of the boundary, 16 three-noded parabolic elements are employed. In Figure 2b, the central node of each element is numbered and denoted by a full circle; also, the computed mean values for the displacements at various nodes are plotted in Figures 2c and 2d. The accuracy of the numerical results for the mean values was gauged against the deterministic ones of Baron & Matthews [91], and found to be entirely satisfactory.

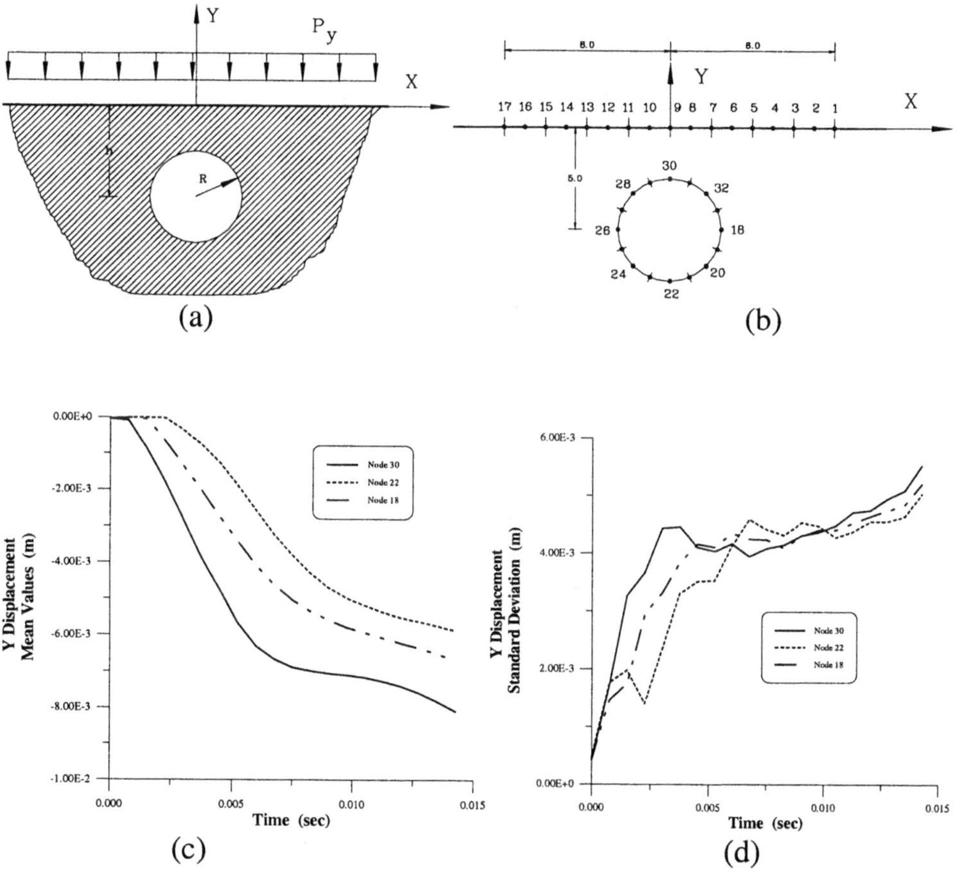

Figure 3. Circular tunnel in a half-space under surface load : (a) the geometry, (b) the discretization, (c) mean values and (d) standard deviations of vertical displacements u_y versus time.

Finally, we plot the standard deviation (s.d.) of the displacements at nodes 1, 21 and 25. We note that for a s.d. of the random parameter γ equal to σ_γ=32%, the s.d. of the displacements are of the order of 15%.

5.4.2. *Circular Unlined Tunnel in a Half-Plane under Surface Load*
An unlined circular tunnel of radius R=2.5m lies at a depth h=5.0 m from the free surface of a half-plane, which in turn is subjected to a uniform surface load P_y= -10 MPa/m, as shown in Figure 3a. The mean material properties of the surrounding soil are the same as before, that is $\lambda=\mu$=8000 Mpa and ρ=2550 kg/m^3, which correspond to mean wave propagation velocities c_{01}= 3068 m/sec and c_{02}= 1770 m/sec. The statistical measures of the vertical displacements at various points of the tunnel surface are evaluated for σ_γ^2=0.10, as in the previous example. The total time span examined is 0.15 sec, divided into 20 equal time steps.

For the discretization of the free surface of the half-plane 8 three-noded parabolic elements are used, each of length of 1 m, as seen in Figure 3b. Similarly, for the discretization of the tunnel surface 8 three-noded parabolic elements are also used. The mean values of the vertical displacements at nodes 18, 22 and 33 are presented in Figure 3c. The values of the s.d. of the displacements (Figure 3d) are found to be of the same order of magnitude as that of their respective mean values. This fact indicates the significant role that the existence of the free surface plays on the dynamic response of the tunnel, as well as the importance the medium randomness has on the response of the system.

6. BEM Formulation Based on Polynomial Chaos

6.1. BACKGROUND

Time harmonic elastic waves that propagate in a three-dimensional random continuum under anti-plane strain conditions are governed by Helmholtz's equation

$$\nabla^2 u(\mathbf{x},\gamma) + k^2(\gamma)u(\mathbf{x},\gamma) = f(\mathbf{x},\gamma) \tag{88}$$

where u is the displacement component in the z-direction and k is the wave number equal to $\omega / c(\gamma)$, with ω the frequency and c the wave speed. Furthermore, $\mathbf{x}$ is the position vector restricted to lie on the x-y plane and γ denotes a random parameter. Note that the common factor $\exp(-i\omega t)$, where t is the time, is implied and that ∇^2 is Laplace's operator. The solution to equation (88) when the forcing function f is Dirac's delta function $f_0 \delta(r)$, where r is the relative distance between source and receiver in the unbounded medium and f_0 is the force magnitude (usually $f_0 = 1.0$), is Green's function and obeys the radiation boundary condition as previously discussed.

For the deterministic, homogeneous elastic medium, equation (88) for the Green's function in cylindrical co-ordinates becomes [92]

$$\frac{d^2 u(r)}{dr^2} + \frac{1}{r}\frac{du(r)}{dr} + k_0^2 u(r) = \delta(r) \tag{89}$$

with k_O the mean wave number. The solution is well known, that is,

$$u(r) = (i/4)\, H_0^{(1)}(k_0 r) \tag{90}$$

where $H_0^{(1)}$ is the Hankel function of first kind, zero order and represents outgoing waves. Also, $i = \sqrt{-1}$. A second fundamental solution is necessary within the context of a boundary integral equation formulation for wave problems. This is given as

$$w(r) = \partial u(r) / \partial n = (du(r)/dr)(dr/dn) = -(i k_0/4) H_1^{(1)}(k_0 r)(dr/dn) \tag{91}$$

where $H_1^{(1)}$ is the Hankel function of first kind, first order and n is the direction of the outward normal at the surface(s) of the medium.

6.2. FORMULATION

Stochasticity here results due to randomness in the wave number, which is defined in terms of a mean plus a fluctuating component as

$$k(\gamma) = k_o + k_1 \gamma \tag{92}$$

In the above, k_1 is a deterministic coefficient and γ is a random variable with zero mean and unit variance σ_γ^2. Thus, the second moment representation of the wave number is

$$\left.\begin{aligned} &< k(\gamma) >=< k_0 >= k_0 \\ &< k^2(\gamma) >=< (k_1\gamma)^2 >= \sigma_\gamma^2 k_1^2 = k_1^2 \end{aligned}\right\} \tag{93}$$

where the expectation operator <> denotes statistical averaging.

The next step is to expand both fundamental solution $u(r,\gamma)$ and forcing function $f(r,\gamma)$ as a series in terms of an orthogonal set of polynomials in γ [24,30]. This implies a separation of variables, since the orthogonal polynomials P_n are weighted by spatially dependent coefficients U_n and F_n in the form

$$\left.\begin{aligned} u(r,\gamma) &= \sum_{n=0}^{N} P_n(\gamma) U_n(r) \\ f(r,\gamma) &= \sum_{n=0}^{N} P_n(\gamma) F_n(r) \end{aligned}\right\} \tag{94}$$

where N is the order of approximation in the random space. The choice of the polynomial basis is dictated by the fact that the expectation is essentially an orthogonality condition, that is

$$< P_n(\gamma) P_m(\gamma) >= \int_\Gamma pdf(\gamma) P_n(\gamma) P_m(\gamma) d\gamma = \delta_{nm} \tag{95}$$

with $pdf(\gamma)$ the probability density function of the random variable γ. Therefore, the selection of the polynomials basis depends on the pdf assumed for the wave number, with Hermite polynomials H_n corresponding to a Gaussian distribution. In such a case, the polynomials are given as

$$\mathcal{P}_n(\bullet) = 2^{-n/2} \bullet^{-1/4} (n!)^{-1/2} \mathcal{H}_n(\bullet) = a_n \mathcal{H}_n(\bullet) \tag{96}$$

The first step in developing the present methodology is substitution of the expansions given by equations (94) into the governing equation of motion (89). Subsequent pre-multiplication by $H_m(\gamma)$ and application of the expectation operator yields the following coupled set of equations:

$$\begin{aligned} &< H_m H_n > \nabla_r^2 U_n + \\ &\left[k_0^2 < H_m H_n > +2k_0 k_1 < \gamma H_m H_n > +k_1^2 < \gamma^2 H_m H_n > \right] U_n =< H_m H_n > F_n \end{aligned} \tag{97}$$

Use of the recurrence relations for Hermite polynomials in conjunction with the orthogonality property [90] finally results in the following coupled system of differential equations of the Helmholtz type, which governs the spatially dependent coefficients $U_n(r)$ of the random fundamental solution:

$$\begin{aligned} &h_{m,n}\nabla^2 U_n + \\ &\left[\left(k_1^2/4\right)h_{m,n+2} + k_0 k_1 h_{m,n+1} + \left(k_0^2 + k_1^2(n+1/2)\right)h_{m,n} + 2n k_0 k_1 h_{m,n+1} + n(n-1) k_1^2 h_{m,n+1}\right]U_n \\ &= h_{m,n} F_n \end{aligned} \tag{98}$$

In the above the "weights" $h_{m,n}$ are given as

$$h_{m,n} =< H_m(\gamma) H_n(\gamma) >= \sqrt{\pi}\, 2^n (n!)\, \delta_{nm} \tag{99}$$

Carrying out the expansion for five terms (that is n = 0,1,2,3,4) yields the following matrix differential equation of the Helmholtz type:

$$\begin{Bmatrix} F_0 \\ F_1 \\ F_2 \\ F_3 \\ F_4 \end{Bmatrix} = \nabla_r^2 \begin{Bmatrix} U_0 \\ U_1 \\ U_2 \\ U_3 \\ U_4 \end{Bmatrix} +$$

$$\begin{bmatrix} k_0+\frac{1}{2}k_1^2 & 2k_0k_1 & 2k_1^2 & 0 & 0 \\ k_0k_1 & k_0^2+\frac{3}{2}k_1^2 & 4k_0k_1 & 6k_1^2 & 0 \\ \frac{1}{4}k_1^2 & k_0k_1 & k_0^2+\frac{5}{2}k_1^2 & 6k_0k_1 & 12k_1^2 \\ 0 & \frac{1}{4}k_1^2 & k_0k_1 & k_0^2+\frac{7}{2}k_1^2 & 8k_0k_1 \\ 0 & 0 & \frac{1}{4}k_1^2 & k_0k_1 & k_0^2+\frac{9}{2}k_1^2 \end{bmatrix} \begin{Bmatrix} U_0 \\ U_1 \\ U_2 \\ U_3 \\ U_4 \end{Bmatrix} \quad (100)$$

where ∇_r is Laplace's operator in cylindrical co-ordinates. We note the following with respect to the above equation: (i) The system matrix is non-symmetric, and (ii) it has been truncated, that is the column corresponding to U_5 and U_6 have been deleted. Thus, an important check on the present method is to ascertain the effect that the $h_{n,m+1}$ and $h_{n,m+2}$ terms multiplying U_n in equation (98) have on the accuracy level of the "polynomial chaos" approximation for fundamental solution $u(r,\gamma)$.

Equation (100) can be written using matrix notation as

$$\frac{d^2}{dr^2}\{U\}+\frac{1}{r}\frac{d}{dr}\{U\}+[K]\{U\}=\{F\} \quad (101)$$

In order to uncouple the above equation, system matrix $[K]$ must be diagonalized. This is achieved by using its eigenvalues $\lfloor \lambda_1,\lambda_2,\ldots,\lambda_n \rfloor$ and the corresponding eigenvector matrix $[\Phi]=\left[\ \{\varphi\}_1,\{\varphi\}_2,\ldots,\{\varphi\}_n\ \right]$, with the eigenvectors arranged column-wise. Although $[K]$ is non-symmetric its coefficients are all positive, real numbers and thus it will posses a complete set of complex eigenvalues and eigenvectors. We note in passing that if $k_1=0$, equations (100) uncouple into five scalar Helmholtz equations which all have the same wave number, namely k_0; they are all equivalent to the governing equation of motion (88) with $k(\gamma)=k_0$.

Next, we observe that matrix $[\Lambda]$ defined as

$$[\Lambda]=[\Phi]^{-1}[K]\ [\Phi] \quad (102)$$

is diagonal with the eigenvalues of $[K]$ as its diagonal elements (note that superscript -1 denotes matrix inversion). Pre-multiplying equation (101) by $[\Phi]^{-1}$ yields

$$[\Phi]^{-1}\frac{d^2}{dr^2}\{U\}+\frac{1}{r}[\Phi]^{-1}\frac{d}{dr}\{U\}+[\Phi]^{-1}[K][\Phi][\Phi]^{-1}\{U\}=[\Phi]^{-1}\{F\} \tag{103}$$

By defining new variables as

$$\{V\}=[\Phi]^{-1}\{U\} \quad \text{and} \quad \{B\}=[\Phi]^{-1}\{F\} \tag{104}$$

we finally get an uncoupled system of Helmholtz equations which is

$$\frac{d^2}{dr^2}\{V\}+\frac{1}{r}\frac{d}{dr}\{V\}+[\Lambda]\{V\}=\{B\} \tag{105}$$

At this stage, we need to examine the right-hand side vector $\{B\}$ which acts as the forcing function. Since randomness is confined to the medium, the original forcing function expansion of equation (94) contains only one term, namely

$$f(r,\gamma)=\delta(r)=P_o(\gamma)F_0(r)=F_0(r) \tag{106}$$

Furthermore, the multiplication given by the second of equations (104) when $\{F\}^T=\lfloor F_0,0,0,0,0\rfloor$ simply yields $\{B\}=\{b\}\delta(r)$, where $\{b\}$ is a constant vector and contains the first column of the inverse of the matrix of eigenvectors $[\Phi]$. Thus, the uncoupled system of scalar Helmholtz equations is of form:

$$\frac{d^2}{dr^2}V_n(r)+\frac{1}{r}\frac{d}{dr}V_n(r)+\lambda_n\dot{V}_n=b_n\delta(r) \tag{107}$$

with $n=0,1,2,\ldots$. The solution for outgoing waves is simply

$$V_n(r) = (i/4) H_0^{(1)}(\lambda_n r) \tag{108}$$

where λ_n is the appropriate wave number corresponding to the n-th term of the "polynomial chaos" expansion.

6.3. RESPONSE STATISTICS

Given the "polynomial chaos" approximation for the stochastic fundamental solution $u(r,\gamma)$ of equation (94), the first two moments are now computed. We note that the same information can be just as easily computed for the second fundamental solution $w(r,\gamma) = \partial u(r,\gamma)/\partial n$. We start with the mean value

$$\begin{aligned} m_u(r) &=< u(r,\gamma) >=< P_n(\gamma) > U_n(r) =< P_n H_o > U_n = \\ &= a_0 h_{0,0} U_0 + a_1 h_{0,1} U_1 + \ldots = \pi^{\frac{1}{4}} U_0(r) + 0 + \ldots = \pi^{\frac{1}{4}} U_0(r) = P_0\, U_0(r) \end{aligned} \tag{109}$$

where coefficients a_n were given in equation (96). With respect to the above solution, we note that: (i) The summation convention is implied for repeated indices (subscripts) so as to avoid using the summation symbol Σ and (ii) since the first Hermite polynomial $H_0(\gamma)=1$ it is possible to introduce it into the above equation so as to take advantage of the orthogonality property given by equation (99). Furthermore, it is obvious here that the mean solution is not equal to the deterministic one which is obtained when randomness is absent and the effective wave number of the problem is k_0. In fact, the former solution exceeds the latter (in absolute value terms), and the physical explanation is that this is due to the interference effect caused by continuous scattering of the propagating signal as it travels through the random medium. This type of behaviour was noticed early on by Karal and Keller [5] and by others [4]. We note that the factor in question is not exactly $\pi^{\frac{1}{4}}$ because the wave number corresponding to U_0 does not coincide with k_0. In fact, all N terms contribute to $u_0(r)$ since the eigenvalues λ_n directly determine the $V_n(r)$ components only, which in turn yield $U_n(r)$ through the algebraic transformation given by equation (104).

Next, we determine the covariance of $u(r,\gamma)$ as

$$\text{cov}_u(r_1,r_2) = \langle\, [u(r_1,\gamma) - m_u(r_1)][u(r_2,\gamma) - m_u(r_2)]\,\rangle =$$
$$= \left\langle \left[P_0U_0(r_1) + P_1U_1(r_1) + \ldots - P_0U_0(r_1)\right]\left[P_0U_0(r_2) + P_1U_1(r_2) + \ldots - P_0U_0(r_2)\right]\right\rangle =$$
$$= \langle P_1P_1\rangle U_1(r_1)U_1(r_2) + \langle P_1P_2\rangle U_1(r_1)U_2(r_2) + \ldots + \langle P_2P_2\rangle U_2(r_1)U_2(r_2) + \ldots \quad (110)$$

Given the orthogonality condition in equation (95), the above result can be written as

$$\text{cov}_u(r_i, r_j) = \sum_{n=1}^{\infty} U_n(r_i)\,U_n(r_j) \quad (111)$$

Finally, it is well known that the variance is $\text{var}_u(r)=\text{cov}_u(r,r)$ and that the standard deviation (s.d.) is the square root of the variance.

6.4. NUMERICAL EXAMPLE

As an example, consider SH wave propagation through a continuous geological medium with a shear wave velocity c_s = 6.28 km/sec and at a transmission frequency ω = 1 Hz = 2π rad/sec. For this case, the reference deterministic wave number is k_0 = 1.0 km^{-1}. Also, the amount of randomness in the wave number $k(\gamma)$ ranges from 1% to 50 % of k_0. It is assumed that randomness of 1% is negligible and the material cannot be distinguished from the reference homogeneous background, while values in the neighbourhood of 5% - 10% are at the limit of the perturbation method. Of course, at 50% randomness we expect a significant departure of the response covariance from the usual results obtained for mild randomness.

With respect to the geometry of the problem we have a unit impulse placed at the source, which is at the origin of the co-ordinate system, while receivers trace a path along a straight line starting from that source. Specifically, forty receiver stations span a total distance of r = 10 km, which implies that the spacing between any two receivers is 0.25 km. For comparison, the reference wavelength of the SH wave at this frequency is $l_0 = 2\pi / k_0 = 6.28$ km.

The first series of results pertains to fundamental solution $u(r,\gamma)$ and, more specifically, to its mean value and variance. Therefore, Figures 4a to 4d plot the mean amplitude, mean phase angle, variance amplitude and variance phase angle of the fundamental solution when $k_1 = 1\,\%$ of k_0. The

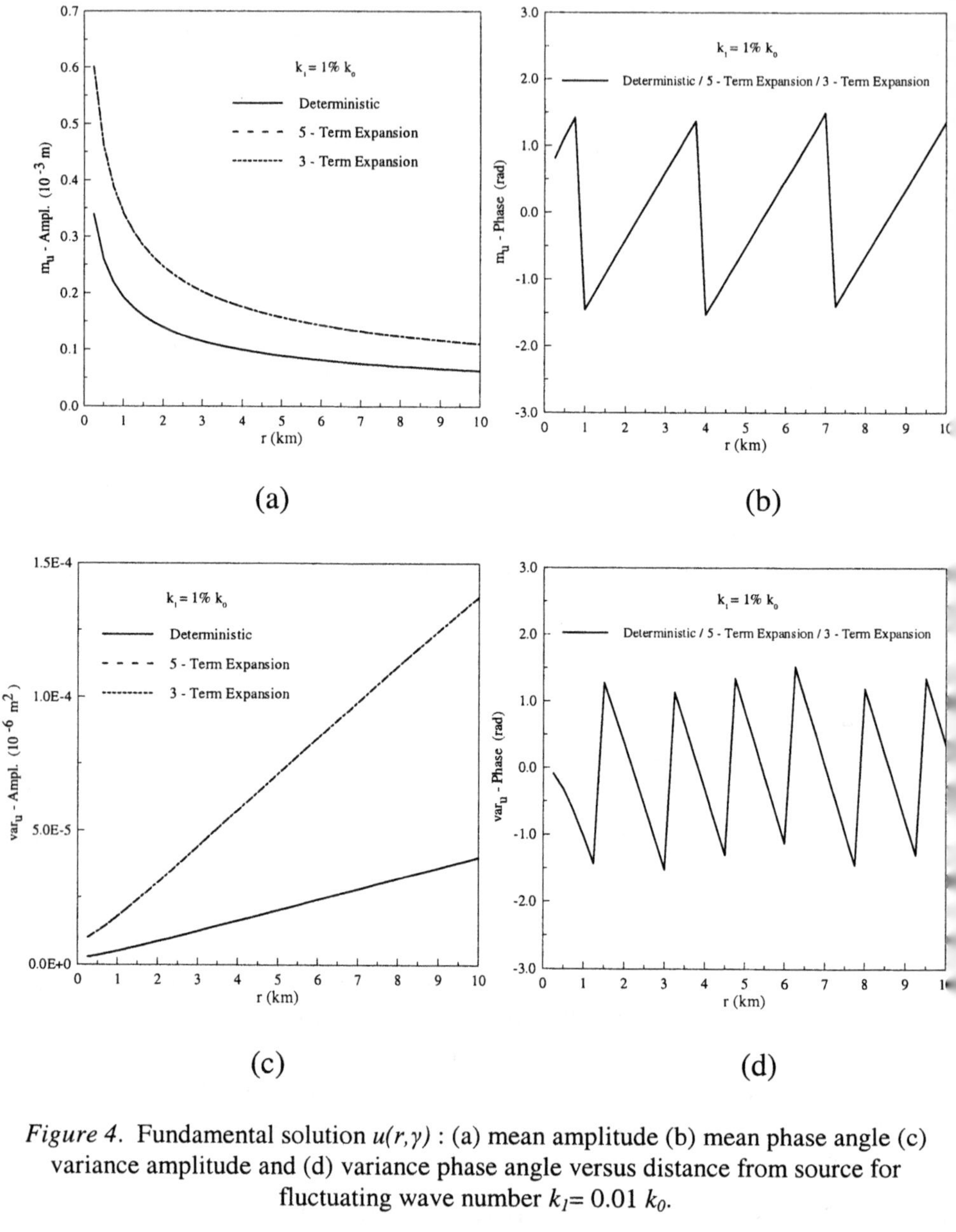

Figure 4. Fundamental solution $u(r,\gamma)$: (a) mean amplitude (b) mean phase angle (c) variance amplitude and (d) variance phase angle versus distance from source for fluctuating wave number $k_1 = 0.01\ k_0$.

results of both five-term (N=5) and three-term (N=3) polynomial chaos expansions are given in each of those figures. For comparison purposes, the deterministic solution, which is obtained from equation (88) when

$k(\gamma) = k_0$, is plotted in conjunction with the mean solutions. Also, the results obtained from the perturbation method with $\sigma_\gamma = 1$ are given along with the variance plots.

We first observe that $u(r,\gamma)$ is singular at the origin where the unit point force is applied. Next, the mean value of $u(r,\gamma)$, which has units of length, follows the basic pattern exhibited by all deterministic fundamental solutions of time-harmonic elastodynamics in that there is pronounced radiation decay in amplitude with increasing distance from the source. As far as the amplitude of the variance is concerned, we observe a gradual increase with distance r. The polynomial chaos expansion method again predicts larger values when compared with the perturbation method. When we look at the standard deviation of $u(r,\gamma)$, this magnification factor is about 1.87. Finally, although not plotted here, for both mean and variance, all methods give identical phase angles. This indicates that the propagating signal remains coherent and does not suffer phase distortion in the presence of small amounts of randomness.

It is interesting to compare here the magnitude of the mean solution of $u(r,\gamma)$ with its corresponding standard deviation. Specifically, as the magnitude of the former drops from about 0.6 to 0.2 (x 10^{-3}) m, that of the latter increases from 0.012 to 0.0135 (x 10^{-3}) m. This indicates that the effect of small randomness is cumulative as its presence becomes more pronounced with increasing distance from the source of the unit force disturbance.

Next, Figures 5a to 5d present the same types of plots as before, only this time they are evaluated at $k_1 = 50\,\%$ of k_0. Here we clearly see the effect of large randomness. At first, the mean value amplitude is no longer a smoothly decreasing function of distance, but significantly overshoots the deterministic amplitude at small distances from the source and subsequently exhibits an oscillatory behaviour. Furthermore, beyond a distance of r = 2 km, the N =3 term polynomial chaos expansion results start to differ from those obtained by the N = 5 term expansion. Again, the N = 3 and N = 5 term expansion results differ past r = 2 km. The effect of large randomness is more striking in the case of the variance of $u(r,\gamma)$. In contrast to the perturbation method, the polynomial chaos expansion no longer predicts a nearly linear increase in the variance amplitude. In fact, the variance attains large values close to the source of the disturbance, decreases with increasing distance from it and at the same time exhibits an oscillatory behaviour.

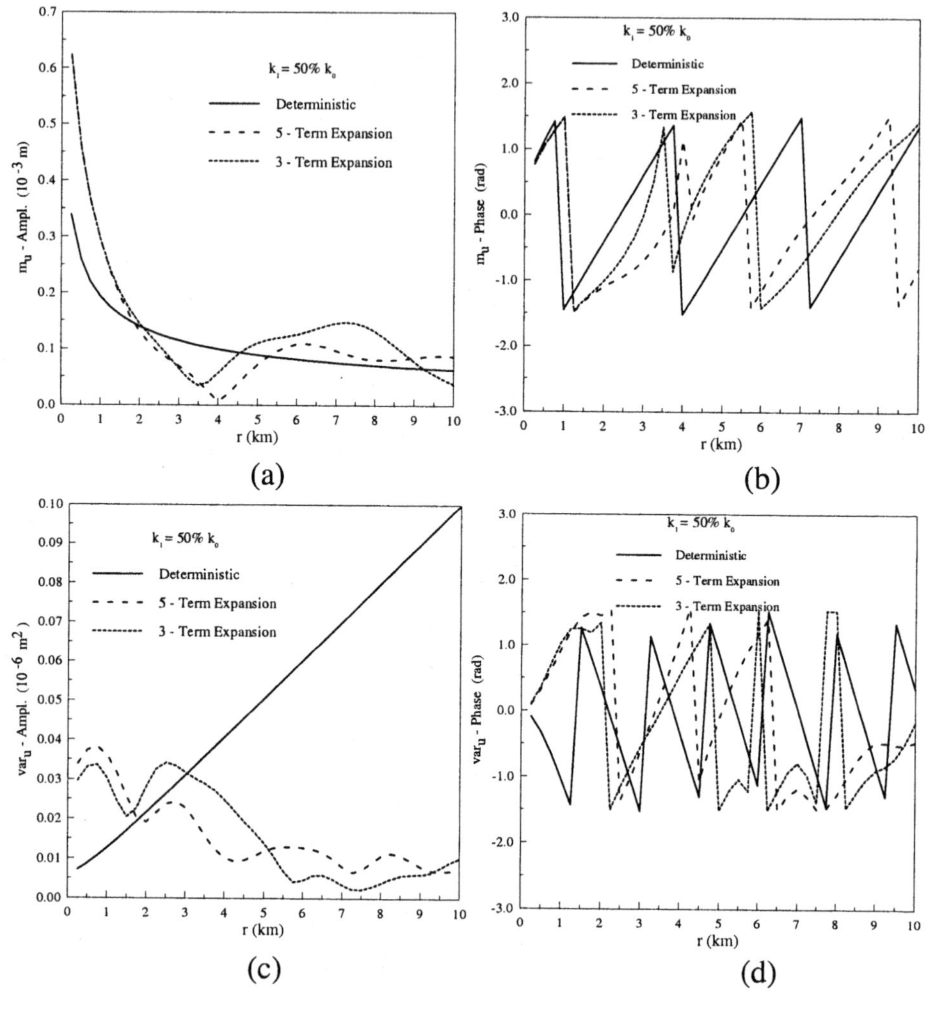

Figure 5. Fundamental solution $u(r,\gamma)$: (a) mean amplitude (b) mean phase angle (c) variance amplitude and (d) variance phase angle versus distance from source for fluctuating wave number $k_1 = 0.50\ k_0$.

It is again interesting to compare the amplitude of the mean value with that of its standard deviation. For a mean amplitude range of 0.7 to 0.0 (x 10^{-3}) m, the s.d. is now between 0.18 to 0.08 (x 10^{-3}) m, while the s.d of the wave number is fixed at $\sigma_k = k_l = 0.50$ km^{-1}. Generally speaking, we see that the perturbation method yields a deterministic solution plus a variance, which at high randomness is essentially a scaled version of what is obtained at low randomness. By contrast, the polynomial chaos expansion method gives a markedly different picture for the mean and variance at high versus low randomness.

7. Conclusions

Boundary integral equation based numerical methodologies are very useful for analysing wave motion problems in continuous media with infinite or semi-infinite boundaries. There is, however, a need to expand these methods to cases involving stochastic media, because depending on the characteristics of the propagating disturbance, it is quite possible for the signal's wavelength to be of dimensions comparable to the spacing of the randomly distributed inhomogeneities. As a result, the propagating signal can be noticeably altered, through mean amplitude increase and phase change, due to continuous scattering from the random irregularities. This point is elaborated through the use of SH wave propagation in a random 3D soil continuum as an example.

This chapter presents a direct boundary element formulation for the mean vector plus covariance matrix solutions of SSI problems involving unlined tunnels under plane-strain conditions subjected to general transient loads. The perturbation method is used for expanding the problem's dependent variables, as well as the fundamental solutions, about their mean values. Substitution of these expansions in the appropriate boundary integral equations, along with conventional numerical integration schemes, yields a compact BEM solution, valid for a wide range of problems involving the dynamic response of stochastic media. The entire methodology is defined in the Laplace transform domain and an efficient inverse transformation algorithm is employed for reconstructing the temporal response.

For large randomness, a general technique is introduced for computing fundamental solutions for time harmonic, scalar wave propagation, based on a series expansion employing Hermite polynomials. In addition, the spatial part of the fundamental solution is obtained from the vector Helmholtz wave equation, which is uncoupled by using the eigenproperties (that is, wavenumbers) of the system matrix. This technique is valid for arbitrary amounts of randomness in the wave number and the total number of terms retained in the expansion can be adjusted so as obtain a desired degree of accuracy. Satisfactory results can be obtained with as little as three terms. The computation of the mean, standard deviation and higher moments of the Green's function is straightforward and does not present any difficulties. Finally, the solution obtained here can also be used for stochastic realisations of SH wave motions in the unbounded continuum.

8. References

1. Sato, H. and Fehler, M.C. (1998) *Seismic Wave Propagation and Scattering in the Heterogeneous Earth*, Springer-Verlag, New York.
2. Adomian, G. (1983) *Stochastic Systems*, Academic Press, New York.
3. Benaroya, H. and Rehak, M. (1987) The Neumann series/Born approximation applied to parametrically excited stochastic systems, *Probabilistic Engineering Mechanics* **2**(2), 74-81.
4. Chernov, L.A. (1962) *Wave Propagation in a Random Medium*, Dover, New York.
5. Karal, F.C. and Keller, J.B. (1964) Elastic, electromagnetic and other waves in a random medium, *Journal of Mathematical Physics* **5**(4), 537-549.
6. Chu, L., Askar A., and Cakmak A.S. (1981) Earthquake waves in a random medium, *International Journal of Numerical and Analytical Methods in Geomechanics* **5**, 79-96.
7. Hryniewicz, Z. (1991) Mean response to distributed dynamic loads across the random layer for anti-plane shear motion, *Acta Mechanica* **90**, 81-89.
8. Kotulski, Z. (1990) Elastic waves in randomly stratified medium. Analytical results, *Acta Mechanica* **83**, 61-75.
9. Kohler, W., Papanikolau, G., and White, B. (1991) Reflection of waves generated by a point source over a randomly layered medium, *Wave Motion* **13**, 53-87.
10. Sobczyk, K. (1976) Elastic wave propagation in a discrete random medium, *Acta Mechanica* **25**, 13-28.
11. Varadan, V.K., Ma, Y., and Varadan, V.V. (1985) Multiple scattering theory of elastic wave propagation in discrete random media, *Journal of the Acoustical Society of America*, **77**, 375-389.
12. Liu, K.C. (1991) Wave scattering in discrete random media by the discontinuous stochastic field method, I. Basic method and general theory, *Journal of Sound and Vibration*, **147**, 301-311.
13. Sobczyk, K. (1985) *Stochastic Wave Propagation*, Elsevier, Amsterdam.
14. Shinozuka, M. (1972) Digital simulation of random processes and its application, *Journal of Sound and Vibration,* **25**, 111-128.

15. Vanmarke, E., Shinozuka, M., Nakagiri, S., Schueller, G.I. and Grigoriu, M. (1986) Random fields and stochastic finite elements, *Structural Safety*, **3**, 143-166.
16. Liu, W.K., Belytschko, T. and Mani, A. (1986) Random field finite elements, *International Journal of Numerical Methods in Engineering*, **23**, 1831-1845.
17. Burczynski, T. (1985) The boundary element method for stochastic potential problems, *Applied Mathematical Modelling*, **9**, 189-194.
18. Manolis, G.D. and Shaw, R.P. (1992) Wave motion in a random hydroacoustic medium using boundary integral/element methods, *Engineering Analysis with Boundary Elements*, **9**, 61-70.
19. Askar, A. and Cakmak, A.S., (1988) Seismic waves in random media, *Probabilistic Engineering Mechanics*, **3**(3), 124-129.
20. Jensen, H. and Iwan, W. D. (1992) Response of systems with uncertain parameters to stochastic excitation, *Journal of Engineering Mechanics of ASCE* **118**(5), 1012-1025.
21. Sun, T.C. (1979) A finite element method for random differential equations with random coefficients, *SIAM Journal of Numerical Analysis*, **16**(6), 1019-1035.
22. Jensen, H. and Iwan, W.D. (1991) Response variability in structural dynamics, *Earthquake Engineering and Structural Dynamics*, **20**, 949-959.
23. Lawrence, M.A. (1987) Basis random variable in finite element analysis, *International Journal of Numerical Methods in Engineering*, **24**, 1849-1863.
24. Spanos, P.D. and Ghanem, R. (1989) Stochastic finite element expansion for random media, *Journal of Engineering Mechanics of ASCE*, **115**(5), 1035-1053.
25. Belyaev, A.K. and Ziegler, F. (1998) Uniaxial waves in randomly heterogeneous elastic media, *Probabilistic Engineering Mechanics* **13**, 27-38.
26. Burczynski, T. (1993) Stochastic boundary element methods: Computational methodology and applications, Chapter 12 in P. Spanos and W.Wu (eds.), *Probabilistic Structural Mechanics, an IUTAM Symposium*, Springer-Verlag, Berlin.
27. Manolis, G.D. (1994) The ground as a random medium, Chapter 15 in G.D. Manolis and T.G. Davies (eds.), *Boundary Element Techniques in Gemechanics*, Elsevier Applied Science, London, pp. 497-533.
28. Manolis, G.D., Davies, T.G. and Beskos, D.E. (1994) Overview of boundary element techniques in geomechanics, Chapter 1 in G.D. Manolis and T.G. Davies (eds.), *Boundary Element Techniques in Gemechanics*, Elsevier Applied Science, London, pp. 1-35.
29. Karakostas, C.Z. and Manolis, G.D. (2000) Dynamic response of unlined tunnels in soil with random properties, *Engineering Structures* **18**(2), 1013-1027.
30. Manolis, G.D. and Pavlou, S. (1999) Fundamental solutions for SH waves in a continuum with large randomness, *Engineering Analysis with Boundary Elements* **23**, 721-736.
31. Vanmarke, E.H. (1977) Random vibration approach to soil dynamics, in *The Use of Probability in Earthquake Engineering*, ASCE Publication, New York, 143-176.
32. Gazetas, G., Debchaudhury, A. and Gasparini, D.A. (1982) Stochastic estimation of the nonlinear response of dams to strong earthquakes, *Soil Dynamics and Earthquake Engineering* **1**(1).

33. Luco, J.E. and Wong, H.L. (1986) Response of a rigid foundation to a spatially random ground motion, *Earthquake Engineering and Structural Dynamics* **14**, 891-908.
34. Pais, A.L. and Kausel, E. (1990) Stochastic response of rigid foundations, *Earthquake Engineering and Structural Dynamics* **19**, 611-622.
35. Hao, H. (1991) Response of multiply supported rigid plate to spatially correlated seismic excitations *Earthquake Engineering and Structural Dynamics* **20**, 821-838.
36. Harichandran, R.S. and Vanmarke, E.H. (1986) Stochastic variation of earthquake ground motion in space and time, *Journal of Engineering Mechanics of ASCE* **112**(2), 154-174.
37. Madhav, M.R. and Ramakrishna, K.S. (1988) Probabilistic prediction of pile group capacity, in P.D. Spanos (ed.) *Probabilistic Methods in Civil Engineering*, ASCE Publication, New York, pp. 5-8.
38. Stickler, D.C. (1991) Scattering from a soft slightly rough surface, *Wave Motion* **13**, 211-221.
39. Ogura, H., Takahashi, N. and Kuwahara, M. (1991) Scattering of waves from a random cylindrical surface, *Wave Motion* **14**, 273-295.
40. Nakagiri, S. and Hisada, T. A. (1980) Note on stochastic finite element method (Part1) - Variation of stress and strain caused by shape fluctuation, *Seisan-Kenkyu Institute of Industrial Sciences of University of Tokyo* **32**(2),. 39-42.
41. Tang, W.H., Halim, I. and Gilbert, R.B. (1988) Reliability of geotechnical systems considering geological anomaly, in P.D. Spanos (ed.) *Probabilistic Methods in Civil Engineering,* ASCE Publication, New York, pp.136-139.
42. Asaoka, A and Athanassiou-Grivas, D. (1982) Spatial variability of the undrained strength of clays, *Journal of Geotechnical Engineering Division of ASCE* **108**(GT5), 743-755.
43. Baecher, G. (1981) Optimal estimators for soil properties, *Journal of Geotechnical Engineering Division of ASCE* **107**(5), 649-653.
44. Wu, T., Potter, J. and Kjekstad, O. (1986) Probabilistic analysis of offshore site exploration, *Journal of Geotechnical Engineering Division of ASCE* **112(**11), 981-1000.
45. Tang, W. and Quek, S.T. (1986) Statistical model of boulder size and fraction, *Journal of Geotechnical Engineering Division of ASCE* **112(**1), 79-90.
46. Tang, W.H. (1988) Updating anomaly statistics-multiple anomaly pieces, *Journal of Engineering Mechanics of ASCE* **114**(6), 1091-1096.
47. Tang, W.H., Sidi, I., and Gilbert, R.B. (1989) Average property in a random two-state medium, *Journal of Engineering Mechanics of ASCE* **115(**1), 131-144.
48. Benaroya, H. (1988) Markov chains and soil constitutive modelling, in P.D. Spanos (ed.) *Probabilistic Methods in Civil Engineering*, ASCE Publication, New York, pp. 49-52.
49. Ronold, K.O. (1990) Probabilistic foundation stability analysis, *International Journal of Numerical and Analytical Methods in Geomechanics* **14**, 279-296.
50. Quek, S.T., Phoon, K.K. and Chow, Y.K. (1991) Pile group settlement: a probabilistic approach, *International Journal of Numerical and Analytical Methods in Geomechanics* **15**, 817-832.

51. Drumm, E.C., Bennett, R.M. and Oakley, G.J. (1990) Probabilistic response of laterally loaded piers by three-point approximation, *International Journal of Numerical and Analytical Methods in Geomechanics* **14**, 499-507.
52. Rosenblueth, E. (1975) Point estimates for probability moments, *Proceedings of National Academy of Sciences (USA)* **72**(10), 3812-3814.
53. Baecher, G.B. and Ingra, T.S. (1981) Stochastic FEM in settlement prediction, *Journal of Geotechnical Engineering Division of ASCE* **107**(GT4), 449-463.
54. Nakagiri, S. and Hisada, T. (1983) A note on stochastic finite element method (Part 6) - An application in problems of uncertain elastic foundation, *Seisan-Kenkyu Institute of Industrial Sciences of University of Tokyo* **35**(1), 20-23.
55. Lin, J.S. (1990) Regional seismic slope failure probability matrices, *Earthquake Engineering and Structural Dynamics* **19**, 911-923.
56. Lahlaf, A.M. and Marciano, E. (1988) Probabilistic slope stability-Case study, in P.D. Spanos (ed.) *Probabilistic Methods in Civil Engineering*, ASCE Publication, New York, pp. 128-131.
57. Resheidat, M.R. (1988) Probabilistic model for stability of slopes, in P.D. Spanos (ed.) *Probabilistic Methods in Civil Engineering*, ASCE Publication, New York, pp.132-135.
58. Yucemen, M.S. and Al-Homoud, A.S. (1988) A probabilistic 3-D short-term stability model, in P.D. Spanos (ed.) *Probabilistic Methods in Civil Engineering*, ASCE Publication, New York, pp. 140-143.
59. Ishii, K. (1987) Stochastic finite element method for slope stability analysis, *Structural Safety* **4**, 111-129.
60. Wong, F.S. (1985) Slope reliability and response surface method, *Journal of Geotechnical Engineering Division of ASCE* **111**(GT1), 32-53.
61. Koppula, S.D. (1988) Pore pressure development: Probability analysis, in P.D. Spanos (ed.) *Probabilistic Methods in Civil Engineering*, ASCE Publication, New York, pp. 57-60.
62. Kavazanjian, E. and Wang, J.N. (1988) Frequency domain site response with pore pressure, in P.D. Spanos (ed.) *Probabilistic Methods in Civil Engineering*, ASCE Publication, New York, pp.1-4.
63. Hoshiya, M. and Ishii, K. (1983) Evaluation of kinematic interaction of soil-foundation systems by a stochastic model, *Soil Dynamics and Earthquake Engineering* **2**(3), 128-134.
64. Hoshiya, M. and Ishii, K. (1984) Deconvolution method between kinematic interaction and dynamic interaction of soil-foundation systems based on observed data, *Soil Dynamics and Earthquake Engineering* **3**(3), 157-164.
65. Harren, S.V. and Fossum, A.F. (1991) Probabilistic analysis of impulsively loaded deep-buried structures, *International Journal for Numerical and Analytical Methods in Geomechanics* **15**, 513-526.
66. Righetti, G. and Harrop-Williams, K. (1988) Finite element analysis of random soil media, *Journal of Geotechnical Engineering Division of ASCE* **114**(GT1), 59-75.
67. Izutani, Y. and Katagiri, F. (1992) Empirical Green's function corrected for source effect, *Earthquake Engineering and Structural Dynamics* **21**, 341-349.

68. Faravelli, L., Kiremidjian, A.S. and Suzuki, S. (1988) A stochastic seismological model in earthquake engineering, in P.D. Spanos (ed.) *Probabilistic Methods in Civil Engineering*, ASCE Publication, New York, pp. 241-244.
69. Anagnos, T. and Kiremidjian, A.S. (1984) Stochastic time-predictable model for earthquake occurrences, *Bulletin of the Seismological Society of America* **74**, 2593-2611.
70. Nishenko, S. and Singh S.K. (1987) Conditional probabilities for the recurrence of large and great earthquakes along the Mexican subduction zone, *Bulletin of the Seismological Society of America* **77**, 2094-2114.
71. Suzuki, S. and Kiremidjian, A.S. (1991) A random slip rate model for earthquake occurrences with Bayesian parameters, *Bulletin of the Seismological Society of America* **81**(3), 781-795.
72. Hoshiya, M. Naruyama, M. and Kurita, H. (1988) Autoregressive model of spatially propagating earthquake ground motion, in P.D. Spanos (ed.) *Probabilistic Methods in Civil Engineering*, ASCE Publication, New York, pp. 257-260.
73. Spanos, P.D. (1983) ARMA algorithms for ocean wave modelling, *Journal of Energy Resources Technology of ASME* **105**, 300-309.
74. Samaras, E., Shinozuka, M. and Tsurui A. (1985) ARMA representation of random processes, *Journal of Engineering Mechanics of ASCE*,**111**(3), 449-461.
75. Mignolet, M.P. and Spanos, P.D. (1987) Recursive simulation of stationary multivariate processes - Part I., *Journal of Applied Mechanics of ASME* **109**, 674-680.
76. Kozin, F. (1988) Autoregressive moving average models of earthquake records, *Probabilistic Engineering Mechanics* **3**(2), 58-63.
77. Cakmak, A.S., Sherif, R.I. and Ellis, G. (1985) Modelling earthquake ground motions in California using parametric time series methods, *Soil Dynamics and Earthquake Engineering* **4**(3), 124-131.
78. Polhemus, N.W. and Cakmak, A.S. (1981) Simulation of earthquake ground motions using autoregressive moving average (ARMA) models, *Earthquake Engineering and Structural Dynamics* **9**, 343-354.
79. Turkstra, C.J., Tallin, A.G., Brahimi, M. and Kim, H.J. (1988) Application of ARMA models for seismic damage prediction, in P.D. Spanos (ed.) *Probabilistic Methods in Civil Engineering*, ASCE Publication, New York, pp. 277-280.
80. Dong, W., Shah, H.C., Bao, A. and Mortgat, C.P. (1984) Utilization of geophysical information in Bayesian seismic hazard model, *Soil Dynamics and Earthquake Engineering* **3**(2), 103-111.
81. Boissonnade, A.C., Dong, W.M., Liu, S.C. and Shah, H.C. (1984) Use of pattern recognition and Bayesian classification for earthquake intensity and damage estimation, *Soil Dynamics and Earthquake Engineering* **3**(3), 145-149.
82. Spanos, P.D. (1983) Digital synthesis of response design spectrum compatible earthquake records for dynamic analyses, *The Shock and Vibration Digest* **15**(3), 21-27.
83. Spanos, P.D. and Vargas-Loli, L.M. (1985) A statistical approach to generation of design spectrum compatible earthquake time histories, *Soil Dynamics and Earthquake Engineering* **4**(1), 2-8.

84. Amini, A and Trifunac, M.D. (1984) Statistical extension of response spectrum superposition. *Soil Dynamics and Earthquake Engineering* **4**(2), 54-63.
85. Barucha - Reid A.T. (1972) *Random Integral Equations*, Academic Press, New York.
86. Hinch, E.J. (1991) *Perturbation Methods*, Cambridge Univ. Press, Cambridge.
87. Nigam, N.C. (1983) *Introduction to Random Vibrations*, MIT Press, Cambridge, Massachusetts.
88. Caughey, T.K. (1963) Derivation and application of the Fokker-Planck equation in discrete non-linear dynamic systems subjected to white random noise, *Journal of the Acoustical Society of America* **35**(11), 1683-1692.
89. Manolis, G.D. and Beskos, D.E. (1988) *Boundary Element Methods in Elastodynamics*, Chapman and Hall, London.
90. Abramowitz, M. and Stegun, I.R. (1972) *Handbook of Mathematical Functions*, Dover, New York.
91. Baron, M.L. and Matthews, A.T. (1961) Diffraction of a pressure wave by a cylindrical cavity in an elastic medium, *Journal Applied Mechanics,* **28**, 347-354.
92. Kausel, E. and Manolis, G.D. (2000) *Wave Motion Problems in Earthquake Engineering*, WIT Press, Southampton.

CHAPTER 5.
SOIL STRUCTURE INTERACTION IN PRACTICE

CONSTANTINE C. SPYRAKOS
Earthquake Engineering Laboratory
Civil Engineering Department
National Technical University Of Athens

1. Introduction

1.1. BRIEF REVIEW OF LITERATURE ON BUILDING STRUCTURES AND SSI

Seismic incidents of recent decades have evoked extensive studies focusing on the effects of Soil-Structure Interaction (SSI). Rodriguez and Montes [1] evaluated the importance of SSI effects on the seismic response and damage of buildings in Mexico City during the 1985 earthquake. A simple structural model was used to conduct a parametric study using a representative record obtained in the soft soil area of Mexico City. The results indicated that in many cases of inelastic response, SSI can be evaluated considering the rigid-base case and the amplified period of the SSI system. A similar procedure can be followed to assess seismic damage in multi-story buildings supported on flexible soils. Literature in the area is rather extensive. This brief introduction gives only a flavour of issues in recent studies on the topic.

Based on an observation on the damage pattern caused by the 1994 Northridge earthquake, that is that the number of severely damaged buildings was reduced in areas where the surface soil experienced some form of non-linear response, Trifunac and Todorovska [2] studied the effects of non-linear soil response. They attempted to quantify the relationship between the density of red-tagged buildings and the severity of shaking, including the density of breaks in water pipes as a variable specifying the level of strain in the soil.

W.S. Hall and G. Oliveto (eds.), Boundary Element Methods For Soil-Structure Interaction, 235–272.

A series of studies was conducted on three reinforced concrete buildings with RC shear walls damaged in the 1995 Hyogo-ken Nanbu earthquake by Hayashi et al [3]. They performed site inspections, including micro-tremor measurements of buildings, evaluated input motions and conducted analyses considering SSI. The results of simulation analyses for the two severely damaged buildings were validated from the actual damage state. Analyses of one slender building with no structural damage, using a 2-D FEM model and taking basemat uplift into consideration lead to the conclusion that uplifting was the main reason it did not suffer any structural damage.

Izuru [4] showed that tuning of the natural period of a building structure with that of the surface ground causes remarkable response amplification of the building structure. He studied the response of a multi-story building for two cases; in the first case the building was lying on a bedrock level and in the second case on the surface ground under which the bedrock lies.

Iida [5] performed a three-dimensional (3-D) non-linear SSI analysis for several types of low- to high-rise buildings during the hypothetical Guerrero earthquake, focusing on the real cause of heavy damage to mid-rise buildings founded at the lakebed zone during the 1985 Michoacan earthquake. The results of the non-linear interaction analysis appear to be the most consistent with the observed damage pattern. On the contrary base-fixed analyses have not been able to explain the building damage pattern in Mexico City, whereas linear SSI analysis has provided only a partial explanation of the damage pattern.

Analytical investigations were carried out to evaluate the effect of base flexibility on a ductile 20-story coupled-shear-wall building designed for Montreal in Canada. Chaallal and Ghlamallah [6] used a two-dimensional model to idealize the structure including the supporting soil. Allowances were made for non-linearities not only in the walls and the coupling beams, but also in the soil and foundation elements modelled with equivalent massless springs allowing for inelastic bearing of the soil and uplift. Results from this study showed that SSI calculations resulted in a lengthening of the period of the building by a maximum of 33% and an increase in the lateral deflections by a maximum of 81%. However, the maximum shear and flexural stress in the walls and the coupling beams were reduced, particularly in lower stories.

Kocak and Mengi [7] have proposed a simple soil-structure interaction model which accommodates not only the interaction between soil and structure, but also the interaction between footings. First they proposed a model for layered soil conditions; then, based on the layered soil model, they developed a finite element model for three-dimensional SSI analysis.

A general coupled boundary element/finite element formulation was presented for the investigation of dynamic soil-structure interaction including non-linearities by Estorff and Firuziaan [8]. This formulation was applied to investigate the transient inelastic response of structures founded on a half-space. The structure and the surrounding soil in the near field were modelled with finite elements, whereas the remaining soil region was discretized with boundary elements. The methodology was also applied to three numerical examples yielding reliable results.

Kellezi [9] investigated alternative methods to analyze structures including SSI. He developed a simple finite element procedure to solve directly in the time domain transient SSI problems. A central feature of the procedure is that local absorbing boundaries are used to render the computational domain finite. These boundaries are local in both time and space and are completely defined by a pair of symmetric stiffness and damping matrices. The validity and accuracy of the procedures were verified with numerical examples.

Wen-Hwa Wu [10] attempted to account for SSI with appropriate fixed-base models. He applied his methodology to determine equivalent fixed-base models of a general multi degree-of-freedom SSI system using simple system identification techniques in the frequency domain. Various fixed-base models were formulated and their accuracy was compared for a five-story shear building resting on soft soil.

Badie et al [11] presented a method to analyze shear wall structures on elastic foundations. The shear walls were modelled using isoparametric quadrilateral plane stress elements and the soil was modelled using a quadratic element that accounted for vertical subgrade reaction and soil shear stiffness. Including soil shear stiffness in the analysis reduced the maximum drift and increased the normal stresses in the wall especially for poor soils. They also showed that ignoring soil deformation underestimates the bending moment in the lintel beams of coupled shear walls as well as the transverse bending moment in core structures.

1.2. BRIEF REVIEW OF LITERATURE ON BRIDGES AND SSI

Saadeghvaziri et al [12] studied the effects of soil-structure interaction for the longitudinal seismic response of bridges. Based on a comprehensive study of three actual multi-span simply supported bridges, they concluded that SSI plays a predominant role in seismic response in the longitudinal direction. Impact forces, deck sliding and SSI affect the development of plastic rotations at the base of the columns.

Foundation behaviour plays a major role in the performance of highway bridges during earthquakes. For many highway bridges, abutments attract a large portion of the seismic force, particularly in the longitudinal direction. After the 1971 San Fernando earthquake, it became quite evident that many abutments had been subjected to large seismic forces. In fact on many bridges, abutment damage was the only damage reported.

Soil-abutment interaction under seismic loads is a highly non-linear phenomenon. This non-linearity plays an important role in the overall structural response (Spyrakos [13], Spyrakos and Vlassis [14], Maragakis et al [15]). As a result there is a definite need to employ a proper methodology to design bridges including the effects of soil-abutment interaction.

Eurocode 8 [16] refers to the non-linearity deriving from soil-structure interaction. However it does not suggest a proper methodology to include soil-structure interaction and non-linear soil behaviour, nor verifications to examine and limit soil failure in the earthquake resistant bridge design. Some guidance is currently provided by Caltrans Bridge Design Aids [17] and AASHTO [18]. Both documents recognise the highly non-linear behaviour that could be caused by large deformations in the backfill at the abutments during seismic excitations.

2. Seismic Design of Building Structures including SSI

2.1. BRIEF INTRODUCTION

This section presents a procedure to design building structures including SSI. The methodology is based on design-oriented literature, e.g. NEHRP [19], EC8 [16]. In order to demonstrate the procedure, a typica multi-story reinforced building structure is designed and comparisons with current practice are presented. In the design procedure the natural period, damping factor and base shear of the flexibly supported structure are computed considering SSI and used to determine the design earthquake forces and the corresponding displacements of the building according to current seismic design codes, for example EC8.

Many studies on the dynamic response of multi degree-of-freedom flexibly supported systems, ATC 40 [20] and FEMA 356 [21], reach the conclusion that, as far as it concerns the design of building structures subject to seismic excitation, soil-structure interaction mainly affects the fundamental period. Therefore, when considering soil-structure interaction only the contribution of the fundamental vibration mode per direction is required. The contribution of these vibration modes must be computed as in the case of fixed structures and the response of the system must be computed with the use of an appropriate superposition rule, such as the square root of the sum of each contribution's maximum response square (SRSS). This issue is further elaborated in Section 2.3.

It should be noted that the suggested approach termed as "simplified dynamic analysis with SSI" is recommended for building structures having a rather uniform distribution of both inertia and stiffness along the height of the building; that is, any change of either stiffness or inertia between successive floors does not exceed 35%.

2.2. DESIGN PROCEDURE

The twelve steps of the simplified dynamic analysis with SSI procedures are presented in this section followed by a numerical example.

Step 1. Computation of the fundamental period of the building along the x or y direction, assuming rigid connection of the footings to the supporting soil, using any appropriate method or the formula [20]

$$T_{x,y} = 0.09 \frac{H}{\sqrt{L}} \sqrt{\frac{H}{H + \rho_{x,y} \cdot L}} \tag{1}$$

where H is the building height, L is the building's length in the direction being analyzed, $\rho_{x,y}$ is ratio of the cross sectional areas of the shear walls along the x or y direction, respectively, to the total cross sectional area of the shear walls and columns.

Step 2. Computation of the ratio

$$v_s / \overline{fh} \tag{2}$$

where v_s is the shear wave velocity of the supporting soil, $\overline{f}$ is the frequency of the rigidly supported structure, $\overline{h}$ is the effective height of the building, which will be taken as 0.7 times the total height. For single story buildings, $\overline{h}$ will be taken as the height of the building.
If $v_s / \overline{fh} \geq 20$ there is no necessity to incorporate the effects of SSI; thus they may be ignored.
If $v_s / \overline{fh} < 20$, the effects of soil-structure interaction are important and the analysis should proceed as described in the following steps.

Step 2 serves as a simple criterion to decide whether SSI should be included in the analysis.

Step 3. Computation of the characteristic foundation radii r_a, and r_m defined by:

$$r_a = \sqrt{\frac{A_0}{\pi}} \quad \text{and} \quad r_m = \sqrt[4]{\frac{4I_0}{\pi}} \tag{3}$$

where, A_o is the area of the foundation, and I_o is the static moment of the foundation about a horizontal centroidal axis normal to the direction in which the structure is analyzed.

Obviously, computation of r_a and r_m is not required for circular foundations, since the actual radius of the foundation is equal to $r_a = r_m$. The computational procedure to arrive at the proper dimensions of r_a and r_m is iterative for every foundation. Estimation of the initial dimensions for a foundation is based on the analysis of the rigidly supported structure.

Step 4. Computation of shear modulus and shear wave velocity for the supporting soil at strain levels that correspond to design spectra.

The initial shear modulus G_o is related to the shear wave velocity at low strains v_{so} and the mass density of the soil ρ with the relationship $G_0 = \rho v_{s0}^2$. Converting mass density ρ to unit weight, γ, leads to an alternative expression $G_0 = \frac{\gamma}{g} v_s^2$, where g is the acceleration of gravity.

Under seismic loads the behaviour of most soils is non-linear, and the shear wave modulus decreases with increasing shear strain. The large strain shear wave velocity v_s and the effective shear modulus G can be estimated on the basis of the anticipated maximum ground acceleration $A_v g$ in accordance with Table I:

TABLE I. Estimation of the effective shear modulus and the large strain shear wave velocity

Seismic Risk Zone	I	II	III	IV
A_v	≤ 0.10	≤ 0.15	≤ 0.20	≥ 0.30
G/G_o	0.81	0.64	0.49	0.42
v_s/v_{so}	0.90	0.80	0.70	0.65

Step 5. Computation of the horizontal k_{yi}, vertical k_{xi} and rocking, k_{ei} stiffnesses of every footing using the formulae

$$
\begin{aligned}
k_{yi} &= \left[\frac{8G_i r_{ai}}{2-v}\right]\left[1+\frac{2}{3}\frac{d_i}{r_{ai}}\right] \\
k_{xi} &= \left[\frac{4G_i r_{ai}}{1-v}\right]\left[1+0.4\frac{d_i}{r_{ai}}\right] \\
k_{ei} &= \left[\frac{8G_i r_{mi}^3}{3(1-v)}\right]\left[1+2\frac{d_i}{r_{mi}}\right]
\end{aligned}
\qquad (4)
$$

in which d_i is the depth of embedment and v the Poisson's ratio of the soil.

Equations (4) are valid for footings placed on relatively uniform soil deposits of substantial depth. For foundations on a soil layer overlaying a much stiffer soil layer or rock, see Figure 1, K_y and K_θ are computed from

$$k_{yi} = \left[\frac{8G_i r_{ai}}{2-\nu}\right]\left[1+\frac{2}{3}\frac{d_i}{r_{ai}}\right]\left[1+\frac{1}{2}\frac{r}{D_s}\right]\left[1+\frac{5}{4}\frac{d}{D_s}\right]$$
$$k_{ei} = \left[\frac{8G_i r_{mi}^3}{3(1-\nu)}\right]\left[1+2\frac{d_i}{r_{mi}}\right]\left[1+\frac{1}{6}\frac{r}{D_s}\right]\left[1+0.7\frac{d_i}{D_s}\right] \quad (5$$

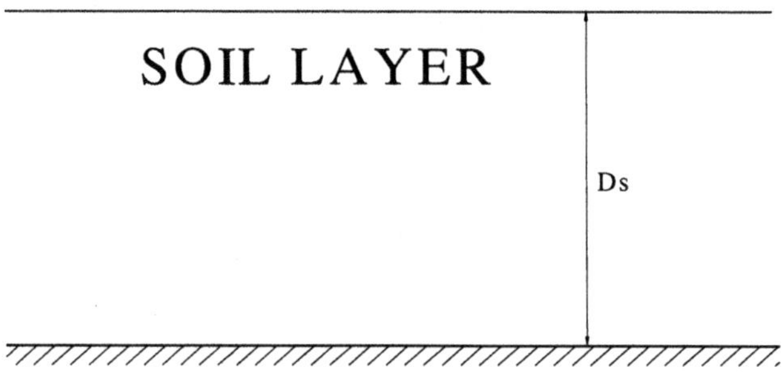

Figure 1. Soil layer over bedrock.

Step 6. Computation of the total stiffness of the foundations made up from individual footings

$$K_y = \Sigma k_{yi} \text{ and } K_\theta = \Sigma k_{xi} y_i^2 + \Sigma k_{ei} \quad (6$$

where y_i is the distance of the i-th footing, from the centre of stiffness of the foundation. In general the contribution of the rocking stiffness k_{ei} is small and can be neglected.

It is evident that, for buildings supported on mat foundations, the total stiffness is equal to the one computed in Step 5, that is $k_y = k_{yi}$, $k_x = k_{xi}$ and $k_e = k_{ei}$.

Step 7. Computation of structural stiffness considering rigid supports using the formula

$$\bar{k} = 4\pi^2 \frac{\overline{W}}{gT^2} \tag{7}$$

where T is the fundamental period of the rigidly supported structure and $\overline{W}$ is the effective gravity load of the building. $\overline{W}$ will be taken as $0.7W$ with the exception of buildings where the gravity load is concentrated at a single level, in such a case $\overline{W}$ will be taken equal to the total dead weight of the building and an appropriate portion of the design live load as defined by seismic codes, e.g., Eurocode 8 [16}.

Step 8. Computation of the effective period of the structure. The effective period $\overline{T}$ is determined from

$$\overline{T} = T\sqrt{\left(1+\frac{\bar{k}}{K_y}\right)\left(1+\frac{K_y\bar{h}^2}{K_\theta}\right)} \tag{8}$$

Alternatively, for buildings supported on mat foundations that rest either on the ground surface or are embedded in such a way that the side wall contact with the soil cannot be considered to remain effective during the design ground motion, the effective period at the building can be determined from

$$\overline{T} = T\sqrt{\left(1+\frac{25dr_a\bar{h}}{v_s^2T^2}\right)\left(1+\frac{1.12r_a\bar{h}^2}{r_m^3}\right)} \tag{9}$$

where r_a and r_m are calculated from equation (3), d is the relative weight density of the structure and the soil given by

$$d = \frac{\overline{W}}{\gamma A_0\bar{h}} \tag{10}$$

For the majority of building structures d can be estimated as equal to 0.15.

Step 9. Computation of the effective damping factor for the foundation, β_o.

The effective soil damping factor β_o accounts for radiation and hysteretic damping in the soil. The variation of β_o in terms of A_v and the ratio $\overline{T}/T$ and $\overline{h}/r$, where $\overline{h}$ is the effective height of building and r is the radius of the foundation, is given in Figure 2.

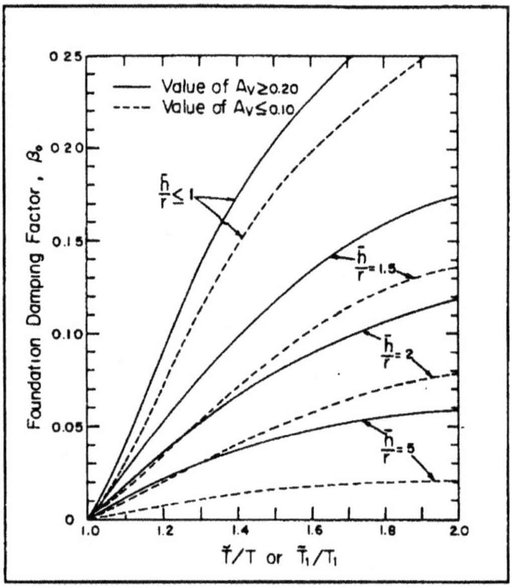

Figure 2. Foundation damping factor.

For buildings supported on point bearing piles and in all other cases where the foundation soil consists of a soft stratum of reasonably uniform properties underlain by a much stiffer rock-like deposit, the factor β_o can be obtained from Figure 2 provided that the ratio $T_s/\overline{T}$ fulfils the inequality

$$\frac{T_s}{\overline{T}} = \frac{4D_s}{V_s\overline{T}} \geq 1 \qquad (11)$$

However, when $\frac{T_s}{\overline{T}} = \frac{4D_s}{V_s\overline{T}} < 1$, then β_o is replaced by β_0' that can be calculated from

$$\beta_0' = \left(\frac{4D_s}{V_s\overline{T}}\right)^2 \beta_0 \tag{12}$$

where D_s is the total depth of the soil stratum, see Figure 1.

Step 10. Computation of the effective damping factor for the structure-foundation system, $\overline{\beta}$, from

$$\overline{\beta} = \beta_0 + \frac{0.05}{\left(\overline{T}/T\right)^3} > 0.05 \tag{13}$$

Step 11. Computation of base shear including the effects of SSI.

The base shear, $\overline{V}$, accounting for SSI is determined from

$$\overline{V} = V - \Delta V \tag{14}$$

where V is the base shear excluding the effects of SSI. The reduction ΔV is computed as follows

$$\Delta V = \left[R_d(T) - R_d\left(\overline{T}\right)\left(\frac{0.05}{\overline{\beta}}\right)^{0.4}\right]\overline{W} \tag{15}$$

where $R_d(T)$ is the seismic design coefficient using the fundamental natural period of the fixed base structure and $R_d(\overline{T})$ is the seismic design coefficient using the fundamental period of the flexibly supported structure. The reduced base shear $\overline{V}$ will always be taken greater than $0.7V$.

Step 12. Determination of design seismic forces.

The distribution over the height of the building of the reduced total seismic force, $\overline{V}$, will be considered to be the same as for the building without SSI, that is the lateral force at level i acting at the centre of mass is given by

$$F_i = (V_0 - V_H)\frac{m_i z_i}{\sum m_i z_i} \quad , i = 1,2,\ldots,\mathrm{N} \tag{16}$$

where z_i is the distance of level i from the base.

2.3. RESPONSE SPECTRUM ANALYSIS WITH SSI

When response spectrum analysis is applied as recommended by seismic codes, the base shear that corresponds to the fundamental mode, V_1, including SSI is given by

$$\overline{V_1} = V_1 - \Delta V_1 \tag{17}$$

The reduction, ΔV_1, is computed as ΔV in the case of the simplified dynamic analysis with SSI. $R_d(T_1)$ and $R_d(\overline{T_1})$ are computed using the fundamental period of the fixed structure, T_1, and the one of the elastically supported structure, $\overline{T_1}$, respectively.

Generally, the effects of SSI concerning the fundamental mode of the structure are determined in almost the same way to the simplified dynamic analysis, except for the effective weight and height of the structure that should be computed in a way so as to express the fundamental mode of an equivalent fixed structure. Specifically, $\overline{W}$ is computed from

$$\overline{W} = \overline{W_1} = \frac{(\Sigma w_i \Phi_i)^2}{\Sigma w_i \Phi_i} \tag{18}$$

and $\overline{h}$ from

$$\bar{h} = \frac{\sum_{i=1}^{n} w_i \Phi_i h_i}{\sum_{i=1}^{n} w_i \Phi_i} \tag{19}$$

where Φ_i represents the first mode at the i-th floor.

In order to avoid overestimation of the role that dynamic SSI plays in design, it is recommended that in no case should the reduced base shear, $\overline{V_1}$, be taken less than $0.7V_1$.

2.4. NUMERICAL EXAMPLE: BUILDING STRUCTURE

A five-story reinforced concrete building with the typical floor plan of Figure 3 is analyzed for the design spectrum given in Figure 4. The foundation of the structure consists of individual footings.

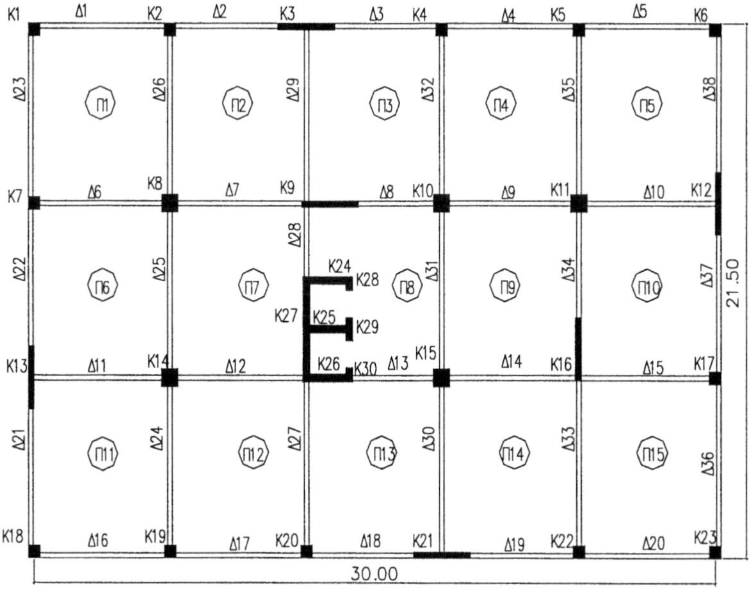

Figure 3. Typical floor plan.

The depth of the upper surface of the foundations from the ground is 1.00 m. The height of the ground floor is 4m, while the height of each one of the other floors is 3 m. The type of concrete is C20/25, the type of steel is S400 and S220 for main reinforcement and stirrups, respectively. The dead load and the live load for the slabs is g = 4 KN/m^2 and q = 2 KN/m^2, respectively. The allowable soil stress is σ_d = 250 KN/m^2 and the shear wave velocity under conditions of low strains is V_{so} = 200 m/s. Poisson's ratio for the soil is v_s = 0.40 and the ductility factor for the structure is q = 3.5.

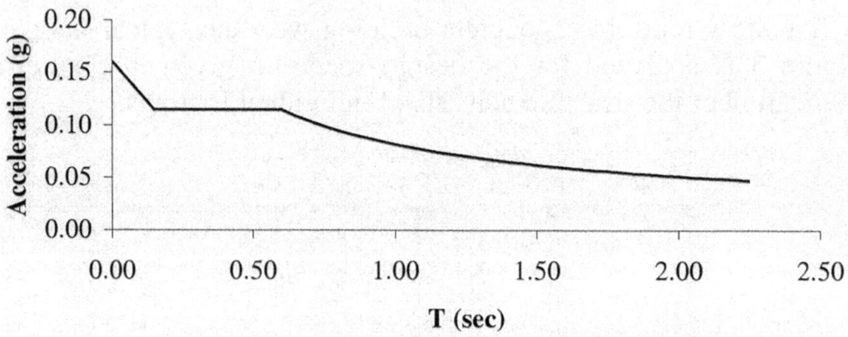

Figure 4. Design spectrum for seismic risk zone II.

The structure is analyzed for the following four different cases.

(a) *Fixed supports.* This case simulates soil conditions for very stiff foundation conditions and is used as a reference condition to assess the more realistic simulation of the foundation-soil system of the other three cases that follow. It should be noted, however, that in practice it is not uncommon to use the fixed base assumption even for soils that cannot be characterized as "rock".

(b) *Elastic supports using one vertical spring k_z and two rotational springs k_e.* The spring stiffnesses are calculated from equations (4) for d_i = 0, i.e., a conservative design assumption, and for low strain level (G_o and v_{so}). This is the most commonly used modelling of a footing in practice. Such a simulation fails to capture the effects of soil softening for large strains, which is the general rule under seismic loading as well as radiation damping in the soil.

(c) *Using spring stiffnesses as in (b), but with magnitudes corresponding to greater strain levels, that is G_s and v_s.* Using v_s and G_s represents a more accurate simulation of soil conditions under seismic loads than the previous two cases. It should be noted that proper practice would involve use of soil springs under low strain conditions for the analysis involving dead and live load combinations and use of reduced soil spring stiffnesses for the seismic load combinations.

(d) *Taking into consideration the effect of soil-structure interaction, as described in Section 2.2.*

2.5. CONCLUSIONS

Results from the analysis for the four cases are presented in Tables II and III. Comparison of the results, as well as remarks drawn from bibliography lead to the following conclusions:

TABLE II. Fundamental period and base shear

	Fundamental period (*s*)		Base shear (*KN*)	
	T_x	T_y	V_x	V_y
Fixed supports	0.21	0.27	3971	3971
Elastic supports	0.52	0.64	3971	3804
Elastic supports (using G_s)	0.54	0.67	3971	3697
Considering SSI	0.33	0.48	2818	2785

For case a), the assumption of elastic supports leads to a substantial increase of the fundamental period, that is 150% in the *x* direction and 140% in the *y* direction of the structure in relation to the one computed assuming fixed supports.

For case b), for seismic risk zones I and II as defined in Table I, soils of category A or B (see Appendix I) and foundations that consist of individual footings, use of a reduced value for the shear modulus leads to a considerably smaller increase of the fundamental period. Therefore, the effect of this reduction on design forces is small. For mat foundations or strip-foundations, as well as flexible tall structures with a large concentrated mass at the top, this effect is obviously greater especially in seismic zones III and IV and for soils of category C or D. In conclusion, the sole reduction of shear modulus in the analysis is not sufficient as the contribution of radiation damping if soil-structure interaction is not taken into consideration.

TABLE III. Reinforcement weight

Analyses	Reinforcement weight (*KN*)
Fixed supports	1150.51
Elastic supports (using G_o)	1134.44
Elastic supports (using G_s)	1125.84
Considering SSI	980.41

For case c), one can observe that soil-structure interaction:

i) Increases considerably the fundamental period, that is 50% in this specific case in relation to the one computed assuming fixed supports.

ii) Greatly reduces the base shear, that is by 30% in both x and y directions for the structure analyzed in comparison to the one computed assuming fixed supports.

iii) Decreases the dimensions of the footings relative to those by the analysis considering rigid supports (Nikolettos [22]).

iv) Significantly reduces the weight of steel reinforcement, that is by 15% in this case, in comparison to the reinforcement given by the analysis with rigid supports.

v) It should be emphasized that at the same time because of SSI, total displacements increase; a fact that must be taken into serious consideration when there is a small gap or contact between adjacent buildings. Inadequate gap can lead to serious damage due to collisions during seismic response (Nikolettos [22]).

3. Seismic Analysis of Bridges including SSI

3.1. BRIEF INTRODUCTION

This section presents seismic design oriented procedures to model and design highway bridges including SSI. Emphasis is placed on modelling the abutment system and the development of procedures that account for non-linear behaviour of the abutments. The first is an iterative design procedure utilizing successive linear analyses. The second is a non-linear static analysis using non-linear springs to account for backfill soil stiffness.

3.2. MODELLING OF THE STRUCTURE AND THE SOIL

3.2.1. *Modelling Backfill Soil Stiffness*

Abutments can be divided into two major categories: a) abutments integrally connected with the superstructure, and b) abutments with bearings. The ones that belong to the first category tend to "move" the soil and dissipate energy in both the longitudinal and the transverse direction. When it is desired to transfer significant forces to the ground, then this type is most appropriate. The possibility of a bridge collapse with integral abutments is relatively small. The abutments with bearings allow for a greater choice as far as the connection with the superstructure is concerned. Designing to avoid possible failures is easier than in the case of the integral abutments, but there is a greater possibility for the superstructure to collapse. Because of the gap between the abutment and the superstructure, the soil can resist considerably higher seismic forces only when the displacements are large enough and the gap is closed. In order to ease the structure, abutments without bearings are apparently more appropriate for the transfer of seismic forces to the soil. It is worth mentioning that one of the methodologies of seismic retrofit of bridges is the conversion of abutments with bearings into abutments rigidly connected with the superstructure (Siros and Spyrakos [23], Lam et al [25] and Priestley et al [30]).

For various abutment configurations and soil conditions, a general form of abutment wall-backfill stiffness equation that considers passive resistance of soil, as recommended by Wilson [24], can be used to estimate the longitudinal stiffness of the end-wall and the transverse stiffness of the wing-wall, that is

$$K_s = \frac{E_S}{(1-\upsilon_s^{\ 2})\cdot I} \tag{20}$$

where K_S is the soil stiffness per unit deflection per unit wall width, E_S is Young's modulus of the backfill soil, v_s is the Poisson's ratio of the backfill soil, and I is a shape factor. Representative values of I are given in Table IV (Lam et al [25]).

TABLE IV. Shape factor for abutment stiffness

[L/B]	Shape Factor [I]
1	0.80
5	1.70
10	2.00
20	2.40

L: Dimension of long side of contact area
B: Dimension of short side of contact area

Expression 20 is used to evaluate the vertical displacement of a uniformly loaded area resting on an elastic half-space, which is available in standard geotechnical references, e.g., (Poulos and Davis [26]). Thus, for a rectangular area with dimensions *a* x *b* (*b* is the shorter dimension) the vertical displacement is given by:

$$\delta_z = \frac{(1-\upsilon_s^{\ 2})}{E_S}\, pbI \tag{21}$$

where *p* is the uniform load per unit area of the rectangle.

Evaluating soil stiffness as described above is just one possible approach to account for translational stiffness of end- and wing-walls. Other models which have received widespread use in estimating foundation stiffness, and are equally as convenient to use, could have also been adopted, e.g., (Wolf [27]).

Expression (20) allows for input of site specific soil parameters and abutment wall configurations. As the length to height ratios for wing-walls are somewhat smaller than end-walls, expression (20) suggests either a lower shape factor I or a higher soil stiffness (K_S) for wing walls as compared to end-walls.

3.2.2. *Modelling Pile Stiffness*

Pile footings are the most commonly used foundation systems to support bridges. A pile foundation can be incorporated in bridge analysis with the aid of several models, including; (i) equivalent cantilever, (ii) uncoupled base spring, and (iii) coupled foundation stiffness matrix. The third model is the most elaborate in representing foundation stiffness for dynamic analyses of the bridge. The main drawback with it relates to the added effort required to develop the coefficients in the stiffness matrix.

In this study a design-oriented procedure is used to evaluate the translational stiffness of the pile-group at the abutments [26]. Translational pile stiffness can be obtained for a combination of bending stiffness of the pile, EI, and the coefficient of variation of soil reaction modulus E_S with depth, f. Proper diagrams are given for example in [25]. There are several simplifying assumptions in this approach. a) The embedment effect has not been taken into account in the procedure, therefore the recommendations are conservative and appropriate for shallow embedment conditions. b) The pile group interaction is neglected for simplicity, a simplification that in special circumstances should not be made.

The relation between the loads at the top of a pile for transverse loading and the displacements is given by

$$[K]\begin{Bmatrix} \delta \\ \theta \end{Bmatrix} = \begin{Bmatrix} P_1 \\ M_1 \end{Bmatrix} \tag{22}$$

where the stiffness matrix is given by

$$[K] = \begin{bmatrix} k_{11} & k_{12} \\ k_{21} & k_{21} \end{bmatrix} \tag{23}$$

In the following we outline the most important steps of the simplified procedure to calculate pile stiffness.

Computation steps.
i) The inertia moment I of the pile is computed. For a circular cross-section I is given by

$$I = \frac{\pi R^4}{4} \tag{24}$$

where R is the radius of the pile.
ii) The factor f is computed from Figures 5 and 6, depending on soil type.

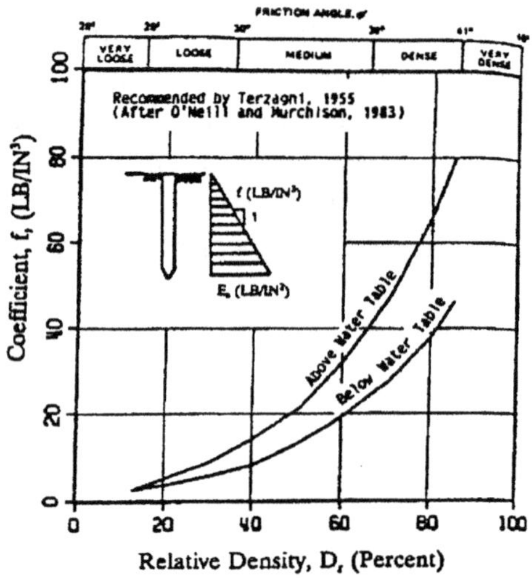

Figure 5. K_I for noncohesive soils.

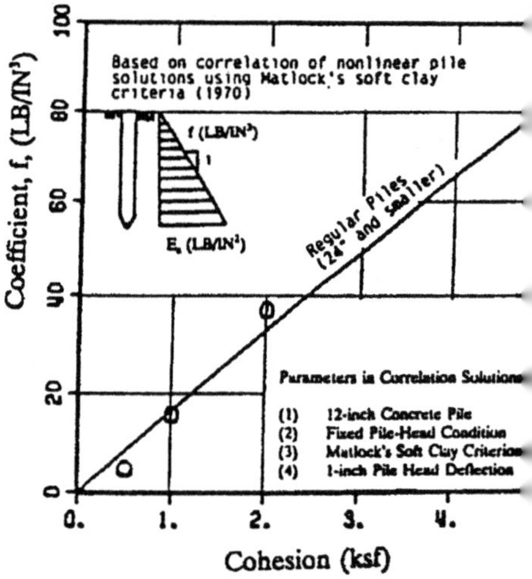

Figure 6. K_I for cohesive soils.

iii) The stiffness k_{11} is computed from Figure 7 knowing EI and f, where E_s is the soil elastic modulus and I is the inertia moment of the pile, as was computed in the first step. Relative diagrams that are used for the computation of the stiffnesses k_{22} and k_{12} are found in the references, e.g., [25]. The diagrams utilize English units, that is, K_I is expressed in lb/in^3, EI in $lb \cdot in^2$ and k_{11} in lb/in. For convenience the conversion from English to S.I. units is given in Table V.

TABLE V. Conversion from English to S.I. units

	English units	S.I. units
K_I	1 (lb/in^3)	271.4 (KN/m^3)
EI	1 $(lb\ in^2)$	2.86 10^{-6} $(KN\ m^2)$
k_{11}	1 (lb/in)	0.175 (KN/m)

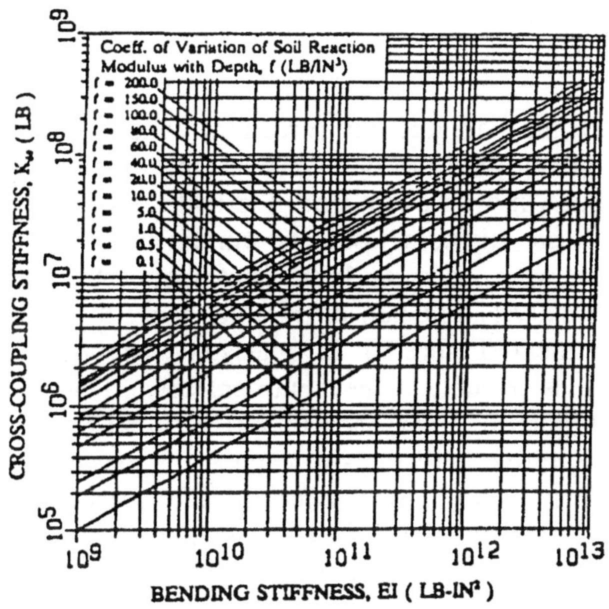

Figure 7. Pile stiffness K_{11}.

iv) If N is the number of piles then the total stiffness for the transverse displacement δ is given by

$$K = N\,k_{11} \tag{25}$$

3.2.3. *Modelling Abutment Stiffness for Linear Iterative Analysis*

The abutment that is used for the analysis is a monolithic type with pile foundation as shown in Figure 8. For simplicity only the translational (longitudinal and transverse) stiffness parameters of the abutment are incorporated in the bridge model for the analysis. Other methods to model the abutment stiffness can be found in the literature, e.g., (Martin and Yan [29]). Proper values of spring constants in the longitudinal and transverse directions are calculated from the backfill soil and pile foundation stiffness according to the following assumptions:

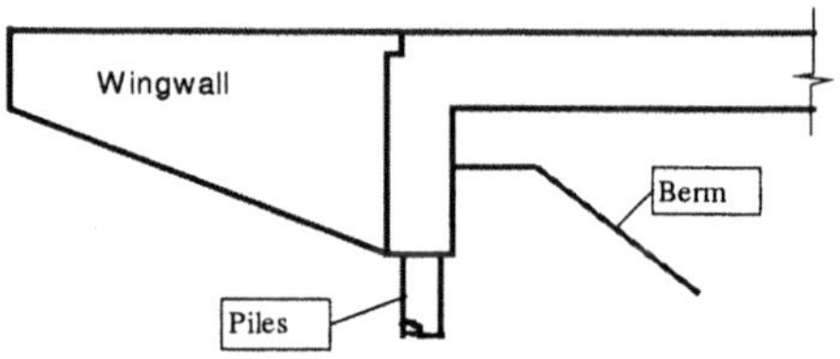

Figure 8. Monolithic abutment.

In the longitudinal direction, when the structure is moving toward the soil, the full passive resistance of the soil is used, but when the structure moves away from the soil no soil resistance is used. The total structure stiffness would be unrealistically high if the full passive resistance were used at both abutments. As an approximation for dynamic analysis, one half of the total backfill soil stiffness is located at each abutment (Figure 9). In quasistatic analyses the full backfill soil resistance is located at the abutment toward which the superstructure moves (Figure 10). The backfill soil stiffness K_{soil} and the pile stiffness K_{pile} are additive until the soil capacity is exceeded, at which point the pile stiffness K_{pile} alone controls the force-deformation behaviour (Priestley et al [30]). In any case, it is important that the total stiffness of the system in the longitudinal direction is determined with the greatest possible accuracy to obtain a realistic evaluation of the system response. In dynamic analyses, the reduction of stiffness at the abutments requires adjustment of the computed resultant forces. When half springs are used, the resulting forces from the analysis should be doubled at each abutment.

In the transverse direction, the flexible wing-walls are not usually fully effective and some judgement is necessary to estimate stiffness realistically. The effective width is taken as the length of the wing-walls multiplied by a factor of 2/3. Also, the soil between the wing walls is more effective (≈100%) than the exterior soil (≈33%). The assumptions are based on several experimental tests and field inspections on abutment response and lead to conservative results for the design of bridge (Caltrans [17]).

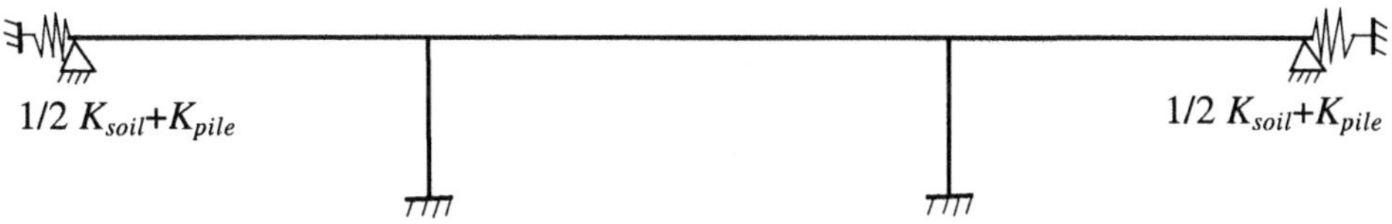

Figure 9. Abutment stiffness for dynamic analyses.

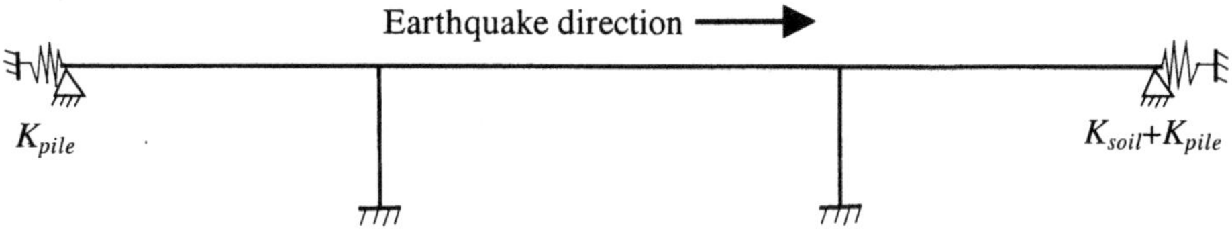

Figure 10. Abutment stiffness for quasistatic analyses.

3.3. ITERATIVE ANALYSIS PROCEDURE

An iterative analysis and design procedure that consists of successive linear dynamic analyses is described. The iterative procedure accounts for the non-linear behaviour of bridges because of backfill soil yielding. The presented procedure is calibrated to Eurocode 8-Part 2 [16] and the Greek Aseismic Code [31] as well as to current bridge design practice. A schematic presentation of the three-step procedure is given in Figure 11.

Step 1. Evaluate the abutment stiffness and the abutment load-displacement characteristics. Assume initial abutment stiffness in the longitudinal and transverse directions. The stiffness should be compatible with the backfill soil stiffness and the foundation type at the abutment. The contribution of the approach slab to abutment stiffness is neglected for simplicity. Soil stiffness and pile foundation stiffness are determined. Load-displacement diagrams for both directions are constructed as shown in Figures 12 and 13.

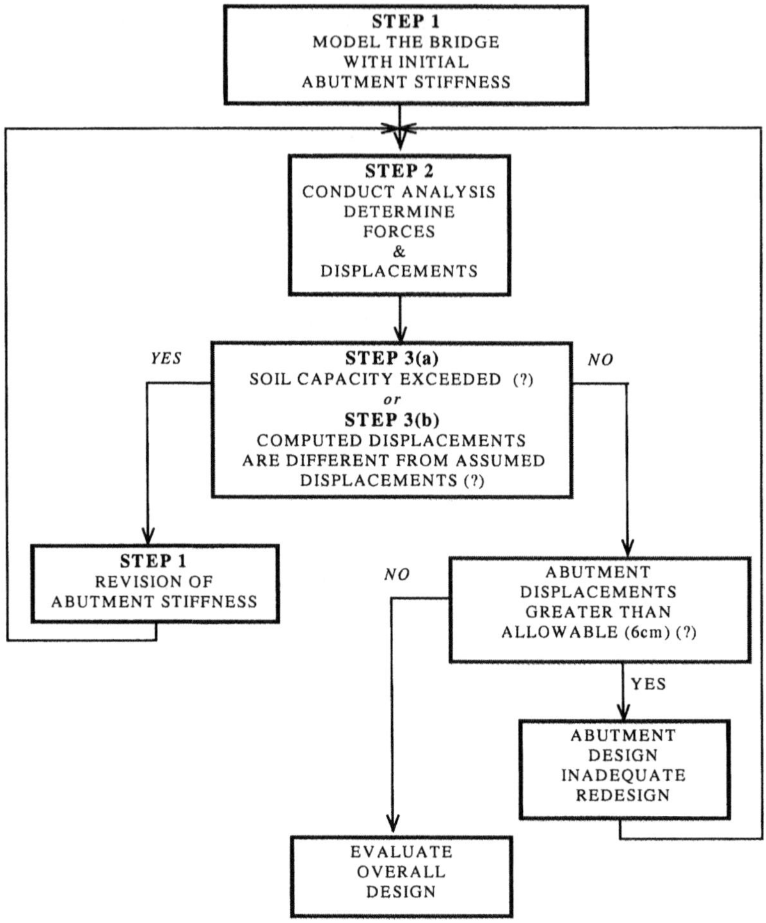

Figure 11. Schematic presentation of iterative analysis procedure.

Step 2. Perform the analysis using the abutment stiffness and conduct linear analysis of the overall bridge to determine forces and displacements. This step is usually repeated as many times as required to arrive at an acceptable solution according to the schematic of Figure 11. Usually three iterations suffice.

Step 3(a). After the first iteration, check that the soil capacity is not exceeded. If the peak soil pressure exceeds soil capacity, the analysis should be repeated with reduced abutment stiffness, using an equivalent linear stiffness (see Trial 2 in Figures 12 and 13) to reflect yielding of the backfill soil. The equivalent linear stiffness for each direction is evaluated on the basis of load-displacement characteristics and assumed displacements.

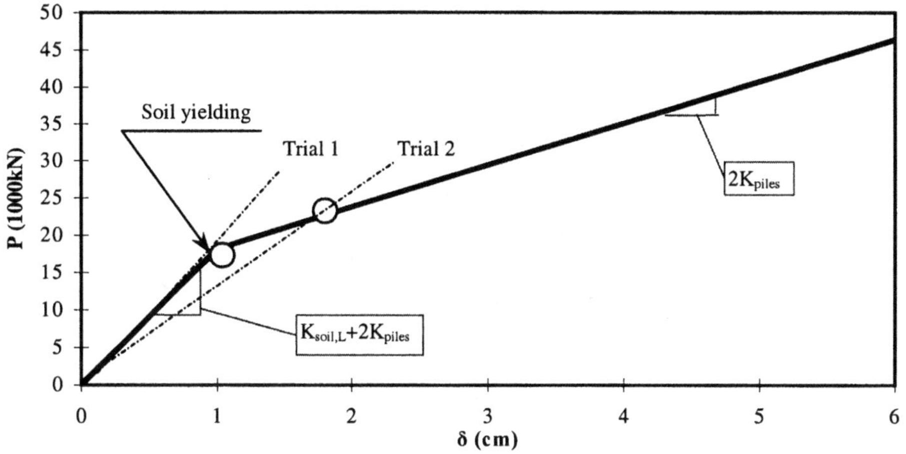

Figure 12. Load-displacement diagram for both abutments in the longitudinal direction.

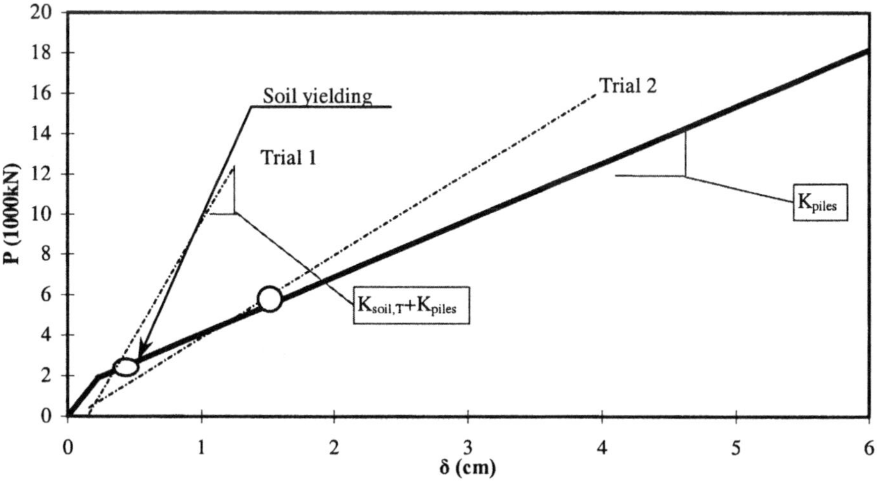

Figure 13. Load-displacement diagram for an abutment in the transverse direction.

Step 3(b). Continue with subsequent iterations and compare for each iteration the displacements with the value assumed for the equivalent linear abutment stiffness. This check is needed to ensure that the assumed abutment stiffness expresses the load-displacement characteristics properly. If the difference in the assumed stiffness between two successive iterations is excessive, the analysis should be repeated with a revised stiffness until convergence is achieved.

Check. Examine for excessive deformations. After the 1971 San Fernando earthquake, field inspections revealed that abutments which moved up to 6cm in the longitudinal direction into the backfill survived with little need for repair. Caltrans [17] and Eurocode 8 [16] suggest that this limit should be maintained. Deformation greater than 6cm in the abutment foundation should be avoided for stability and structural integrity.

3.4. MODELLING ABUTMENT STIFFNESS FOR NON-LINEAR ANALYSIS

Instead of conducting the iterative procedure to account for backfill soil yielding at the abutments, either a non-linear static analysis or a non-linear time-domain dynamic analysis can be implemented. In the following the static non-linear analysis is presented.

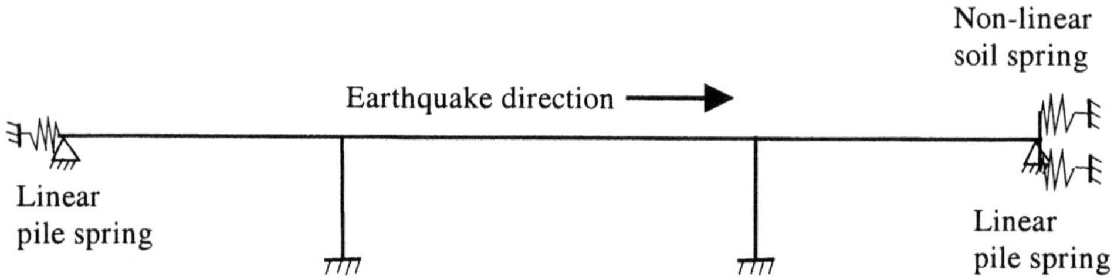

Figure 14. Abutment stiffness for non-linear static analysis.

Two springs are used to model abutment stiffness toward which the structure moves (Figure 14). The first is a non-linear spring, representing the backfill soil stiffness with constant K_{soil} and yield limit at the point where the "failure" soil pressure is reached (Figure 15). The second is a linear spring representing the pile foundation stiffness with constant K_{pile} (Figure 16). At the opposite abutment only the second spring is set.

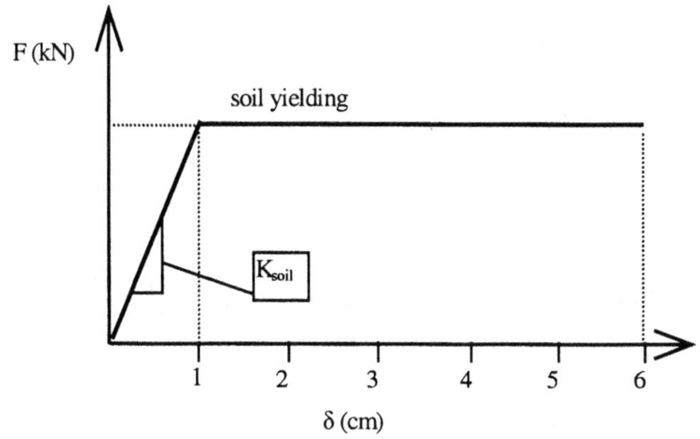

Figure 15. Non-linear soil spring.

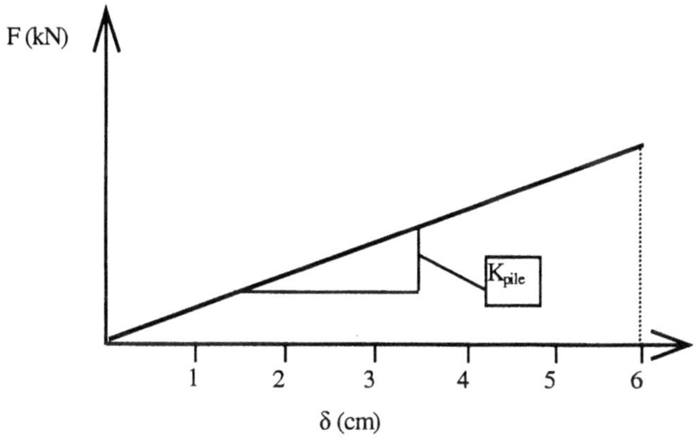

Figure 16. Linear pile foundation spring.

3.5. BRIDGE EXAMPLE

The two procedures (Sections 3.3 and 3.4) are demonstrated with a representative example. Consider a 115 m long three-span bridge with a pre-stressed concrete box girder deck in monolithic connection with bents and abutments. There are three circular columns at each bent founded on spread footings. The width of the bridge is 25 m; geometric characteristics are shown in Figure 17. The geometry of a 2.5 m tall abutment wall is shown in Figure 18. Spectra of the Greek seismic code are used for the analysis for a bridge built in seismic zone III

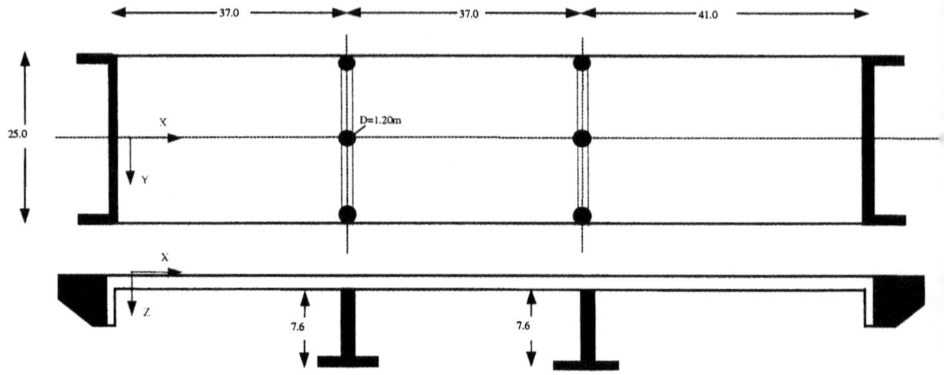

Superstructure: Prestressed concrete box girder $A = 11.427\ m^2$ $I_{yy} = 4.54\ m^4$ $I_{zz} = 565.76\ m^4$ $I_t = 1.00\ m^4$	Bents: Three circular columns, $D = 1.40\ m$ $A = 1.54\ m^2$ $I_{yy} = I_{xx} = 0.188\ m^4$ $I_t = 0.38\ m^4$ Foundation: Footing	Abutments: Height: 2.50 m Foundation: 23 piles Pile diameter $D = 0.50\ m$

Figure 17. Bridge Plan and Elevation.

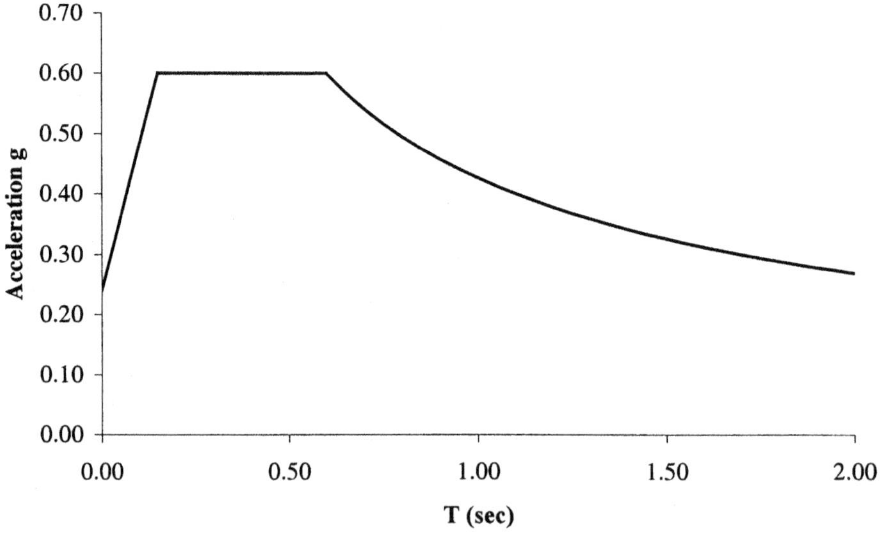

Figure 18. Design Spectrum for Seismic Risk Zone III.

characterised by a peak ground acceleration of $A = 0.24\ g$ (Figure 18). A behaviour factor $q = 1.00$ is adopted to facilitate the parametric studies in order to evaluate the effects of SSI. Detailed calculations of abutment stiffness can be found in Karantzikis [28].

Beam finite elements are used in order to construct the model according to basic rules of finite element analysis for structures (Spyrakos [32]). The inertia characteristics of the box girder are used to model the superstructure (Figure 19). The bents are modelled with beam elements, whereas the rigid connections between them and the superstructure are modelled with rigid elements (Spyrakos [33]). The soil properties to model the foundations at the bents are given in Table VI.

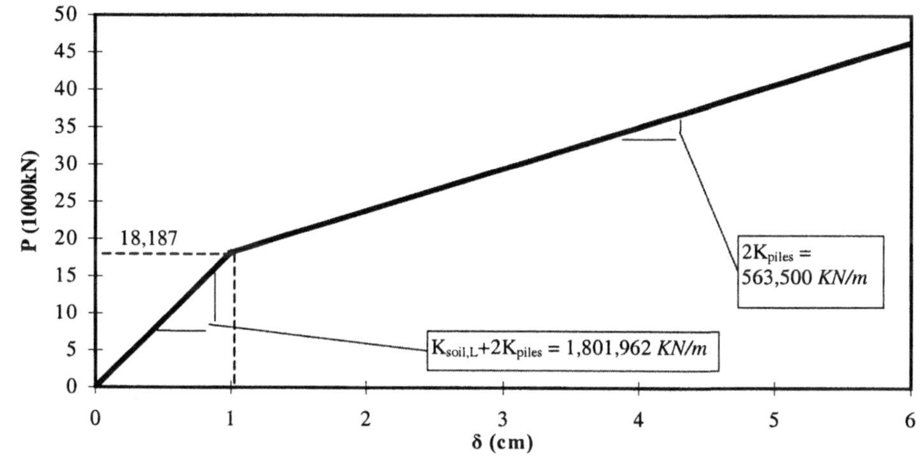

Figure 19. Force-displacement diagram for both abutments in the longitudinal direction.

Translational and rotational springs are used. The vertical displacement and the rotations about the longitudinal axis of the bridge *x-x* and the vertical axis *z-z* at the abutments are constrained. For the specific geometry of the abutment, the stiffness of the soil and the piles is modelled by placing translational springs in the longitudinal and the transverse direction of the bridge and a rotational spring for the rotation about axis *y-y*.

According to Table I for zone III the soil shear modulus is reduced to $G = 0.46G_o$. The elastic modulus is computed from the relation

$$E_s = 0.46[2(1+\nu)G_0] = 86{,}940\,KPa \tag{26}$$

In Table VI, σ_α is the ultimate limit stress of the soil for seismic excitation and δ_{max} is the maximum permissible displacement of the abutment for earthquake movements.

TABLE VI. Soil properties

G_o	70,000 KPa
φ	30°
ν	0.35
E_s	86,940 KPa
σ_α	200 KPa = 200 KN/m^2
δ_{max}	6 cm

3.5.1. *Stiffness Computation*

Computation of pile stiffness. For transverse translation the stiffness of the piles is given by

$$K_{piles} = (23 \text{ piles})(12{,}250\ KN/m/\text{pile}) = 281{,}750\ KN/m \tag{27}$$

Computation of abutment stiffness in the longitudinal direction. The abutment is $L = 25$ m long and $B = 2.5$ m high, so the ratio

$$\frac{L}{B} = \frac{25.0}{2.5} = 10 \tag{28}$$

From Table IV the shape factor of the abutment is I = 2.0. Using expression (20) gives

$$K_s = \frac{E_s}{(1-\nu^2)I} = 49{,}538\ KN/m \tag{29}$$

Consequently, the soil stiffness is K_s = 49,538 *KN/m* per meter of the abutment length. The total soil stiffness is

$$K_{soil} = (49{,}538\ KN/m)(25\ m) = 1{,}238{,}450\ KN/m \tag{30}$$

Assuming that for each abutment only half of the stiffness is active and adding the stiffness of the 23 piles results in

$$K_{abutment} = 0.5K_{soil} + K_{piles} = 900{,}975\ KN/m \tag{31}$$

Computation of abutment stiffness in the transverse direction. The soil stiffness K_s is different in the transverse direction because the geometry of the wing-walls and consequently the ratio *L/B* and the shape factor *I* have different values for this direction. In fact, the ratio *L/B* for the wing-walls is usually smaller than the ratio for the abutment, therefore the stiffness K_s is greater. The wing-wall is 4.0 m long and 2.5 m high, thus the ratio is *L/B* = 1.6 and the shape factor is *I* = 1.0, thus the stiffness is given by

$$K_s = \frac{E_s^{'}}{\left(1-\nu^2\right)I} = 99{,}077\ KN/m \tag{32}$$

Because of the shape of the wing-wall, it is assumed that only 2/3 of its length is active. Therefore, the soil stiffness is

$$K_{soil} = (99{,}077)(2/3)(4.0\ m) = 264{,}205\ KN/m \tag{33}$$

It is also assumed that one wing-wall and only a third of the other wing-wall contribute to the total abutment stiffness. Consequently, the abutment stiffness in the transverse direction is

$$K_{abutment} = (1+1/3)K_{soil} + K_{piles} = 634{,}023\ KN/m \tag{34}$$

Force-displacement diagram in the longitudinal direction. Since σ_α is 200 *KPa*, the maximum force that the soil can resist without failure is given by

$$P_{s,max} = \sigma_\alpha LB = 12{,}500\ KN \tag{35}$$

The equivalent force applied to the system of the two abutments is estimated to be

$$P_{\max} = \frac{K_{soil} + 2K_{piles}}{K_{soil}} P_{s,\max} = 18{,}187\ KN \tag{36}$$

The force P_{max} refers to the displacement $\delta = P_{max}/K_{tot} = 1.00cm$, where $K_{tot} = K_{soil} + 2K_{piles}$. Figure 19 is drawn according to the computation of the forces in the abutments and the relative displacements in the longitudinal direction.

Force-displacement diagram in the transverse direction. The maximum force that the soil can resist in the transverse direction, since $\sigma_\alpha = 200$ KPa is

$$P_{s,\max} = \sigma_a \frac{2}{3}\left(100\% A_{\text{int}} + 33\% A_{ext}\right) = 1{,}252\ KN \tag{37}$$

The relative force applied to the whole abutment is calculated as

$$P_{\max} = \frac{(1+1/3)K_{soil} + K_{piles}}{(1+1/3)K_{soil}} P_{s,\max} = 2{,}253\ KN \tag{38}$$

P_{max} refers to displacement and $\delta = P_{max}/K_{abutment} = 0.355\ cm$.

Figure 20 is drawn based on the computation of the forces in the abutments and the relative displacements in the transverse direction.

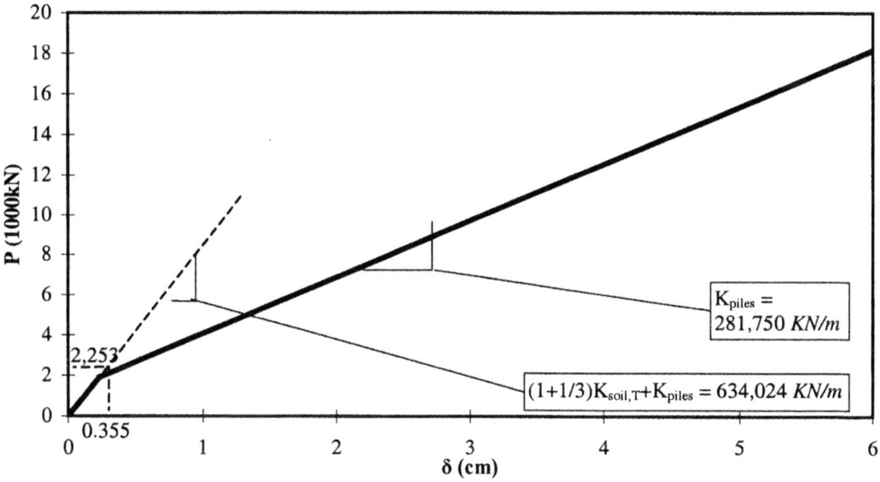

Figure 20. Force-displacement diagram for an abutment in the transverse direction.

Verifications in the longitudinal direction. The displacement of the bridge, and consequently of the two abutments in the longitudinal direction obtained from the analysis is δ_x = 1.323 *cm* > 1.00 *cm*. Apparently, the soil has failed and the analysis must be repeated with reduced stiffness for the abutment-soil system in the longitudinal direction.

Verifications in the transverse direction. The analysis has shown that the displacement of the abutment in the transverse direction is δ_y = 1.753 *cm* > 0.355 *cm*. Apparently, stresses have exceeded the ultimate limit stress of the soil and therefore a reiteration of the analysis is necessary with a reduced value for the stiffness in the transverse direction.

Analysis with reduced stiffnesses. For the second iteration it is assumed that the displacements of the abutments are δ_x = 1.60 *cm* and δ_y = 3.00 *cm* in the longitudinal and the transverse direction, respectively. The equivalent linear stiffness for each direction is computed from the corresponding force-displacement diagrams. The displacements obtained from the second analysis are δ_x = 1.64 $cm \approx 1.60\, cm$ and δ_y = 3.03 $cm \approx 3.00\, cm$. The convergence is satisfactory and there is no need for a third iteration.

3.5.2. *Parametric Studies*

Parametric studies are conducted for three different soils (loose, medium and dense). Results from the analysis are presented in Tables VII to IX in which G_o indicates the shear modulus of the soil for small strains.

TABLE VII. Poor soil, $G_o = 70000$ *Kpa*

Longitudinal Earthquake	Bent 2 M_y (KNm)	Bent 3 M_y (KNm)	Abutment 1 displacement $\delta_x\,(cm)$
Analysis without SSI	-2138	-2003	1.323
Proposed analysis	-2825	-2528	1.640
	+32%	+26%	+24%
Transverse Earthquake	**Bent 2 M_x (KNm)**	**Bent 3 M_x (KNm)**	**Abutment 1 displacement $\delta_y\,(cm)$**
Analysis without SSI	3148	3223	1.753
Proposed analysis	4976	5062	3.030
	+58%	+57%	+73%

TABLE VIII. Medium soil, $G_o = 140000$ *Kpa*

Transverse Earthquake	Bent 2 M_x (KNm)	Bent 3 M_x (KNm)	Abutment 1 displacement $\delta_y\,(cm)$
Analysis without SSI	2101	2181	0.928
Proposed analysis	2876	2957	1.430
	+37%	+36%	+54%

TABLE IX. Dense soil, $G_o = 280000$ *Kpa*

Transverse Earthquake	Bent 2 M_x (KNm)	Bent 3 M_x (KNm)	Abutment 1 displacement $\delta_y\,(cm)$
Analysis without SSI	1651	1735	0.588
Proposed analysis	2263	2346	0.920
	+37%	+35%	+56%

Note that the Tables contain:
i) One linear analysis with the original stiffness of the abutments assuming that the applied forces do not exceed the maximum resistance of the surrounding soil (first line of the Tables).
ii) The suggested analysis procedure consisting successive linear analyses, taking into consideration the non-linear behaviour of the abutment-soil system (second line of the Tables).

3.6. REMARKS AND CONCLUSIONS

Results with the proposed procedures, which consider the abutment non-linearity caused by backfill soil yielding are compared with the results from an analysis that ignores it. The comparison clearly demonstrates that SSI plays a major role in bridge seismic response.

Specifically, in the longitudinal direction the soil fails only for poor soil conditions where great changes in the internal forces in the piers (+(26-32)%) and the displacements of the bridge (+24%) occur. The soil failure results in a reduction of the abutment stiffness in the longitudinal direction and consequently in an increase of the seismic forces and the displacements of the piers. Also in the transverse direction the soil fails in all three cases, therefore the bridge design must follow the iterative analysis procedure, which yields greater forces in the piers (+(36-58)%) as well as greater displacements (+(54-73)%).

In conclusion two procedures to consider non-linear soil-abutment interaction under seismic loads have been presented, the first using iterative linear dynamic response analysis and the second using non-linear static analysis. The procedures are relatively simple and easy to apply for bridge design. However, one of the greatest uncertainties in applying these procedures is the determination of an appropriate value of the soil shear modulus, G. Parametric studies demonstrate that, if the bridge is analysed with the proposed methodology instead of a simple procedure that ignores backfill stiffness reduction, the calculated forces and moments at the piers are greater by 25% to 60% and the displacements by 25% to 75%, depending on soil properties.

Incorporation of abutment stiffness in design and retrofit analysis of highway bridges leads to a more reliable estimation of the overall seismic load level and distribution of seismic loads among bents and abutments. More importantly, it leads to better estimation of displacements.

4. References

1. Rodriguez, Mario E.; Montes, Roberto, (2000), Seismic Response and Damage Analysis of Buildings Supported on Flexible Soils, *Earthquake Engineering and Structural Dynamics*, **29**, 5, 647-665.
2. Trifunac, M.D.; Todorovska, M.I. (1999), Reduction of Structural Damage by Non-Linear Soil Response, *Journal of Structural Engineering*, **125**, 1, 89-97.
3. Hayashi, Yasuhiro; Tamura, Kazuo; Mori, Masafumi; Takahashi, Ikuo, (1999), Simulation Analyses of Buildings Damaged in the 1995 Kobe, Japan, Earthquake Considering Soil-Structure Interaction, *Earthquake Engineering and Structural Dynamics*, **28**, 4, 371-391.
4. Izuru, Takewaki, (1998), Remarkable Response Amplification of Building Frames due to Resonance with the Surface Ground, *Soil Dynamics and Earthquake Engineering*, **17**, 4, 211-218.
5. Iida, Masahiro, (1998), Three-Dimensional Non-Linear Soil-Building Interaction Analysis in the Lakebed Zone of Mexico City during the Hypothetical Guerrero Earthquake, *Earthquake Engineering and Structural Dynamics*, **27**, 12, 1483-1502.
6. Chaallal, O.; Ghlamallah, N. (1998), Seismic Response of Flexible Supported Coupled Shear Walls, *Journal of Structural Engineering*, **122**, 10, 1187-1197.
7. Kocak, S.; Mengi, Y. (2000), A Simple Soil-Structure Interaction Model, *Applied Mathematical Modelling*, **24**, 8-9, 607-635.
8. Estorff, O.; Firuziaan, M. (2000), Coupled BEM/FEM Approach for Non-Linear Soil/Structure Interaction, *Engineering Analysis with Boundary Elements*, **24**, 10, 715-725.
9. Kellezi, L. (2000), Local Transmitting Boundaries for Transient Elastic Analysis, *Soil Dynamics and Earthquake Engineering*, **19**, 7, 533-547.
10. Wen-Hwa Wu, (1997), Equivalent Fixed-Base Models for Soil-Structure Interaction Systems, *Soil Dynamics and Earthquake Engineering*, **16**, 5, 323-336.
11. Badie, S.S.; Salmon, D.C.; Beshara, A.W. (1997), Analysis of Shear Wall Structures on Elastic Foundations, *Computers and Structures*, **65**, 2, 213-224.
12. Saadeghvaziri, M.A.; Yazdani-Motiagh, A.R.; Rashidi, S. (2000), Effects of Soil-Structure Interaction on Longitudinal Seismic Response of MSSS Bridges, *Soil Dynamics and Earthquake Engineering*, **20**, 1-4, 231-242.
13. Spyrakos, C.C. (1992), Seismic Behaviour of Bridge Piers including Soil-Structure Interaction, *Computers and Structures*, **4**, 2, 373-384.
14. Vlassis, A.G.; Spyrakos, C.C. (2001), Seismically Isolated Bridge Piers on Shallow Soil Stratum with Soil-Structure Interaction, *Computers and Structures*, **79**, 2847-2861.
15. Maragakis, E.A., G. Thornton, M. Saiidi, R. Siddharthan, (1989), A Simple Non-Linear Model for the Investigation of the Effects of the Gap Closure at the Abutment Joints of Short Bridges, *Earthquake Engineering and Structural Dynamics*, **18**, 1163-1178.
16. Eurocode 8. Design Provisions for Earthquake Resistant structures. Part-2: Bridges.

17. Caltrans, (1989), Bridge Design Aids Manual, State of California, Department of Transportation, Division of Structures.
18. AASHTO (1992), Standard Specifications for Highway Bridges, Fifteen edition, Washington, D.C.
19. NEHRP 1991, Recommended Provisions for the Development of Seismic Regulations for New Buildings, Part 2. Washington, D.C.
20. ATC 40, 1996, Seismic Evaluation and Retrofit of Concrete Buildings, Cal. Seismic Commission, Report SSC 96-01.
21. FEMA 356, 2000, Prestandard and Commentary for the Seismic Rehabilitation of Buildings, Washington, D.C.
22. Nikolettos, 1999, Soil Structure Interaction for Building Structures, Master Thesis, Department of Civil Engineering, National Technical University of Athens (NTUA), Athens.
23. Siros, K.A.; Spyrakos, C.C. (1995), Creep Analysis of Hybrid Integral Bridges, *Transportation Research Record*, 1476, 147-155.
24. Wilson, J.C., (1988), Stiffness of Non-Skewed Monolithic Bridge Abutments for Seismic Analysis, *Earthquake Engineering and Structural Dynamics*,
25. Lam, I.P., G.R. Martin, R. Imbsen, Modelling Bridge Foundations for Seismic Design and Retrofitting, *Transportation Research Record*, 1290.
26. Poulos, H.G., E.H. Davis, (1974), *Elastic Solutions for Soil and Rock Mechanics,* Wiley, New York.
27. Wolf, J.P. (1994), *Foundation Vibration Analysis using Simple Physical Models,* Prentice Hall, Englewwod Cliffs, NJ, USA.
28. Karantzikis , M.I., (1997), Seismic Analysis and Design of Integral Bridges with Soil-Abutment Interaction, Master Thesis, Department of Civil Engineering, National Technical University of Athens (NTUA), Athens.
29. Martin, G.R. and Yan L., (1995), Modelling Passive Earth Pressure for Bridge Abutments; Earth-induced Movements and Seismic Remediation of Existing Foundations and Abutments, *AICE Geotech. Special Publ.*, 55, 1-16.
30. Priestley, M.J.N., F. Seible, G.M. Calvi, (1996), *Seismic Design and Retrofit of Bridges,* John Wiley & Sons, INC, New York.
31. Greek Aseismic Code (2001), Earthquake Planning and Protection Organization, Athens, Greece.
32. Spyrakos C.C., (1990), *Finite Element Modeling in Engineering Practice*, Algor Publishing Div., Pittsburgh, PA, USA.
33. Spyrakos C.C., and Raftoyiannis J., (1995), *Linear and Non-linear Finite Element Analysis in Engineering Practice*, Algor Publishing Div., Pittsburgh, PA, USA.

5. Appendix

CLASS	DESCRIPTION
A	- Rock or semi-rock formations extending in wide area and large depth provided that they are not strongly weathered. - Layers of dense granular material with little percentage of silty-clayey mixtures having thickness less than 70m. - Layers of stiff overconsolidated clay with thickness less than 70 *m*.
B	- Strongly weathered rocks or soils, which can be considered as granular materials in terms of their mechanical properties. - Layers of granular material of medium density with thickness larger than 5 *m* or of high density with thickness over 70 *m*. - Layers of stiff overconsolidated clay with thickness over 70 *m*.
C	- Layers of granular material of low relative density with thickness over 5 *m* or of medium density with thickness over 70 *m*. - Silty-clayey soils of low strength with thickness over 5 *m*.
D	- Soft clays of high plasticity index ($I_p > 60$) with total thickness over 12 *m*.

PART 2

RELATED TOPICS AND APPLICATIONS

CHAPTER 6.
BEM TECHNIQUES IN NONLOCAL ELASTICITY

C. POLIZZOTTO
Dipartimento di Ingegneria Strutturale e Geotecnica
Università di Palermo,
Viale delle Scienze, 90128 Palermo (Italy)

1. Introduction

BEM techniques for stress analysis in nonlocal elasticity are of interest for soil-structure interaction problems. Many classical models of soil foundations are nonlocal, or weakly nonlocal, (i.e. of differential type) as for example the Pasternak model [1]. Furthermore, no stress singularities occur in a nonlocal elastic medium, or soil, in the presence of cracks, or faults [2,3], which makes it possible to apply the classical stress-based failure criteria.

The nonlocal theories of continua have attracted more and more attention for their ability to provide effective remedies to some specific drawbacks of the material constitutive behaviour, as for example the localization phenomena in plasticity and damage mechanics with loss of ellipticity in the related boundary-value problems, and singularities in the stress response caused by sharp crack tips in elastic media. In the framework of BEM techniques, little work has been devoted to the so-called gradient plasticity [4], whereas no work has been devoted to nonlocal elasticity, to the author's knowledge. The present paper aims to fill this gap.

The nonlocal elasticity theory can be traced back to Kröner [5] who formulated a continuum theory for elastic media with long range cohesive forces. Eringen and coworkers [2,3,6,7] provided a simplified theory for linear homogeneous isotropic nonlocal elastic solids, which differs from the classical one in the stress-strain constitutive relation only, with the elastic moduli being some simple functions of the Euclidean distance between the strain and the stress points. The latter authors addressed many problems that lead to stress singularities in local elasticity (such as, typically, the

W.S. Hall and G. Oliveto (eds.), Boundary Element Methods For Soil-Structure Interaction, 275–296.

crack-tip problems) and showed that these singularities disappear with the nonlocal treatment.

The continuum boundary-value problem of nonlocal elasticity was addressed in [8-10]. It was proved that the solution, if it exists, is unique and that the so-called fundamental solution exists only for certain constitutive models. Extensions of the classical variational principles of elasticity theory to nonlocal elasticity were given in [11]. In the latter work, a nonlocal finite element method (NL-FEM) was also formulated, which is characterized by solving an equation system formally the same as for the standard FEM, but with a stiffness matrix that reflects all the nonlocality features of the problem. It is worth noting that the standard FEM may also be effectively utilized through a suitable iterative procedure of the type local-prediction/nonlocal-correction, in which the problem's nonlocality features are stored in a node vector of fictitious body forces to be updated at every iteration. The latter procedure may be consistently derived from a variational statement as the minimum total potential energy principle [11]; it turns out to be quite similar to that developed in this paper for the application of BEM techniques.

The plan of the present chapter is as follows. In Section 2 the nonlocal (linear) elastic material model is presented and discussed with particular reference to the Eringen model. In Section 3, a nonlocal hyperelastic material model is considered and cast within a thermodynamic framework in which the nonlocal material behaviour is energetically interpreted by means of the so-called nonlocality residual. The latter intervenes in any local energy balance equation, but not in those of global type. The first and second principles of thermodynamics lead to the state equations (i.e. the nonlocal stress-strain relation), as well as to the constitutive equations for the nonlocality residual. Section 4 is devoted to the continuum boundary-value problem for a nonlocal hyperelastic material under static loads and small displacements. It is shown that the solution, if any, is unique, provided the material is endowed with a convex free energy potential, as assumed. The latter problem is then transformed to take the form of a (fictitious) linear isotropic elastic one of local type with an (unknown) initial (correction) strain field. This is aimed at injecting the required nonlocality features into the model by satisfying suitable consistency domain equations, also provided herein. Section 5 is devoted to the extension to nonlocal elasticity of the Hu-Washizu principle of classical elasticity theory; namely, two versions are given for this stationarity principle, one for the original boundary-value problem, another for the transformed one. In Section 6, a boundary/domain stationarity principle is given as one characterizing the transformed boundary-value problem. In Section 7, the latter principle is employed for the formulation of a SGBEM (Symmetric Galerkin BE-

M) and a related iterative procedure of the type local-prediction/nonlocal-correction is proposed, in which the prediction phase is achieved by a S-GBEM technique. The use of standard (nonsymmetric collocation) BEM technique is discussed in Section 7. Conclusions are drawn in Section 8.

The following notation is used throughout this chapter. As a rule, the compact notation is used, with vectors and tensors denoted by bold-face symbols. The *dot* and *colon* products indicate the simple and double index contraction, respectively. For instance, for the vectors $\mathbf{u} = \{u_i\}$, $\mathbf{n} = \{n_i\}$, the second order tensors $\boldsymbol{\sigma} = \{\sigma_{ij}\}$ and $\boldsymbol{\varepsilon} = \{\varepsilon_{ij}\}$ and the fourth-order tensor $\mathbf{D} = \{D_{ijkh}\}$, one can write: $\mathbf{u} \cdot \mathbf{n} = u_i n_i$; $\boldsymbol{\sigma} : \boldsymbol{\varepsilon} = \sigma_{ij}\varepsilon_{ji}$, $\boldsymbol{\varepsilon} : \mathbf{D} : \boldsymbol{\varepsilon} = \varepsilon_{ij} D_{jihk}\varepsilon_{kh}$, $(\boldsymbol{\sigma} \cdot \mathbf{n})_i = n_j \sigma_{ji}$, $(\mathbf{D} : \boldsymbol{\varepsilon})_{ij} = D_{ijkh}\varepsilon_{hk}$, where the repeated index summation rule holds. Latin subscripts denote components with respect to a Cartesian orthogonal co-ordinate system (x_1, x_2, x_3). The symbol 'div' is the divergence operator, i.e. $\text{div}\boldsymbol{\sigma} = \{\sigma_{ji,j}\}$, where $(\cdot)_{,i} = \partial(\cdot)/\partial x_i$; also the symbol ∇^s denotes the symmetric part of the gradient operator, i.e. $\nabla^s \mathbf{u} = \{(u_{i,j} + u_{j,i})/2\}$. Upper dot denotes time derivative, e.g. $\dot{u} = \partial u/\partial t$. Other symbols will be defined in the text where they appear for the first time.

2. Nonlocal Elasticity

Eringen and co-workers developed a simplified nonlocal theory for linear homogeneous isotropic elastic solids [2,3,6,7], which is referred to here as the Eringen model (see also [11]). According to the latter model, the long range forces arising in a homogeneous isotropic elastic material as a consequence of a strain field $\boldsymbol{\varepsilon}(\mathbf{x}) = \{\varepsilon_{ij}(\mathbf{x})\}$ are described by the stress field $\boldsymbol{\sigma}(\mathbf{x}) = \{\sigma_{ij}(\mathbf{x})\}$ expressed as

$$\boldsymbol{\sigma}(\mathbf{x}) = \int_V a(\mathbf{x}, \mathbf{x}')\, \mathbf{D} : \boldsymbol{\varepsilon}(\mathbf{x}')\, \mathrm{d}V' \qquad \forall \mathbf{x} \in V \tag{1}$$

where: $\mathrm{d}V' = \mathrm{d}V(\mathbf{x}')$, $\mathbf{D} = \{D_{ijhk}\}$ is the elastic moduli tensor of the classical isotropic elasticity, that is

$$D_{ijhk} = \lambda \delta_{ij}\delta_{hk} + \mu\left(\delta_{ih}\delta_{jk} + \delta_{ik}\delta_{jh}\right), \tag{2}$$

where λ and μ are the Lamé constants and δ_{ij} the Kronecker symbol; also, $\mathbf{x} = (x_1, x_2, x_3)$ denotes Cartesian orthogonal co-ordinates to which the indicial notation is referred to. The scalar function $a(\mathbf{x}, \mathbf{x}')$ is the *attenuation function*, which relates the source point $\mathbf{x}'$ to the field point $\mathbf{x}$ through the Euclidean distance $r = |\mathbf{x} - \mathbf{x}'|$, that is $a(\mathbf{x}, \mathbf{x}') = \bar{a}(r)$, where $\bar{a}(r)$ is non-negative and decays more or less rapidly with increasing r; i.e. $\bar{a}(r) \to 0$ for $r \to \infty$, but in practice $\bar{a} \simeq 0$ for $r \geq R$, R being finite (*influence distance*).

Equation (1) interprets the material properties as long as the geometry of the domain occupied by the material is in its natural state.

The material stress response $\sigma(\mathbf{x})$ in equation (1) is referred to as the *nonlocal stress*, to signify that it is a functional of the local strain $\varepsilon(\mathbf{x})$, expressed as a weighted value of the strain field $\varepsilon(\mathbf{x})$ in V. Note that $\sigma(\mathbf{x})$ as given by equation (1) exhibits the same regularity degree of $a(\mathbf{x},\mathbf{x}')$ as function of $\mathbf{x}$. Typical choices for $\bar{a}(r)$ are the following:

$$\bar{a}(r) = k\exp(-r^2/\ell^2) \qquad \text{(error function)} \tag{3a}$$

$$\bar{a}(r) = k\left\langle 1 - r^2/r_0^2\right\rangle^2 \qquad \text{(bell shape)} \tag{3b}$$

$$\bar{a}(r) = k\left\langle 1 - r/r_0\right\rangle \qquad \text{(conical shape)} \tag{3c}$$

where k is a constant, $\langle\cdot\rangle$ is the Macauley symbol, i.e. $\langle x\rangle = (x+|x|)/2$ for any value of the scalar x, ℓ is the internal length scale of the material, $r_0 = m_0\ell$, with $m_0 =$ integer > 1. The constant k is determined by imposing the condition

$$\int_{V_\infty} \bar{a}(r)\,\mathrm{d}V(r) = 1. \tag{4}$$

If $\ell \to 0$, by equation (4) $\bar{a}(r)$ tends to become a Dirac delta, i.e. $\bar{a}(r) \to \delta_D(r)$ and thus the nonlocal elasticity model becomes a local one; moreover, equation (4) guarantees that in the case of uniform strain in the infinite domain, V_∞, the nonlocal elasticity model provides a uniform nonlocal stress (see [11] for further comments on this point).

Often in the literature [8, 9, 10, 12] the constitutive equation of nonlocal elasticity is set in the form

$$\sigma(\mathbf{x}) = \xi_1\mathbf{D} : \varepsilon(\mathbf{x}) + \xi_2\int_V a(\mathbf{x},\mathbf{x}')\,\mathbf{D} : \varepsilon(\mathbf{x}')\,\mathrm{d}V' \qquad \forall\mathbf{x}\in V \tag{5}$$

where ξ_1 and ξ_2 are nonnegative material constants. Assuming $\xi_1+\xi_2 = 1$, equation (5) can be interpreted as the constitutive equation for a two-phase elastic material, with phase 1 (of volume fraction ξ_1) complying with local elasticity and phase 2 (or volume fraction ξ_2) complying with nonlocal elasticity. It appears that the local fraction possesses an effective stabilizing effect on the model and that fundamental solutions may exist only for $\xi_1 > 0$ [9]. Note that equation (5) can be set in a form similar to equation (1), but with the attenuation function $\bar{a}(r)$ replaced by

$$\bar{a}^*(r) = \xi_1\delta_D(r) + \xi_2\bar{a}(r). \tag{6}$$

Two alternative expressions can be given for equation (1), namely

$$\mathbf{e} = \mathcal{R}(\boldsymbol{\varepsilon}), \qquad \boldsymbol{\sigma} = \mathbf{D} : \mathbf{e} \tag{7a}$$

$$\boldsymbol{\sigma} = \mathcal{R}(\mathbf{s}), \qquad \mathbf{s} = \mathbf{D} : \boldsymbol{\varepsilon} \tag{7b}$$

where $\mathcal{R}$ is the integral operator

$$\mathcal{R}(\cdot)\Big|_{\mathbf{x}} = \int_V a(\mathbf{x}, \mathbf{x}')\,(\cdot)(\mathbf{x}')\,\mathrm{d}V' \qquad \forall \mathbf{x} \in V \tag{8}$$

which transforms the local field $(\cdot)$ into the corresponding nonlocal one. In equation (7*a*), Hooke's law operates at the nonlocal level by relating the (nonlocal) stress $\boldsymbol{\sigma}$ to the (fictitious) nonlocal strain $\mathbf{e}$; in equation (7*b*), Hooke's law operates at the local level by relating the (fictitious) local stress $\mathbf{s}$ to the (local) strain $\boldsymbol{\varepsilon}$. Both formulations of equation (7*a*) and equation (7*b*) are in turn employed in practice according to convenience.

As pointed out in [8-11], a fundamental hypothesis, assumed here, is that the strain energy stored in V in any nontrivial strain state, say $W[\boldsymbol{\varepsilon}]$, is positive definite, that is, the inequality

$$W[\boldsymbol{\varepsilon}] = \int_V \int_V a(\mathbf{x}, \mathbf{x}')\,\boldsymbol{\varepsilon}(\mathbf{x}) : \mathbf{D} : \boldsymbol{\varepsilon}(\mathbf{x}')\,\mathrm{d}V'\,\mathrm{d}V > 0 \tag{9}$$

is satisfied for any nontrivial $\boldsymbol{\varepsilon}(\mathbf{x})$ in V, and thus, as a consequence, the identity

$$\int_V a(\mathbf{x}, \mathbf{x}')\,\boldsymbol{\varepsilon}(\mathbf{x}')\,\mathrm{d}V' = \mathbf{0} \qquad \forall \mathbf{x} \in V \tag{10}$$

implies that the field $\boldsymbol{\varepsilon}(\mathbf{x})$ is vanishing (almost everywhere) in V. The above conditions can be achieved by operating in the (L_2) space of the square summable functions —as it is the rule in this chapter— and by assuming that the eigenvalue integral equation

$$\phi(\mathbf{x}) = \alpha \int_V a(\mathbf{x}, \mathbf{x}')\,\phi(\mathbf{x}')\,\mathrm{d}V' \tag{11}$$

is nondegenerate, admits positive eigenvalues $(\alpha_1, \alpha_2, \ldots)$ and the related (orthonormal) eigensolutions $(\phi_1, \phi_2, \ldots)$ constitute a complete set of functions in L_2 [11, 13].

3. Thermodynamic Framework

As shown in [11], the above material model can be cast in a suitable thermodynamic framework. For completeness, this point is pursued here by assuming that the material is nonlocal hyperelastic at constant temperature. By hypothesis, there exists a free energy density potential of the form

$$\psi = \psi(\varepsilon, \mathcal{R}(\varepsilon)) \tag{12}$$

which is a C^2-continuous convex function with respect to its arguments. $\mathcal{R}$ is the nonlocality integral operator (8), which by hypothesis is self-adjoint, that is it satisfies the Green's identity:

$$\int_V \mathbf{s} : \mathcal{R}(\varepsilon)\, \mathrm{d}V = \int_V \mathcal{R}(\mathbf{s}) : \varepsilon\, \mathrm{d}V \tag{13}$$

for any pair of tensors $(\mathbf{s}, \varepsilon)$ of L_2.

Considering that only reversible deformation processes are allowed in this material, the relevant Clausius-Duhem inequality takes on the form of an equality, i.e.

$$\sigma : \dot{\varepsilon} - \dot{\psi} + P = 0 \qquad \text{in } V \tag{14}$$

where P is the *nonlocality residual* [14]. The latter accounts for the nonlocality effects, by which some diffusion processes with energy transfers occur in the material; more precisely, P is the power per unit volume transmitted to the generic particle from all other particles in V. Because the latter diffusion processes do not extend outside of the boundary ∂V, the following insulation condition is satisfied for P, i.e. [14]:

$$\int_V P\, \mathrm{d}V = 0. \tag{15}$$

Then, integrating equation (14) over V and expanding $\dot{\psi}$ gives

$$\int_V \left[\sigma : \dot{\varepsilon} - \frac{\partial \psi}{\partial \varepsilon} : \dot{\varepsilon} - \frac{\partial \psi}{\partial \mathcal{R}(\varepsilon)} : \mathcal{R}(\dot{\varepsilon})\right] \mathrm{d}V = 0 \tag{16}$$

which, by equation (13), turns out to be equivalent to

$$\int_V \left[\sigma - \frac{\partial \psi}{\partial \varepsilon} - \mathcal{R}\left(\frac{\partial \psi}{\partial \mathcal{R}(\varepsilon)}\right)\right] : \dot{\varepsilon}\, \mathrm{d}V = 0. \tag{17}$$

Considering that equation (17) must be satisfied for any elastic deformation process, hence for any choice of the strain rate field $\dot{\varepsilon}$, and that the latter condition can be satisfied only if the square bracketed expression vanishes (almost everywhere) in V, one obtains the state equation

$$\sigma = \frac{\partial \psi}{\partial \varepsilon} + \mathcal{R}\left(\frac{\partial \psi}{\partial \mathcal{R}(\varepsilon)}\right) \qquad \text{in } V \tag{18}$$

which is the relevant stress-strain relation. Also, equation (14) gives the constitutive equation for P, i.e.

$$P = \frac{\partial\psi}{\partial\mathcal{R}(\varepsilon)} : \mathcal{R}(\dot{\varepsilon}) - \mathcal{R}\left(\frac{\partial\psi}{\partial\mathcal{R}(\varepsilon)}\right) : \dot{\varepsilon} \qquad \text{in } V \tag{19}$$

which satisfies equation (15).

On choosing ψ as

$$\psi = \frac{1}{2}\varepsilon : \mathbf{D}_1 : \varepsilon + \frac{1}{2}\varepsilon : \mathbf{D}_2 : \mathcal{R}(\varepsilon) \tag{20}$$

where $\mathbf{D}_1$, $\mathbf{D}_2$ are suitable (constant) elastic moduli tensors, applying equation (18) gives

$$\sigma = \mathbf{D}_1 : \varepsilon + \mathbf{D}_2 : \mathcal{R}(\varepsilon) \qquad \text{in } V. \tag{21}$$

For $\mathbf{D}_1 = \xi_1\mathbf{D}$, $\mathbf{D}_2 = \xi_2\mathbf{D}$, equation (21) coincides with equation (5), and thus for $\xi_1 = 0$ and $\xi_2 = 1$, with the Eringen model. Therefore, the Eringen model belongs to a class of nonlocal hyperelastic material models.

4. Boundary-value Problem

A solid body of (open) domain V is made of a nonlocal hyperelastic material endowed with a free energy density $\psi = \psi(\varepsilon, \mathcal{R}(\varepsilon)) > 0$, which is C^2-continuous and convex with respect to its arguments, with $\psi(\mathbf{0}, \mathcal{R}(\mathbf{0})) = 0$. Its strain energy $W[\varepsilon]$ turns out to be positive definite, since in fact one can write:

$$\begin{aligned} W[\varepsilon] &= \int_V \int_0^{\varepsilon} \sigma(\varepsilon) : \mathrm{d}\varepsilon\, \mathrm{d}V \\ &= \int_V \int_0^{\varepsilon} \left[\frac{\partial\psi}{\partial\varepsilon} + \mathcal{R}\left(\frac{\partial\psi}{\partial\mathcal{R}(\varepsilon)}\right)\right] : \mathrm{d}\varepsilon\, \mathrm{d}V \\ &= \int_V \int_0^{\varepsilon} \left[\frac{\partial\psi}{\partial\varepsilon} : \mathrm{d}\varepsilon + \frac{\partial\psi}{\partial\mathcal{R}(\varepsilon)} : \mathcal{R}(\mathrm{d}\varepsilon)\right] \mathrm{d}V \\ &= \int_V \psi(\varepsilon, \mathcal{R}(\varepsilon))\, \mathrm{d}V > 0, \end{aligned} \tag{22}$$

which holds for any nontrivial strain field ε. The last integral in equation (22) coincides with that on the right hand side of equation (9) for $\psi = \frac{1}{2}\varepsilon : \mathbf{D} : \mathcal{R}(\varepsilon)$, that is for the Eringen model.

The body considered is subjected to external actions which are volume forces $\bar{\mathbf{b}}(\mathbf{x})$ in V, surface tractions $\bar{\mathbf{t}}(\mathbf{x})$ on the portion S_t of the boundary surface $S = \partial V$, and imposed displacements $\bar{\mathbf{u}}(\mathbf{x})$ on $S_u = S \backslash S_t$. By hypothesis, all these given fields are sufficiently continuous in their respective domains. Imposed thermal-like strains in V are not considered for the sake of simplicity, but such external actions could be easily included with small changes.

The body's response to the above static external actions can be obtained by solving the following equation system:

$$\operatorname{div}\boldsymbol{\sigma} + \bar{\mathbf{b}} = \mathbf{0} \quad \text{in } V \qquad \text{(field equilibrium)} \tag{23a}$$

$$\boldsymbol{\sigma}\cdot\mathbf{n} = \bar{\mathbf{t}} \quad \text{on } S_t \qquad \text{(boundary equilibrium)} \tag{23b}$$

$$\boldsymbol{\varepsilon} = \nabla^s \mathbf{u} \quad \text{in } V \qquad \text{(field compatibility)} \tag{23c}$$

$$\mathbf{u} = \bar{\mathbf{u}} \quad \text{on } S_u \qquad \text{(boundary compatibility)} \tag{23d}$$

$$\boldsymbol{\sigma} = \frac{\partial\psi}{\partial\boldsymbol{\varepsilon}} + \mathcal{R}\left(\frac{\partial\psi}{\partial\mathcal{R}(\boldsymbol{\varepsilon})}\right) \quad \text{in } V \qquad \text{(nonlocal hyperelastic model)} \tag{23e}$$

For a given stress field $\boldsymbol{\sigma}$, equation (23e) constitutes a Fredholm integral equation which, by hypothesis, provides a unique related strain field $\boldsymbol{\varepsilon}$. For the Eringen model, equation (23e) reduces to

$$\boldsymbol{\sigma} = \mathbf{D} : \mathcal{R}(\boldsymbol{\varepsilon}) \qquad \text{in } V. \tag{23f}$$

The above equation set constitutes an analysis problem for (nonlinear) nonlocal elasticity. It can be proved that the solution to the above problem, if it exists, is unique. In fact, to set up a reduction to absurdity argument, assume that there exist two such solutions, say $\mathbf{u}'$, $\boldsymbol{\varepsilon}'$, $\boldsymbol{\sigma}'$ and $\mathbf{u}''$, $\boldsymbol{\varepsilon}''$, $\boldsymbol{\sigma}''$. By the virtual work principle one can write:

$$\int_V (\boldsymbol{\sigma}' - \boldsymbol{\sigma}'') : (\boldsymbol{\varepsilon}' - \boldsymbol{\varepsilon}'')\,\mathrm{d}V = 0 \tag{24}$$

which, using equation (23e), becomes

$$\int_V \left[\frac{\partial\psi}{\partial\boldsymbol{\varepsilon}'} - \frac{\partial\psi}{\partial\boldsymbol{\varepsilon}''} + \mathcal{R}\left(\frac{\partial\psi}{\partial\mathcal{R}(\boldsymbol{\varepsilon}')}\right) - \mathcal{R}\left(\frac{\partial\psi}{\partial\mathcal{R}(\boldsymbol{\varepsilon}'')}\right)\right] : (\boldsymbol{\varepsilon}' - \boldsymbol{\varepsilon}'')\,\mathrm{d}V = 0$$

and thus, by equation (13),

$$\int_V \left[\left(\frac{\partial\psi}{\partial\boldsymbol{\varepsilon}'} - \frac{\partial\psi}{\partial\boldsymbol{\varepsilon}''}\right) : (\boldsymbol{\varepsilon}' - \boldsymbol{\varepsilon}'') + \left(\frac{\partial\psi}{\partial\mathcal{R}(\boldsymbol{\varepsilon}')} - \frac{\partial\psi}{\partial\mathcal{R}(\boldsymbol{\varepsilon}'')}\right) : (\mathcal{R}(\boldsymbol{\varepsilon}') - \mathcal{R}(\boldsymbol{\varepsilon}''))\right]\mathrm{d}V = 0. \tag{25}$$

Since ψ is convex, the square-bracketed integrand expression in equation (25) is nonnegative, and thus from equation (25) it follows that

$$\left(\frac{\partial\psi}{\partial\varepsilon'}-\frac{\partial\psi}{\partial\varepsilon''}\right):(\varepsilon'-\varepsilon'')+\left(\frac{\partial\psi}{\partial\mathcal{R}(\varepsilon')}-\frac{\partial\psi}{\partial\mathcal{R}(\varepsilon'')}\right):(\mathcal{R}(\varepsilon')-\mathcal{R}(\varepsilon''))=0 \text{ in } V \quad (26)$$

which can be satisfied if, and only if, $\varepsilon' = \varepsilon''$ in V and thus $\mathbf{u}' = \mathbf{u}''$ and $\sigma' = \sigma''$ by (23e). That is, the solution (if any) is unique, as previously stated.

The nonlocal elasticity problem (23$a-e$) admits variational principles which are extensions to nonlocal elasticity of analogous principles of local elasticity, namely the total potential energy principle, the complementary energy principle, Hu-Washizu's, etc. In [11], these extensions were achieved for the case of linear nonlocal elasticity. The same might be carried out for a nonlocal hyperelastic material, but here only principles suitable for BEM formulations are considered, that is, the Hu-Washizu principle and the boundary min-max principle. This will be carried out in the next Section.

Due to nonlinearity, or (in the linear case) either to the lack of fundamental solutions, or to their complexity, BEM approaches to problem (23$a-e$) can work only if based on classical fundamental solutions. For this purpose, problem (23$a-e$) must be transformed into another problem, which is linear isotropic elastic of local type, but with an (unknown) initial strain carrying the nonlocality (and the nonlinearity as well, if any). The latter strain is referred to as the *(nonlocality) correction strain* and is required to satisfy some appropriate domain *consistency* equations, by which the nonlocality (and nonlinearity) features are injected into the problem. The above *fictitious linear problem* is the same as (23$a-e$) except for equation (23e); that is:

$$\operatorname{div}\sigma + \bar{\mathbf{b}} = \mathbf{0} \quad \text{in } V \qquad \text{(field equilibrium)} \qquad (27a)$$

$$\sigma\cdot\mathbf{n} = \bar{\mathbf{t}} \quad \text{on } S_t \qquad \text{(boundary equilibrium)} \qquad (27b)$$

$$\varepsilon = \nabla^s\mathbf{u} \quad \text{in } V \qquad \text{(field compatibility)} \qquad (27c)$$

$$\mathbf{u} = \bar{\mathbf{u}} \quad \text{on } S_u \qquad \text{(boundary compatibility)} \qquad (27d)$$

$$\sigma = \mathbf{D}:(\varepsilon-\varepsilon^c) \quad \text{in } V \qquad \text{(Hooke's law).} \qquad (27e)$$

Equations (27$a-d$) are exactly the same as equations(23$a-d$), but they have been rewritten for convenience. In equation (27e), $\mathbf{D}$ is the elasticity tensor (2) and ε^c denotes the (unknown) correction strain. For ε^c treated as a fixed field in V, problem (27$a-e$) is a linear elastic problem solvable by classical BEM techniques.

The consistency equations which must be appended to equations (27a–e) can be written as follows:

$$\widehat{\varepsilon} = \varepsilon \qquad \text{in } V \tag{28a}$$

$$\varepsilon^c = \widehat{\varepsilon} - \mathbf{D}^{-1} : \left[\frac{\partial \psi}{\partial \widehat{\varepsilon}} + \mathcal{R}\left(\frac{\partial \psi}{\partial \mathcal{R}(\widehat{\varepsilon})}\right)\right] \qquad \text{in } V, \tag{28b}$$

where $\widehat{\varepsilon}$ denotes some *auxiliary* strain field, but with equation (28b) simplified to

$$\varepsilon^c = \widehat{\varepsilon} - \mathcal{R}(\widehat{\varepsilon}) \qquad \text{in } V \tag{28c}$$

in the common case of the Eringen model. Obviously, if equations (28a, b) are satisfied, it follows that the correction strain field is such that the related stress σ given by Hooke's law (27e) is also related to the compatible strain field ε through the nonlocal constitutive equation (23e).

Equations (27$a-e$) and (28a, b) lend themselves to an iterative analysis procedure of the type local-prediction/nonlocal-correction, in which the prediction is obtained by solving equations (27$a - e$) with ε^c taken fixed. Then, the solution so found can be introduced into equations (28a, b) to obtain a new (probably better approximated) value for ε^c and then another iteration can be made.

5. Hu-Washizu Principle Extended to Nonlocal Elasticity

In this Section the classical Hu-Washizu principle of linear elasticity [15] is first extended to nonlocal nonlinear elasticity. Then, the extended principle is transformed into one for local linear elasticity with initial correction strains to account for nonlocality and nonlinearity.

5.1. NONLOCAL HYPERELASTIC MATERIAL

The functional

$$\begin{aligned} H &= \int_V [\psi(\varepsilon, \mathcal{R}(\varepsilon)) + \sigma : (\nabla^s \mathbf{u} - \varepsilon) - \bar{\mathbf{b}} \cdot \mathbf{u}] \, \mathrm{d}V \\ &- \int_{S_u} \mathbf{t} \cdot (\mathbf{u} - \bar{\mathbf{u}}) \, \mathrm{d}S - \int_{S_t} \bar{\mathbf{t}} \cdot \mathbf{u} \, \mathrm{d}S, \end{aligned} \tag{29}$$

in which all the unknown variables are free, is to be made stationary. For this purpose, making use of equation (13) and the divergence theorem, the first variation of H can be written as follows:

$$\delta H = \int_V \delta\varepsilon : \left[\frac{\partial \psi}{\partial \varepsilon} + \mathcal{R}\left(\frac{\partial \psi}{\partial \mathcal{R}(\varepsilon)}\right) - \sigma\right] \mathrm{d}V + \int_V \delta\sigma : [\nabla^s \mathbf{u} - \varepsilon] \, \mathrm{d}V +$$

$$- \int_V \delta \mathbf{u} \cdot [\operatorname{div}\boldsymbol{\sigma} + \bar{\mathbf{b}}] \, \mathrm{d}V + \int_{S_t} \delta \mathbf{u} \cdot [\boldsymbol{\sigma} \cdot \mathbf{n} - \bar{\mathbf{t}}] \, \mathrm{d}S$$

$$+ \int_{S_u} \delta \mathbf{u} \cdot [\boldsymbol{\sigma} \cdot \mathbf{n} - \mathbf{t}] \, \mathrm{d}V - \int_{S_u} \delta \mathbf{t} \cdot [\mathbf{u} - \bar{\mathbf{u}}] \, \mathrm{d}S. \tag{30}$$

Because δH must vanish for arbitrary choices of the variations in the L_2 space, it follows that all the square-bracketed fields must vanish in their respective domains, and thus that the governing equations ($23a-e$) must be satisfied. The converse is also true, that is, if equations ($23a-e$) are satisfied, δH vanishes for arbitrary variations and thus H is stationary. Therefore, the conclusion is that the solution (if any) to the nonlinear nonlocal elasticity problem ($23a-e$) is provided by the stationarity conditions of the functional (29) and conversely.

However, the above variational principle is not suitable for BEM formulations because of nonlinearity, or, in the linear case, because the fundamental solutions either do not exist, or are most probably cumbersome. Instead the fundamental solutions of local linear isotropic elasticity can effectively be employed to address the transformed problem ($27a-e$) with appended domain equations ($28a, b$). The latter problem too admits a Hu-Washizu stationarity principle which is a transformation of that given above.

5.2. LINEAR LOCAL ELASTICITY WITH CORRECTION STRAIN

Let one consider the functional

$$H^* = H_1 + H_2 \tag{31}$$

where

$$H_1 = \int_V \left[\frac{1}{2} \boldsymbol{\varepsilon} : \mathbf{D} : \boldsymbol{\varepsilon} + \boldsymbol{\sigma} : (\nabla^s \mathbf{u} - \boldsymbol{\varepsilon}) - \boldsymbol{\varepsilon} : \mathbf{D} : \boldsymbol{\varepsilon}^c - \bar{\mathbf{b}} \cdot \mathbf{u} \right] \mathrm{d}V$$

$$- \int_{S_u} \mathbf{t} \cdot (\mathbf{u} - \bar{\mathbf{u}}) \, \mathrm{d}S - \int_{S_t} \bar{\mathbf{t}} \cdot \mathbf{u} \, \mathrm{d}S, \tag{32a}$$

$$H_2 = \int_V \left[\psi(\widehat{\boldsymbol{\varepsilon}}, \mathcal{R}(\widehat{\boldsymbol{\varepsilon}})) - \frac{1}{2} \widehat{\boldsymbol{\varepsilon}} : \mathbf{D} : \widehat{\boldsymbol{\varepsilon}} + \widehat{\boldsymbol{\varepsilon}} : \mathbf{D} : \boldsymbol{\varepsilon}^c \right] \mathrm{d}V. \tag{32b}$$

Here $\mathbf{D}$ denotes the moduli tensor (2), whereas $\boldsymbol{\varepsilon}$, $\boldsymbol{\varepsilon}^c$ and $\widehat{\boldsymbol{\varepsilon}}$ denote independent strain fields in V. Note that, for $\boldsymbol{\varepsilon}^c$ taken fixed, H_1 represents the Hu-Washizu functional of linear (local) elasticity with an initial strain field $\boldsymbol{\varepsilon}^c$. The symbols adopted here are aimed at giving $\widehat{\boldsymbol{\varepsilon}}$ the meaning of auxiliary strain and $\boldsymbol{\varepsilon}^c$ the meaning of correction strain.

Following the procedure used for equation (29), the first variation of H^* can be written as follows:

$$\begin{aligned}\delta H^* &= \int_V \delta\boldsymbol{\varepsilon} : [\mathbf{D} : (\boldsymbol{\varepsilon} - \boldsymbol{\varepsilon}^c) - \boldsymbol{\sigma}]\,\mathrm{d}V + \int_V \delta\boldsymbol{\sigma} : [\nabla^s \mathbf{u} - \boldsymbol{\varepsilon}]\,\mathrm{d}V \\ &- \int_V \delta\mathbf{u} \cdot [\mathrm{div}\,\boldsymbol{\sigma} + \bar{\mathbf{b}}]\,\mathrm{d}V + \int_{S_t} \delta\mathbf{u} \cdot [\boldsymbol{\sigma} \cdot \mathbf{n} - \bar{\mathbf{t}}]\,\mathrm{d}S \\ &+ \int_{S_u} \delta\mathbf{u} \cdot [\boldsymbol{\sigma} \cdot \mathbf{n} - \mathbf{t}]\,\mathrm{d}S - \int_{S_u} \delta\mathbf{t} \cdot [\mathbf{u} - \bar{\mathbf{u}}]\,\mathrm{d}S \\ &+ \int_V \delta\boldsymbol{\varepsilon}^c : \mathbf{D} : [\widehat{\boldsymbol{\varepsilon}} - \boldsymbol{\varepsilon}]\,\mathrm{d}V \\ &+ \int_V \delta\widehat{\boldsymbol{\varepsilon}} : \left[\frac{\partial\psi}{\partial\widehat{\boldsymbol{\varepsilon}}} + \mathcal{R}\left(\frac{\partial\psi}{\partial\mathcal{R}(\widehat{\boldsymbol{\varepsilon}})}\right) - \mathbf{D} : (\widehat{\boldsymbol{\varepsilon}} - \boldsymbol{\varepsilon}^c)\right]\mathrm{d}V. \end{aligned} \tag{33}$$

In order that $\delta H^* = 0$ for arbitrary choices of the variations in L_2, all the square-bracketed fields of equation (33) must vanish in their respective domains, which implies that all the governing equations $(27a - e)$ together with the consistency equations $(28a, b)$ are satisfied.

In other words, in the stationarity conditions: *i)* the auxiliary strain field $\widehat{\boldsymbol{\varepsilon}}$ coincides with the compatible field, $\boldsymbol{\varepsilon}$, and *ii)* the correction strain field, $\boldsymbol{\varepsilon}^c$, is such that the stress field, $\boldsymbol{\sigma}$, related to $\boldsymbol{\varepsilon} - \boldsymbol{\varepsilon}^c$ through Hooke's law of the fictitious linear elastic material, is related to the compatible strain field, $\boldsymbol{\varepsilon}$, through the pertinent (nonlinear) nonlocal constitutive equation $(23e)$.

The latter result is of value for numerical computations. One would in fact like to reach the stationarity condition through an iterative procedure in which $\boldsymbol{\varepsilon}^c$ and $\widehat{\boldsymbol{\varepsilon}}$ are taken constant at every iteration. Then, H_2 drops out from H^* in equation (31) because it is constant and the stationarity problem reduces to the stationarity of H_1 with $\boldsymbol{\varepsilon}^c$ taken fixed. Since the first variation of H_1 with $\boldsymbol{\varepsilon}^c$ fixed coincides with equation (33) (where the last two integrals vanish because $\delta\boldsymbol{\varepsilon}^c = \delta\widehat{\boldsymbol{\varepsilon}} = \mathbf{0}$ identically), one obtains that the relevant stationarity equations are $(27a - e)$.

Therefore, denoting the iteration sequence by $n = 1, 2, \ldots$, and the known fixed values of $\widehat{\boldsymbol{\varepsilon}}$ and $\boldsymbol{\varepsilon}^c$ by $\widehat{\boldsymbol{\varepsilon}}_{(n-1)} = \boldsymbol{\varepsilon}_{(n-1)}$, $\boldsymbol{\varepsilon}^c_{(n-1)}$, solving equations $(27a - e)$ with $\boldsymbol{\varepsilon}^c = \boldsymbol{\varepsilon}^c_{(n-1)}$ gives the n-th iteration solution, say $\mathbf{u}_{(n)}$, $\boldsymbol{\varepsilon}_{(n)}$, $\boldsymbol{\sigma}_{(n)}$, after which one can write, using equations $(28a, b, c)$:

$$\widehat{\boldsymbol{\varepsilon}}_{(n)} = \boldsymbol{\varepsilon}_{(n)} \qquad \text{in } V \tag{34a}$$

$$\boldsymbol{\varepsilon}^c_{(n)} = \widehat{\boldsymbol{\varepsilon}}_{(n)} - \mathbf{D}^{-1} : \left[\frac{\partial\psi}{\partial\widehat{\boldsymbol{\varepsilon}}_{(n)}} + \mathcal{R}\left(\frac{\partial\psi}{\partial\mathcal{R}(\widehat{\boldsymbol{\varepsilon}}_{(n)})}\right)\right] \qquad \text{in } V, \tag{34b}$$

or, instead of equation $(34b)$,

$$\boldsymbol{\varepsilon}^c_{(n)} = \widehat{\boldsymbol{\varepsilon}}_{(n)} - \mathcal{R}(\widehat{\boldsymbol{\varepsilon}}_{(n)}) \tag{34c}$$

for the Eringen model. The above n-th iteration values, which are likely to constitute a better approximation of the unknown fields, can be employed in the next iteration, and so forth. The first iteration solution starts by solving equations $(27a-e)$ with $\boldsymbol{\varepsilon}^c = \boldsymbol{\varepsilon}^c_{(o)} \equiv \mathbf{0}$. Whenever the solution to equations $(27a-e)$ exhibits singularities, equation $(34a)$ will be applied by taking $\widehat{\boldsymbol{\varepsilon}}_{(n)}$ equal to a suitably regularized $\boldsymbol{\varepsilon}_{(n)}$ field.

This iteration procedure can be implemented numerically by either BEM or FEM techniques. The Hu-Washizu principle under discussion here is known [16] to be a suitable means for symmetric Galerkin (SG) BEM formulations. In fact, coming back to the fictitious linear local problem $(27a-e)$, let the unknown fields satisfy rigorously all the field equations, i.e. $(27a)$, $(27c)$ and $(27e)$, but the boundary equations $(27b)$ and $(27d)$ only in some weak form. Then the stationarity conditions of H^* with $\boldsymbol{\varepsilon}^c$ and $\widehat{\boldsymbol{\varepsilon}}$ held fixed reduce to (see equation (33)) the following global boundary conditions:

$$\int_{S_u} \delta\mathbf{u}\cdot(\mathbf{t}-\bar{\mathbf{t}})\,\mathrm{d}S = 0, \qquad \int_{S_t} \delta\mathbf{t}\cdot(\mathbf{u}-\bar{\mathbf{u}})\,\mathrm{d}S = 0 \tag{35}$$

where $\mathbf{t} = \boldsymbol{\sigma}\cdot\mathbf{n}$ by definition. The latter approach can be achieved by making use of the Somigliana formulae [17–19] to represent the unknown fields $\mathbf{u}$ and $\mathbf{t} = \boldsymbol{\sigma}\cdot\mathbf{n}$ and by approximating $\delta\mathbf{u}$ and $\delta\mathbf{t}$ by a suitable boundary element (BE) discretization. However, as pointed out in [20, 21], a better implementation of the same approach can be obtained through the so-called *boundary min-max principle*, or something else equivalent to it. This point will be pursued in next Sections.

6. A Boundary/Domain Stationarity Principle

This principle is analogous to the boundary min-max principle [20, 21] and also rooted in the use of classical fundamental solutions. The functional to consider is:

$$\chi = \chi_1 + \chi_2 \tag{36}$$

where

$$\begin{aligned}
\chi_1 \;=\;& \frac{1}{2}\int_{S_u}\int_{S_u} \mathbf{t}\cdot\mathbf{G}_{uu}\cdot\mathbf{t}\,\mathrm{d}S_u\mathrm{d}S_u - \int_{S_u}\int_{S_t} \mathbf{t}\cdot\mathbf{G}_{ut}\cdot\mathbf{u}\,\mathrm{d}S_t\mathrm{d}S_u \\
&+ \frac{1}{2}\int_{S_t}\int_{S_t} \mathbf{u}\cdot\mathbf{G}_{tt}\cdot\mathbf{u}\,\mathrm{d}S_t\mathrm{d}S_t + \frac{1}{2}\int_V\int_V \boldsymbol{\varepsilon}^c : \mathbf{G}_{\sigma\sigma} : \boldsymbol{\varepsilon}^c\,\mathrm{d}V\mathrm{d}V \\
&+ \int_{S_u}\int_V \mathbf{t}\cdot\mathbf{G}_{u\sigma}\cdot\boldsymbol{\varepsilon}^c\,\mathrm{d}V\mathrm{d}S_u - \int_{S_t}\int_V \mathbf{u}\cdot\mathbf{G}_{t\sigma} : \boldsymbol{\varepsilon}^c\,\mathrm{d}V\mathrm{d}S_t \\
&+ \int_{S_u} \mathbf{t}\cdot(\mathbf{q}_u - \bar{\mathbf{u}})\,\mathrm{d}S_u - \int_{S_t} \mathbf{u}\cdot(\mathbf{q}_t - \bar{\mathbf{t}})\,\mathrm{d}S_t + \int_V \mathbf{q}_\sigma : \boldsymbol{\varepsilon}^c\,\mathrm{d}V,
\end{aligned} \tag{37}$$

$$\chi_2 = \int_V \left[\frac{1}{2}(\hat{\varepsilon}-\varepsilon^c):\mathbf{D}:(\hat{\varepsilon}-\varepsilon^c) - \psi(\hat{\varepsilon},\mathcal{R}(\hat{\varepsilon}))\right] \mathrm{d}V. \tag{38}$$

Here, the following notation has been employed. The two-point tensor-valued functions, $\mathbf{G}_{ab} = \mathbf{G}_{ab}(\mathbf{x},\mathbf{y})$, $(a,b = u,t,\sigma)$ denote Green-type influence functions for the infinite domain having the same properties as the fictitious linear local elastic material with the moduli considered before, equation (2). As shown in [16, 22, 23], $\mathbf{G}_{ab}(\mathbf{x},\mathbf{y})$ provides the effects produced at point $\mathbf{x}$ due to a singular action applied at point $\mathbf{y}$. The effects are specified by the first subscript, i.e. *displacement* for $a = u$, *traction* (on an area element of normal $\mathbf{n}(\mathbf{x})$) for $a = t$, *stress* for $a = \sigma$; whereas the cause is specified by the second subscript through a conjugacy rule in the virtual work principle sense, i.e. unit *force* for $b = u$, unit *relative displacement* for $b = t$ and unit *imposed strain* for $b = \sigma$. The following symmetry conditions hold [16, 22, 23]:

$$\mathbf{G}_{ab}(\mathbf{x},\mathbf{y}) = \mathbf{G}^T_{ba}(\mathbf{y},\mathbf{x}) \quad \forall\, \mathbf{x},\mathbf{y} \neq \mathbf{x}, \quad \forall\,(a,b = u,t,\sigma) \tag{39}$$

which are a consequence of Maxwell's theorem. The $\mathbf{G}_{ab}(\mathbf{x},\mathbf{y})$ functions collect the classical fundamental solutions, which are well known in the literature [17-19].

It should be noted that the symmetric matrix of two-point tensor-valued functions $\mathbf{G}_{ab}(\mathbf{x},\mathbf{y})$, i.e. the matrix

$$\left[\begin{array}{c:cc} \mathbf{G}_{uu} & \mathbf{G}_{ut} & \mathbf{G}_{u\sigma} \\ \hdashline \mathbf{G}_{tu} & \mathbf{G}_{tt} & \mathbf{G}_{t\sigma} \\ \mathbf{G}_{\sigma u} & \mathbf{G}_{\sigma t} & \mathbf{G}_{\sigma\sigma} \end{array}\right] \tag{40}$$

contains entries which become singular as $\mathbf{y} \to \mathbf{x}$; more precisely, those in the first principal block are weakly singular, those in the second principal block are hypersingular, whereas those out of the diagonal blocks are singular. This implies that an integral having $\mathbf{G}_{ab}(\mathbf{x},\mathbf{y})$ as kernel must be treated with due care in the singular and hypersingular cases [22, 23].

The quantities $\mathbf{q}_a$, $(a = u,t,\sigma)$, appearing in equation (37) represent the effects produced in the infinite domain, V_∞, in which the body V is embedded, due to the given loads, that is, the volume force $\bar{\mathbf{b}}$ upon V, the surface traction $\bar{\mathbf{t}}$ (as single layer source $\bar{\mathbf{t}}$) applied on the interface S_u, and the imposed displacement $\bar{\mathbf{u}}$ (as double layer source $-\bar{\mathbf{u}}$) applied on the interface S_u.
These effects are expressed as follows [16, 22, 23]:

$$\mathbf{q}_u = \int_V \mathbf{G}_{uu}\cdot\bar{\mathbf{b}}\,\mathrm{d}V + \int_{S_t} \mathbf{G}_{uu}\cdot\bar{\mathbf{t}}\,\mathrm{d}S_t - \int_{S_u} \mathbf{G}_{ut}\cdot\bar{\mathbf{u}}\,\mathrm{d}S_u \tag{41a}$$

$$\mathbf{q}_t = \int_V \mathbf{G}_{tu} \cdot \bar{\mathbf{b}} \, \mathrm{d}V + \int_{S_t} \mathbf{G}_{tu} \cdot \bar{\mathbf{t}} \, \mathrm{d}S_t - \int_{S_u} \mathbf{G}_{tt} \cdot \bar{\mathbf{u}} \, \mathrm{d}S_u \tag{41b}$$

$$\mathbf{q}_\sigma = \int_V \mathbf{G}_{\sigma u} \cdot \bar{\mathbf{b}} \, \mathrm{d}V + \int_{S_t} \mathbf{G}_{\sigma u} \cdot \bar{\mathbf{t}} \, \mathrm{d}S_t - \int_{S_u} \mathbf{G}_{\sigma t} \cdot \bar{\mathbf{u}} \, \mathrm{d}S_u . \tag{41c}$$

The first variation of χ of equation (36) has the following expression:

$$\begin{aligned}
\delta\chi &= \int_{S_u} \delta\mathbf{t} \cdot \left[\int_{S_u} \mathbf{G}_{uu} \cdot \mathbf{t} \, \mathrm{d}S_u - \int_{S_t} \mathbf{G}_{ut} \cdot \mathbf{u} \, \mathrm{d}S_t \right. \\
&+ \left. \int_V \mathbf{G}_{u\sigma} : \boldsymbol{\varepsilon}^c \, \mathrm{d}V + \mathbf{q}_u - \bar{\mathbf{u}} \right] \mathrm{d}S_u \\
&- \int_{S_t} \delta\mathbf{u} \cdot \left[\int_{S_u} \mathbf{G}_{tu} \cdot \mathbf{t} \, \mathrm{d}S_u - \int_{S_t} \mathbf{G}_{tt} \cdot \mathbf{u} \, \mathrm{d}S_t \right. \\
&+ \left. \int_V \mathbf{G}_{t\sigma} : \boldsymbol{\varepsilon}^c \, \mathrm{d}V + \mathbf{q}_t - \bar{\mathbf{t}} \right] \mathrm{d}S_t \\
&+ \int_V \delta\boldsymbol{\varepsilon}^c : \left[\int_{S_u} \mathbf{G}_{\sigma u} \cdot \mathbf{t} \, \mathrm{d}S_u - \int_{S_t} \mathbf{G}_{\sigma t} \cdot \mathbf{u} \, \mathrm{d}S_t \right. \\
&+ \left. \int_V \mathbf{G}_{\sigma\sigma} : \boldsymbol{\varepsilon}^c \, \mathrm{d}V + \mathbf{q}_\sigma - \mathbf{D} : (\hat{\varepsilon} - \boldsymbol{\varepsilon}^c) \right] \mathrm{d}V \\
&+ \int_V \delta\hat{\varepsilon} : \left[\mathbf{D} : (\hat{\varepsilon} - \boldsymbol{\varepsilon}^c) - \frac{\partial\psi}{\partial\hat{\varepsilon}} - \mathcal{R}\left(\frac{\partial\psi}{\partial\mathcal{R}(\hat{\varepsilon})} \right) \right] \mathrm{d}V.
\end{aligned} \tag{42}$$

Then, the stationarity conditions for χ read:

$$\begin{aligned}
\mathbf{u}|_{S_u} \equiv \int_{S_u} \mathbf{G}_{uu} \cdot \mathbf{t} \, \mathrm{d}S_u - \int_{S_t} \mathbf{G}_{ut} \cdot \mathbf{u} \, \mathrm{d}S_t \\
+ \int_V \mathbf{G}_{u\sigma} : \boldsymbol{\varepsilon}^c \, \mathrm{d}V + \mathbf{q}_u = \bar{\mathbf{u}} \qquad \text{on } S_u
\end{aligned} \tag{43a}$$

$$\begin{aligned}
\mathbf{t}|_{S_t} \equiv \int_{S_u} \mathbf{G}_{tu} \cdot \mathbf{t} \, \mathrm{d}S_u - \int_{S_t} \mathbf{G}_{tt} \cdot \mathbf{u} \, \mathrm{d}S_t \\
+ \int_V \mathbf{G}_{t\sigma} : \boldsymbol{\varepsilon}^c \, \mathrm{d}V + \mathbf{q}_t = \bar{\mathbf{t}} \qquad \text{on } S_t
\end{aligned} \tag{43b}$$

$$\begin{aligned}
\boldsymbol{\sigma}|_V \equiv \int_{S_u} \mathbf{G}_{\sigma u} \cdot \mathbf{t} \, \mathrm{d}S_u - \int_{S_t} \mathbf{G}_{\sigma t} \cdot \mathbf{u} \, \mathrm{d}S_t \\
+ \int_V \mathbf{G}_{\sigma\sigma} : \boldsymbol{\varepsilon}^c \, \mathrm{d}V + \mathbf{q}_\sigma = \mathbf{D} : (\hat{\varepsilon} - \boldsymbol{\varepsilon}^c) \ \text{ in } V
\end{aligned} \tag{43c}$$

$$\mathbf{D} : (\hat{\varepsilon} - \boldsymbol{\varepsilon}^c) = \frac{\partial\psi}{\partial\hat{\varepsilon}} + \mathcal{R}\left(\frac{\partial\psi}{\partial\mathcal{R}(\hat{\varepsilon})} \right) \quad \text{in } V. \tag{44}$$

Equations ($43a, b$) are the usual boundary integral equations constituting the bases of the SGBEM [16, 21-23]. More precisely, the left-hand members of equations ($43a, b$) are the Somigliana formulae [17-19] for the displacements and tractions, respectively enforced on S_u and S_t. (This implies some care in computing $\mathbf{q}_u$ and $\mathbf{q}_t$, since these contain integral terms becoming singular for $\mathbf{x} \to \mathbf{y}$, see equations ($41a, b$); that is, these integrals must be computed as Cauchy principal values (the related free terms are here intended as incomporated in the symbols $\mathbf{q}_u$ and $\mathbf{q}_t$, respectively.) The left-hand side of equation ($43c$) is the Somigliana formula for the stress $\boldsymbol{\sigma}$, which thus turns out to be related to the displacement $\mathbf{u}$ given by equation ($43a$) through Hooke's law, i.e. $\boldsymbol{\sigma} = \mathbf{D} : (\boldsymbol{\varepsilon} - \boldsymbol{\varepsilon}^c)$, where $\boldsymbol{\varepsilon}$ is the compatible strain field corresponding to this $\mathbf{u}$. Therefore, equation ($43c$) implies that, in the stationarity conditions, the auxiliary strain, $\hat{\boldsymbol{\varepsilon}}$, and the compatible strain, $\boldsymbol{\varepsilon}$, coincide with each other, i.e.

$$\hat{\boldsymbol{\varepsilon}} = \boldsymbol{\varepsilon} \quad \text{in } V, \tag{45}$$

and that thus, by equation (44), the correction strain, $\boldsymbol{\varepsilon}^c$, is such that the stress $\boldsymbol{\sigma}$ is related to the compatible strain, $\boldsymbol{\varepsilon} = \hat{\boldsymbol{\varepsilon}}$, through the (nonlinear) nonlocal constitutive equation. It can be concluded that the stationarity conditions of χ coincide with the equations governing the transformed problem, ($27a - e$), together with the consistency equations ($28a, c$). Therefore, the solution (if any) to problem ($23a - e$) is equivalent to the stationarity conditions of the boundary/domain functional χ and conversely.

7. Symmetric Galerkin BEM Technique

The boundary/domain stationarity principle of Section 6 can be effectively employed to solve the boundary-value problem ($23a - e$) via an iterative procedure of the type local-prediction/nonlocal-correction equivalent to that sketched in Sections 4 and 5.2. For this purpose, let $n = 1, 2, \ldots$ be the iteration sequence and let $\hat{\boldsymbol{\varepsilon}}_{(n-1)} = \boldsymbol{\varepsilon}_{(n-1)}$ and $\boldsymbol{\varepsilon}^c_{(n-1)}$ be known fixed values of $\hat{\boldsymbol{\varepsilon}}$ and $\boldsymbol{\varepsilon}^c$. Then, the functional χ of equation (36), with the fields $\hat{\boldsymbol{\varepsilon}}$ and $\boldsymbol{\varepsilon}^c$ taken fixed at the values mentioned, reduces to χ_1, whereas the first variation $\delta\chi$ of equation (42) simplifies by losing the two integral expressions with the variations $\delta\boldsymbol{\varepsilon}^c$ and $\delta\hat{\boldsymbol{\varepsilon}}$ being both identically vanishing in V. In other words, χ reduces to the functional χ_1 of equation (37), which for $\boldsymbol{\varepsilon}^c = \boldsymbol{\varepsilon}^c_{(n-1)}$ fixed identifies with the functional relating to the boundary min-max principle [20, 21] for a linear local elastic problem with an imposed initial strain $\boldsymbol{\varepsilon}^c_{(n-1)}$ among the external actions. Dropping the constant terms and referring to the n-th iteration, χ_1 can be written as:

$$\chi_{1(n)} = \frac{1}{2}\int_{S_u}\int_{S_u} \mathbf{t} \cdot \mathbf{G}_{uu} \cdot \mathbf{t}\, \mathrm{d}S_u \mathrm{d}S_u - \int_{S_u}\int_{S_t} \mathbf{t} \cdot \mathbf{G}_{ut} \cdot \mathbf{u}\, \mathrm{d}S_t \mathrm{d}S_u$$

$$
\begin{aligned}
&+ \frac{1}{2}\int_{S_t}\int_{S_t} \mathbf{u}\cdot\mathbf{G}_{tt}\cdot\mathbf{u}\,\mathrm{d}S_t\mathrm{d}S_t \\
&+ \int_{S_u}\int_V \mathbf{t}\cdot\mathbf{G}_{u\sigma}\cdot\varepsilon^c_{(n-1)}\,\mathrm{d}V\mathrm{d}S_u - \int_{S_t}\int_V \mathbf{u}\cdot\mathbf{G}_{t\sigma}:\varepsilon^c_{(n-1)}\,\mathrm{d}V\mathrm{d}S_t \\
&+ \int_{S_u}\mathbf{t}\cdot(\mathbf{q}_u-\bar{\mathbf{u}})\,\mathrm{d}S_u - \int_{S_t}\mathbf{u}\cdot(\mathbf{q}_t-\bar{\mathbf{t}})\,\mathrm{d}S_t. \qquad (46)
\end{aligned}
$$

The n-th local prediction for the boundary-value problem $(23a - c)$ can be obtained as the set $\left(\mathbf{t}|_{S_u}, \mathbf{u}|_{S_t}\right)$ which solves the problem:

$$
\min_{(\mathbf{t})}\max_{(\mathbf{u})} \chi_{1(n)} = \chi_{1(n)}[\mathbf{t},\mathbf{u}] \qquad (47)
$$

subject to suitable continuity conditions on $\mathbf{t}$ and $\mathbf{u}$ in L_2. Once the solution to equation (47) has been obtained, say $\mathbf{t}_{(n)}, \mathbf{u}_{(n)}$, by the Somigliana formulae one obtains the related displacement field, i.e.

$$
\begin{aligned}
\mathbf{u}_{(n)}(\mathbf{x}) &= \int_{S_u}\mathbf{G}_{uu}(\mathbf{x},\mathbf{y})\cdot\mathbf{t}_{(n)}(\mathbf{y})\,\mathrm{d}S(\mathbf{y}) - \int_{S_t}\mathbf{G}_{ut}(\mathbf{x},\mathbf{y})\cdot\mathbf{u}_{(n)}(\mathbf{y})\,\mathrm{d}S(\mathbf{y}) \\
&+ \int_V \mathbf{G}_{u\sigma}(\mathbf{x},\mathbf{y}):\varepsilon^c_{(n-1)}(\mathbf{y})\,\mathrm{d}V(\mathbf{y}) + \mathbf{q}_u(\mathbf{x}) \quad \forall\mathbf{x}\in V. \qquad (48)
\end{aligned}
$$

Then, the nonlocal correction consists in obtaining an updated value of ε^c, which is achieved by setting $\hat{\varepsilon}_{(n)} = \varepsilon_{(n)}$ and thus writing for $\varepsilon^c_{(n)}$ the *correction formula*:

$$
\varepsilon^c_{(n)} = \varepsilon_{(n)} - \mathbf{D}^{-1} : \left[\frac{\partial\psi}{\partial\varepsilon_{(n)}} + \mathcal{R}\left(\frac{\partial\psi}{\partial\mathcal{R}(\varepsilon_{(n)})}\right)\right] \quad \text{in } V \qquad (49)
$$

where $\varepsilon_{(n)}$ is derived from $\mathbf{u}_{(n)}$ of equation (48). Whenever $\varepsilon_{(n)}$ is singular, $\varepsilon_{(n)}$ can be suitably regularized for use in equation (49). Note that, for the Eringen model, equation (49) reduces to

$$
\varepsilon^c_{(n)} = \varepsilon_{(n)} - \mathcal{R}(\varepsilon_{(n)}) \quad \text{in } V. \qquad (50)
$$

For numerical computation purposes, it is convenient to discretize the problem by BEs. To this end, following standard procedures of BEM techniques, let the unknown fields of equation (46) be approximated as:

$$
\mathbf{t}(\mathbf{x}) = \mathbf{N}_t(\mathbf{x})\,\mathbf{X}_t \quad \text{in } S_u, \qquad \mathbf{u}(\mathbf{x}) = \mathbf{N}_u(\mathbf{x})\,\mathbf{X}_u \quad \text{in } S_t \qquad (51)
$$

where $\mathbf{N}_t(\mathbf{x})$ and $\mathbf{N}_u(\mathbf{x})$ are matrices of suitable shape functions. Substituting equation (51) in equation (46) gives

$$
\begin{aligned}
\chi^h_{1(n)} &= \frac{1}{2}\mathbf{X}^T_{t(n)}\mathbf{B}_{uu}\mathbf{X}_{t(n)} - \mathbf{X}^T_{t(n)}\mathbf{B}_{ut}\mathbf{X}_{u(n)} + \frac{1}{2}\mathbf{X}^T_{u(n)}\mathbf{B}_{tt}\mathbf{X}_{u(n)} \\
&+ \mathbf{X}^T_{t(n)}\left(\mathbf{L}_u + \mathbf{L}^c_{u(n-1)}\right) - \mathbf{X}^T_{u(n)}\left(\mathbf{L}_t + \mathbf{L}^c_{t(n-1)}\right). \qquad (52)
\end{aligned}
$$

Here the following notation holds:

$$\mathbf{B}_{uu} = \int_{S_u}\int_{S_u} \mathbf{N}_t^T \cdot \mathbf{G}_{uu} \cdot \mathbf{N}_t \, \mathrm{d}S\mathrm{d}S = \mathbf{B}_{uu}^T \tag{53a}$$

$$\mathbf{B}_{ut} = \int_{S_u}\int_{S_t} \mathbf{N}_t^T \cdot \mathbf{G}_{ut} \cdot \mathbf{N}_u \, \mathrm{d}S\mathrm{d}S = \mathbf{B}_{tu}^T \tag{53b}$$

$$\mathbf{B}_{tt} = \int_{S_t}\int_{S_t} \mathbf{N}_u^T \cdot \mathbf{G}_{tt} \cdot \mathbf{N}_u \, \mathrm{d}S\mathrm{d}S = \mathbf{B}_{tt}^T \tag{53c}$$

$$\mathbf{L}_u = \int_{S_u} \mathbf{N}_t^T \cdot (\mathbf{q}_u - \bar{\mathbf{u}}) \, \mathrm{d}S \tag{54a}$$

$$\mathbf{L}_t = \int_{S_t} \mathbf{N}_u^T \cdot (\mathbf{q}_t - \bar{\mathbf{t}}) \, \mathrm{d}S \tag{54b}$$

$$\mathbf{L}_{u(n-1)}^c = \int_{S_u}\int_V \mathbf{N}_t^T \cdot \mathbf{G}_{u\sigma} : \varepsilon_{(n-1)}^c \, \mathrm{d}V\mathrm{d}S \tag{55a}$$

$$\mathbf{L}_{t(n-1)}^c = \int_{S_t}\int_V \mathbf{N}_u^T \cdot \mathbf{G}_{t\sigma} : \varepsilon_{(n-1)}^c \, \mathrm{d}V\mathrm{d}S. \tag{55b}$$

In a more compact form, equation (52) can be written as

$$\chi_{1(n)}^h = \frac{1}{2}\mathbf{X}_{(n)}^T \mathbf{B} \mathbf{X}_{(n)} - \mathbf{X}_{(n)}^T \left(\mathbf{L} + \mathbf{L}_{(n-1)}^c\right), \tag{56}$$

where the following definitions hold:

$$\mathbf{B} = \begin{bmatrix} \mathbf{B}_{uu} & -\mathbf{B}_{ut} \\ -\mathbf{B}_{tu} & \mathbf{B}_{tt} \end{bmatrix}, \qquad \mathbf{X}_{(n)} = \begin{bmatrix} \mathbf{X}_{t(n)} \\ \mathbf{X}_{u(n)} \end{bmatrix} \tag{57a}$$

$$\mathbf{L} = \begin{bmatrix} -\mathbf{L}_u \\ \mathbf{L}_t \end{bmatrix}, \qquad \mathbf{L}_{(n)}^c = \begin{bmatrix} -\mathbf{L}_{u(n)}^c \\ \mathbf{L}_{t(n)}^c \end{bmatrix}. \tag{57b}$$

It can be proved [20, 21] that $\mathbf{B}_{uu}$ is positive definite and $\mathbf{B}_{tt}$ negative definite (the latter condition being valid for $S_u \neq \emptyset$, otherwise $\mathbf{B}_{tt}$ is negative semidefinite).

The discrete local prediction in the $n-$th iteration is given by the min-max solution to equation (56) and thus by the solution to the equation system

$$\mathbf{B}\,\mathbf{X}_{(n)} = \mathbf{L} + \mathbf{L}^c_{(n-1)} \tag{58}$$

Obviously, the solution $\mathbf{X}_{(n)}$ is the discrete counterpart of $\mathbf{t}_{(n)}$, $\mathbf{u}_{(n)}$ for equation (47). The related nonlocal correction can be obtained by equation (49) with $\varepsilon_{(n)}$ computed using a specific Somigliana formula, i.e.

$$\begin{aligned}\varepsilon_{(n)}(\mathbf{x}) &= \left(\int_{S_u} \mathbf{G}_{\varepsilon u}(\mathbf{x},\mathbf{y})\cdot\mathbf{N}_t(\mathbf{y})\mathrm{d}S(\mathbf{y})\right)\mathbf{X}_{t(n)} \\ &- \left(\int_{S_t} \mathbf{G}_{\varepsilon t}(\mathbf{x},\mathbf{y})\cdot\mathbf{N}_u(\mathbf{y})\,\mathrm{d}S(\mathbf{y})\right)\mathbf{X}_{u(n)} \\ &+ \int_V \mathbf{G}_{\varepsilon\sigma}(\mathbf{x},\mathbf{y}) : \varepsilon^c_{(n-1)}(\mathbf{y})\,\mathrm{d}V(\mathbf{y}) + \mathbf{q}_u(\mathbf{x}).\end{aligned} \tag{59}$$

Here $\mathbf{G}_{\varepsilon b}$, $(b = u, t, \sigma)$ is derived from $\mathbf{G}_{ub}$ by application of the gradient operator ∇^s.

8. Nonsymmetric Collocation BEM Technique

The SGBEM has been considered so far. However, since the computer programs for the numerical implementation of SGBEM are less developed than those related to the standard (nonsymmetric collocation) BEM, it is of interest here to show that the iterative procedure described in the preceding sections can be implemented also by making use of the standard BEM technique.

For this purpose, the boundary integral equation (43a) alone is to be applied on the whole boundary $S = \partial V$ by collocation at a discrete set of nodes, say $\mathbf{x}_j$, $(j = 1, 2, ..., m = m_u + m_t)$. Thus, using the interpolation formulas (51), assuming that the vectors $\mathbf{X}_u$ and $\mathbf{X}_t$ collect m_u nodal values of $\mathbf{u}$ in S_u and m_t nodal values of $\mathbf{t}$ in S_t, one can write:

$$\left(\int_{S_u} \mathbf{G}_{uu}(\mathbf{x}_j,\mathbf{y})\cdot\mathbf{N}_t(\mathbf{y})\mathrm{d}S(\mathbf{y})\right)\mathbf{X}_{t(n)} - \left(\int_{S_t} \mathbf{G}_{ut}(\mathbf{x}_j,\mathbf{y})\cdot\mathbf{N}_u(\mathbf{y})\mathrm{d}S(\mathbf{y})\right)\mathbf{X}_{u(n)}$$

$$= \mathbf{U}_j - \int_V \mathbf{G}_{u\sigma}(\mathbf{x}_j,\mathbf{y}) : \varepsilon^c_{(n-1)}(\mathbf{y})\,\mathrm{d}V(\mathbf{y}), \qquad \forall \mathbf{x}_j \in S_u, \tag{60a}$$

$$\left(\int_{S_u} \mathbf{G}_{uu}(\mathbf{x}_j,\mathbf{y})\cdot\mathbf{N}_t(\mathbf{y})\mathrm{d}S(\mathbf{y})\right)\mathbf{X}_{t(n)} - \left(\oint_{S_t} \left[\mathbf{G}_{ut}(\mathbf{x}_j,\mathbf{y}) + \frac{1}{2}\mathbf{I}\right]\cdot\mathbf{N}_u(\mathbf{y})\mathrm{d}S(\mathbf{y})\right)\mathbf{X}_{u(n)}$$

$$= \mathbf{U}_j - \int_V \mathbf{G}_{u\sigma}(\mathbf{x}_j,\mathbf{y}) : \varepsilon^c_{(n-1)}(\mathbf{y})\,\mathrm{d}V(\mathbf{y}), \qquad \forall \mathbf{x}_j \in S_t. \tag{60b}$$

Here, the following notation holds:

$$\mathbf{U}_j = \bar{\mathbf{u}}(\mathbf{x}_j) - \mathbf{q}_u(\mathbf{x}_j) \qquad \text{for } \mathbf{x}_j \in S_u \tag{61a}$$

$$\mathbf{U}_j = -\mathbf{q}_u(\mathbf{x}_j) \qquad \text{for } \mathbf{x}_j \in S_t; \tag{61b}$$

that is, remembering equation (41a),

$$\mathbf{U}_j = -\int_V \mathbf{G}_{uu}(\mathbf{x}_j,\mathbf{y}) \cdot \bar{\mathbf{b}}(\mathbf{y})\mathrm{d}V(\mathbf{y}) - \int_{S_t} \mathbf{G}_{uu}(\mathbf{x}_j,\mathbf{y}) \cdot \bar{\mathbf{t}}(\mathbf{y})\mathrm{d}S(\mathbf{y})$$

$$+ \oint_{S_u} \left[\mathbf{G}_{ut}(\mathbf{x}_j,\mathbf{y}) + \frac{1}{2}\mathbf{I}\right] \cdot \bar{\mathbf{u}}(\mathbf{y})\,\mathrm{d}S(\mathbf{y}) \quad \forall \mathbf{x}_j \in S_u \tag{62a}$$

$$\mathbf{U}_j = -\int_V \mathbf{G}_{uu}(\mathbf{x}_j,\mathbf{y}) \cdot \bar{\mathbf{b}}(\mathbf{y})\mathrm{d}V(\mathbf{y}) - \int_{S_t} \mathbf{G}_{uu}(\mathbf{x}_j,\mathbf{y}) \cdot \bar{\mathbf{t}}(\mathbf{y})\mathrm{d}S(\mathbf{y})$$

$$+ \int_{S_u} \mathbf{G}_{ut}(\mathbf{x}_j,\mathbf{y}) \cdot \bar{\mathbf{u}}(\mathbf{y})\,\mathrm{d}S(\mathbf{y}) \quad \forall \mathbf{x}_j \in S_t. \tag{62b}$$

The symbol $\oint$ indicates Cauchy principal value. Also, a smooth boundary surface has been assumed.

Since the unknown vectors $\mathbf{X}_{u(n)}$ and $\mathbf{X}_{t(n)}$ contain in total $m = m_u + m_t$ entries and correspondingly there are m collocation points, the (nonsymmetric) linear equation system (60a, b) can be solved to obtain $\mathbf{X}_{u(n)}$, $\mathbf{X}_{t(n)}$. These vectors represent the discrete local prediction in the $n-$th iteration. The nonlocal correction can then be obtained using equation (58) and equation (49), or equation (50).

9. Conclusions

We have here described boundary element method (BEM) approaches to continuum mechanics problems with nonlocal hyperelastic materials. Because of nonlinearity or (in the linear case) because the fundamental solutions either do not exist, or if exist are probably unsuitable for computational purposes, these proposed BEM approaches are characterized by the systematic use of classical fundamental solutions, typically the Kelvin solution of linear isotropic (local) elasticity. Two BEM techniques have been employed, that is the symmetric Galerkin (SG) BEM technique and the s-tandard (nonsymmetric collocation) BEM technique. The former technique has been supported by adequate variational formulations.

The solution strategy consists in an iterative procedure of the type local-prediction/nonlocal-correction. The proposed BEM techniques are suitable for application in the local-prediction phase of every iteration. The nonlocal-correction phase consists in updating a (fictitious) initial strain which has the role of injecting the nonlocality features into the material model.

All the developments reported in this paper are novel and no numerical applications are available at the time being. The proposed methods are believed to constitute an effective tool for the advancement of research in a number of engineering problems, including those related to soil-structure interaction.

10. References

1. Kerr, A.D.: On the formal development of elastic foundation models, *Ingenieur-Archiv*, **54** (1984), 455–464.
2. Eringen, A.C., Speziale, C.G., Kim, B.S.: Crack-tip problem in nonlocal elasticity, *J. Mech. Phys. Solids*, **25** (1977), 339–355.
3. Eringen, A.C.: Line crack subject to shear, *Int. Jour. Fracture*, **14** (1978), 367–379.
4. Maier, G., Miccoli, S., Novati, G., Perego, U.: Symmetric Galerkin BEM in plasticity and gradient-plasticity, *Comput. Mech.*, **17** (1995), 115–129.
5. Kröner, E.: Elasticity theory of materials with long range cohesive forces, *Int. Jour. Solids Struct.*, **3** (1967), 731–742.
6. Eringen, A.C., Kim, B.S.: Stress concentration at the tip of a crack, *Mech. Res. Comm.*, **1** (1974), 233–237.
7. Eringen, A.C.: Line crack subjected to anti-plane shear, *Engng. Fracture Mech.*, **12** (1979), 211–219.
8. Rogula, D.: Introduction to nonlocal theory of material media, in D. Rogula (ed.), *Theory of Material Media*, Springer-Verlag, Berlin, 1982, pp. 124–222.
9. Sztyren, M.: On solvable nonlocal boundary-value problems, in D. Rogula (ed.), *Theory of Material Media*, Springer-Verlag, Berlin, 1982, pp. 223–278.
10. Altan, S.B.: Uniqueness of the initial-value problem in nonlocal elastic solids, *Int. Jour. Solids Struct.*, **25** (1989), 1271–1278.
11. Polizzotto, C.: Nonlocal elasticity and related variational principles, *Int. Jour. Solids Struct.*, **38** (2001), 7359–7380.
12. Eringen, A.C.: Theory of nonlocal elasticity and some applications, *Res Mechanica*, **21** (1987), 313–342.
13. Tricomi, F.G.: *Integral Equations*, Dover Publications, New York, 1985.
14. Edelen, D.G.B., Laws, N.: On the thermodynamics of systems with nonlocality, *Arch. Rat. Mech. Anal.*, **43** (1971), 24–35.
15. Washizu, K.: *Variational methods in Elasticity and Plasticity*, Third Edition, Pergamon Press, Oxford and New York, 1982.
16. Polizzotto, C.: An energy approach to the boundary element method. Part I: Elastic solids; Part II: Elastic-plastic solids, *Comput. Meths. Appl. Mech. Engng.*, **69** (1988), 167–184, 263–276.
17. Banerjee, P.K.: *The Boundary Element Methods in Engineering*, McGraw-Hill, London, 1994.
18. Brebbia, C.A., Telles, J.C.F., Wrobel, L.C.: *Boundary Element Techniques*, Springer-Verlag, Berlin, 1984.
19. Bonnet, M.: *Equationes Intégrales et Eléments de Frontière*, CNRS Editions, Eyrolles, Paris, 1995.
20. Polizzotto, C.: A consistent formulation of the BEM within elastoplasticity, in T.A. Cruse (Ed.), *Advanced Boundary Element Methods*, Springer-Verlag, Berlin, 1988, pp. 315–324.
21. Polizzotto, C.: A boundary min-max principle as a tool for BEM formulations, *Engng. Anal. with Boundary Elements*, **8** (1991), 89–93.

22. Maier, G., Polizzotto, C.: A Galerkin approach to BE elastoplastic analysis, *Comput. Meths. Appl. Mech. Engng.* , **60** (1987), 175–194.
23. Bonnet, M., Maier, G., Polizzotto, C.: Symmetric Galerkin boundary element methods, *Appl. Mech. Rev.* , **51** (1998), 669–704.

CHAPTER 7.
BEM FOR CRACK DYNAMICS

M.H. ALIABADI
Department of Engineering
Queen Mary, University of London, E1 4NS, UK

Abstract
In this Chapter the modern application of the boundary element method to crack problems in dynamic fracture mechanics is reviewed. Dual boundary formulations are presented for the time domain, Laplace transform and dual reciprocity methods. The three approaches are applied to mixed mode two-dimensional and three-dimensional crack problems.

Keywords : Boundary element method, fracture, crack propagation, dynamic stress intensity factor

1. Introduction

The aim of dynamic fracture mechanics is to analyze the growth, arrest and branching of moving cracks in structures subjected to dynamic loads. The stress field in the vicinity of the crack is usually characterized by dynamic stress intensity factors (DSIF) which are generally functions of time. Structures with arbitrary shape and time-dependent boundary conditions need to be analyzed by numerical methods. One of the earliest studies of the transient problem can be found in the paper by Baker (1962). Later Achenbach and Nuismer (1971) extended Baker's work to include incident waves of arbitrary stress profile. A review of state of the art techniques in computational dynamic fracture mechanics can be found in Aliabadi (1994) where different modelling approaches such as the Finite Element Method, the Boundary Element Method and the Finite Volume Method are described. The Boundary Element Method (BEM) of analysis is better suited to crack problems than the more established Finite Element Method

W.S. Hall and G. Oliveto (eds.), Boundary Element Methods For Soil-Structure Interaction, 297–314.

because the crack and crack propagation modelling are simpler. Some reviews of boundary element methods for the numerical solution of elastodynamic problems are given by Beskos (1998) and Dominguez and Gallego (1992). Solutions in elastodynamics using the BEM are usually obtained by either the time domain method, Laplace or Fourier transforms or the dual reciprocity method.

The time domain method was used by Nishimura, Guo and Kobayashi (1987) to solve crack problems. The boundary integral equations in that formulation contain hypersingular integrals, which were regularized using integration by parts twice. Constant spatial and linear temporal shape functions were used for the approximations. The method was applied for stationary and growing straight cracks in two-dimensional, and plane cracks in three-dimensional infinite domains. Zhang and Gross (1997) used the two-state conservation integral of elastodynamics, which leads to non-hypersingular traction boundary integral equations. The unknown quantities in that approach are the crack opening displacements and their derivatives. Similar time and space discretizations were used. The method was applied to penny-shaped and square cracks in infinite domains. Hirose (1991) used the formulation based on the traction equation. Piecewise linear temporal functions were used and the displacements of the crack were interpolated using the analytical solution of the static problem. The method was applied for both stationary and growing penny-shaped cracks.

The Laplace transform method was used by Sládek and Sládek (1986) who analyzed a penny-shaped crack in an infinite elastic body under a harmonic and an impact load using the traction equation and the displacement discontinuity method. A polar coordinate system was assumed and a linear variation of displacements along the radius. They used the displacement equation to analyze symmetric problems, which require discretization of a part of the body only, a rectangular plate with edge cracks and a thick walled tube with radial cracks.

Application of the indirect displacement discontinuity method to dynamic crack problems was developed by Wen, Aliabadi and Rooke (1995a, 1996a,1998ab).

Recently Fedelinski, Aliabadi and Rooke (1993, 1994, 1995, 1996ab, 1997), Wen, Aliabadi and Rooke (1998c) and Wen, Aliabadi and Young (1999ac) have developed time domain method, Laplace transform method and dual reciprocity method in dual boundary element analysis for two-dimensional and three–dimensional dynamic fracture mechanics problems respectively. By using the displacement equation and traction equation on both crack surfaces, a non–singular system of equations is obtained for the coincident nodes on the crack surfaces.

In this chapter, dynamic formulations of the dual boundary element method (DBEM) in the time domain, in the Laplace transform domain and using the dual reciprocity technique are reviewed for analysis of two and three–dimensional cracked structures subjected to dynamic loading.

2. Time Domain Method (TDM)

In this section, a formulation of the dual boundary element method for three–dimensional dynamic crack problems is presented in the time domain and the numerical procedure for the boundary integral equations is described.

Consider a body which is not subjected to body forces and which has zero initial displacements and velocities. The displacement at point $\mathbf{x}'$ on the boundary Γ can be written as

$$c_{ij}u_j(\mathbf{x}',t) = \int_0^t \left[\int_\Gamma U_{ij}(\mathbf{x}',\mathbf{x},t-\tau)t_j(\mathbf{x},\tau)d\Gamma(\mathbf{x})\right]d\tau$$

$$-\int_0^t \left[⨍_\Gamma T_{ij}(\mathbf{x}',\mathbf{x},t-\tau)u_j(\mathbf{x},\tau)d\Gamma(\mathbf{x})\right]d\tau,\quad i,j=1,2,3 \qquad (1)$$

where $U_{ij}(\mathbf{x}',\mathbf{x},t-\tau)$ and $T_{ij}(\mathbf{x}',\mathbf{x},t-\tau)$ are displacement and traction fundamental solutions of elastodynamics (see for example Dominguez (1993)); $u_j(\mathbf{x},\tau), t_j(\mathbf{x},\tau)$ are displacements and tractions respectively on the boundary; the coefficient $c_{ij}(\mathbf{x}')$ depends on the geometry at $\mathbf{x}'$ and $⨍$ stands for a Cauchy principal value integral.

If $\mathbf{x}'$ is assumed to be on a smooth boundary, the traction boundary integral equation can be written as:

$$\frac{1}{2}t_j(\mathbf{x}',t) = n_i(\mathbf{x}')\left\{\int_0^t \left[⨍_\Gamma U_{kij}(\mathbf{x}',\mathbf{x},t-\tau)t_k(\mathbf{x},\tau)d\Gamma(\mathbf{x})\right]d\tau\right.$$

$$\left.-\int_0^t \left[⨎_\Gamma T_{kij}(\mathbf{x}',\mathbf{x},t-\tau)u_k(\mathbf{x},\tau)d\Gamma(\mathbf{x})\right]d\tau\right\},\quad i,j,k=1,2,3 \qquad (2)$$

where $n_i(\mathbf{x}')$ is the unit outward normal at the collocation point, $U_{kij}(\mathbf{x}',t,\mathbf{x},\tau)$ and $T_{kij}(\mathbf{x}',t,\mathbf{x},\tau)$ are other fundamental solutions of elastodynamics which contain derivatives of T_{ij} and U_{ij} respectively and the symbol $⨎$ stands for a Hadamard principal value integral.

The boundary Γ is divided into M quadratic elements (continuous, semi-continuous and discontinuous elements) with eight nodes per element, and observation time t is divided into N time steps ($N\Delta t$). For the spatial distributions, quadratic shape functions are used for both $u_j(\mathbf{x}, \tau)$ and $t_j(\mathbf{x}, \tau)$, and the values of displacement and traction at time step $n\Delta t (n = 1, 2, ...N)$ can be written as:

$$t_i(\mathbf{x}, \tau) = \sum_{n=1}^{N} \phi_n(\tau) t_i^n(\mathbf{x})$$

$$u_i(\mathbf{x}, \tau) = \sum_{n=1}^{N} \left(M_1(\tau) u_i^n(\mathbf{x}) + M_2(\tau) u_i^{n-1}(\mathbf{x}) \right) \tag{3}$$

where

$$\phi_n(\tau) = H(\tau - [n-1]\Delta t) - H(\tau - n\Delta t)$$

$M_1(\tau)$ and $M_2(\tau)$ are temporal linear interpolation functions

$$M_1(\tau) = \frac{\tau - (n-1)\Delta t}{\Delta t} \phi_n(\tau); \quad M_2(\tau) = \frac{n\Delta t - \tau}{\Delta t} \phi_n(\tau)$$

$H(\tau)$ stands for the Heaviside step function, and $t_i^n(\mathbf{x})$ and $u_i^n(\mathbf{x})$ are the traction and displacement at time $t = n\Delta t$ at boundary point $\mathbf{x}$. The time functions in the fundamental solution are simple enough to carry out the time integration analytically. Considering the above approximation, the equations at observation time $t (= N\Delta t)$ are

$$c_{ij} u_j^N(\mathbf{x}') = \sum_{n=1}^{N} \int_\Gamma \int_{(n-1)\Delta t}^{n\Delta t} U_{ij}(\mathbf{x}', \mathbf{x}, t - \tau) t_j^n(\mathbf{x}) d\tau d\Gamma$$

$$- \sum_{n=1}^{N} \int_\Gamma \int_{(n-1)\Delta t}^{n\Delta t} T_{ij}(\mathbf{x}', \mathbf{x}, t - \tau) \left(M_1(\tau) u_i^n(\mathbf{x}) + M_2(\tau) u_i^{n-1}(\mathbf{x}) \right) d\tau d\Gamma \tag{4}$$

and

$$\frac{1}{2} t_j^N(\mathbf{x}') = n_i(\mathbf{x}') \sum_{n=1}^{N} \int_\Gamma \int_{(n-1)\Delta t}^{n\Delta t} U_{kij}(\mathbf{x}', \mathbf{x}, t - \tau) t_k^n(\mathbf{x}) d\tau d\Gamma$$

$$- n_i(\mathbf{x}') \sum_{n=1}^{N} \int_\Gamma \int_{(n-1)\Delta t}^{n\Delta t} T_{kij}(\mathbf{x}', \mathbf{x}, t - \tau) \times$$

$$\times \left(M_1(\tau) u_k^n(\mathbf{x}) + M_2(\tau) u_k^{n-1}(\mathbf{x}) \right) d\tau d\Gamma \tag{5}$$

Displacement, traction and position on the element m are represented in terms of quadratic shape functions. For example in three-dimensional problems, we have

$$u_i^m(\mathbf{x}) = \sum_{p=1}^{8} N_p(\xi,\eta) u_i^{nmp} \quad t_i^m(\mathbf{x}) = \sum_{p=1}^{8} N_p(\xi,\eta) t_i^{nmp} \tag{6}$$

and

$$x_i^m = \sum_{p=1}^{8} N_p(\xi,\eta) x_i^{mp} \tag{7}$$

where u_i^{nmp}, t_i^{nmp} are values of displacement and traction at the node p of element m at the time $n\Delta t$, x_i^{mp} denotes the node coordinate. Because of the presence of the $\dot{\delta}(t-\tau-r/c)$ function and $\ddot{\delta}(t-\tau-r/c)$ in fundamental solutions T_{ij} and T_{kij}, the following temporal integrations can be approximated as (See Wen et al (1999a):

$$\int_{(n-1)\Delta t}^{n\Delta t} \dot{\delta}\left(t-\tau-\frac{r}{c}\right) f(\mathbf{x},\tau) d\tau \simeq \frac{f^n(\mathbf{x}) - f^{n-1}(\mathbf{x})}{\Delta t} \phi_n(t) \tag{8}$$

$$\int_{(n-1)\Delta t}^{n\Delta t} \ddot{\delta}\left(t-\tau-\frac{r}{c}\right) f(\mathbf{x},\tau) d\tau \simeq \frac{f^n(\mathbf{x}) - 2f^{n-1}(\mathbf{x}) + f^{n-2}(\mathbf{x})}{(\Delta t)^2} \phi_n(t) \tag{9}$$

The use of these approximations enables the displacement and traction boundary integral equations to be discretized as

$$c_{ij} u_j^N(\mathbf{x}^\beta) = \sum_{n=1}^{N} \sum_{m=1}^{M} \sum_{p=1}^{8} t_j^{nmp} \int_{-1}^{1} \int_{-1}^{1} U_{ij1}^{N-n+1} N_p J^m d\xi d\eta$$
$$- \sum_{n=1}^{N} \sum_{m=1}^{M} \sum_{p=1}^{8} u_j^{nmp} \int_{-1}^{1} \int_{-1}^{1} \left(T_{ij1}^{N-n+1} + T_{ij2}^{N-n}\right) N_p J^m d\xi d\eta \tag{10}$$
$$\beta = 1, 2, ..., B_1$$

and

$$\frac{1}{2} t_j^N(\mathbf{x}^\beta) = n_i \sum_{n=1}^{N} \sum_{m=1}^{M} \sum_{p=1}^{8} t_k^{nmp} \int_{-1}^{1} \int_{-1}^{1} \left(U_{kij1}^{N-n+1} + U_{kij2}^{N-n}\right) N_p J^m d\xi d\eta$$
$$-n_i \sum_{n=1}^{N} \sum_{m=1}^{M} \sum_{p=1}^{8} u_k^{nmp} \int_{-1}^{1} \int_{-1}^{1} \left(T_{kij1}^{N-n+1} + T_{kij2}^{N-n} + T_{kij3}^{N-n-1}\right) N_p J^m d\xi d\eta \tag{11}$$

$$\beta = 1, 2, ..., B_2$$

where $U_{ij\alpha}^{N-n+1}, U_{kij\alpha}^{N-n+1}, T_{ij\alpha}^{N-n+1}$ and $T_{kij\alpha}^{N-n+1}$ can be found in Fedelinski et al(1995) and Wen et al(1999a), B_1 and B_2 are the numbers of collocation points for which the displacement and traction equations are applied $B_1 + B_2 = B$ is the total number of nodes and J^m is the Jacobian. Since the time dependent fundamental solutions have the same form as in the static case as r tends to zero, the integrals with singularities can be evaluated in the same way as static ones.

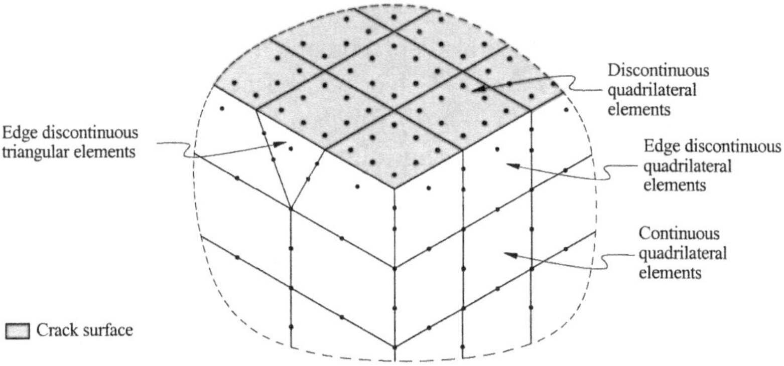

Figure 1. Discretisation strategy.

The displacement and traction boundary integral equations (10) and (11) are discretized with three different types of elements as shown in Figure 1, quadratic continuous elements on the outer boundary except the element at the junction with an edge crack, quadratic discontinuous elements on the crack surfaces and semi–discontinuous elements on the outer boundary at the junction with an edge crack. A set of discretized boundary equations from equations (10) and (11) can be written in matrix form at the time step N as

$$\mathbf{H}^1\mathbf{u}^N - \mathbf{G}^1\mathbf{t}^N = -\sum_{n=1}^{N-1}\left(\mathbf{H}^{n+1}\mathbf{u}^{N-n} - \mathbf{G}^{n+1}\mathbf{t}^{N-n}\right) \tag{12}$$

where $\mathbf{H}^{n+1}$ and $\mathbf{G}^{n+1}$ contain the integrations of fundamental solutions in equations (10) and (11), $\mathbf{u}$ and $\mathbf{t}$ are displacement and traction at the nodes. In equation (12), all terms on the right are known from solution for previous time steps $0 \leq t \leq (N-1)\Delta t$. The values for the first two steps are determined from the initial boundary conditions. Putting the unknowns from the current time step $(N-1)\Delta t \leq t \leq N\Delta t$ on the left, equation (12) can be rearranged as:

$$\mathbf{A}^1\mathbf{x}^N = \mathbf{y}^N - \sum_{n=1}^{N-1}\left(\mathbf{H}^{n+1}\mathbf{u}^{N-n} - \mathbf{G}^{n+1}\mathbf{t}^{N-n}\right) \tag{13}$$

and all the quantities on the right hand side are known. The unknowns of displacement and traction on the boundary at time $t(=N\Delta t)$ can be solved from equation (13). In each time step only the matrices $\mathbf{H}^N$ and $\mathbf{U}^N$ are computed, but all previous matrices $\mathbf{H}^1$ to $\mathbf{H}^{N-1}$ and $\mathbf{G}^1$ to $\mathbf{G}^{N-1}$ must be retained.

3. Laplace Transform Method (LTM)

The Laplace transform method is an efficient way to treat elastodynamic problems. The Laplace transform of a function $f(\mathbf{x},t)$ is defined as

$$\mathcal{L}[f(\mathbf{x},t)] = \tilde{f}(\mathbf{x},s) = \int_0^\infty f(\mathbf{x},t)e^{-st}\mathrm{d}t \tag{14}$$

where s is a Laplace transform parameter. Taking Laplace transforms, the displacement boundary integral equation (1) becomes:

$$c_{ij}(\mathbf{x}')\tilde{u}_j(\mathbf{x}',s) + \int\!\!\!\!\!-_\Gamma \tilde{T}_{ij}(\mathbf{x}',\mathbf{x},s)\tilde{u}_k(\mathbf{x},s)d\Gamma = \int_\Gamma \tilde{U}_{ij}(\mathbf{x}',\mathbf{x},s)\tilde{t}_k(\mathbf{x},s)d\Gamma \tag{15}$$

and the traction equation (2) can be arranged as

$$\frac{1}{2}\tilde{t}_j(\mathbf{x}',s) + n_i(\mathbf{x}')\int\!\!\!\!\!=_\Gamma \tilde{T}_{kij}(\mathbf{x}',\mathbf{x},s)\tilde{u}_j(\mathbf{x},s)d\Gamma =$$

$$n_i(\mathbf{x}')\int\!\!\!\!\!-_\Gamma \tilde{U}_{kij}(\mathbf{x}',\mathbf{x},s)\tilde{t}_j(\mathbf{x},s)d\Gamma \tag{16}$$

where $\tilde{U}_{ij}(\mathbf{x}',\mathbf{x},s)$, $\tilde{T}_{ij}(\mathbf{x}',\mathbf{x},s)$, $\tilde{U}_{kij}(\mathbf{x}',\mathbf{x},s)$ and $\tilde{T}_{kij}(\mathbf{x}',\mathbf{x},s)$ are the Laplace transformed fundamental solutions of elastodynamic fundamental solutions in the time domain in equations (1) and (2).

The displacement and traction on the element m can be approximated as in equations (6) and (7). Then, the displacement and traction boundary integral equations can be discretized as

$$c_{ij}\tilde{u}_j(\mathbf{x}^\beta,s) + \sum_{m=1}^{M}\sum_{p=1}^{8}\tilde{u}_j^{mp}(s)\int_{-1}^{1}\int_{-1}^{1}\tilde{T}_{ij}(\mathbf{x}^\beta,\mathbf{x},s)N_pJ^m d\xi d\eta$$

$$= \sum_{m=1}^{M} \sum_{p=1}^{8} \tilde{t}_j^{mp}(s) \int_{-1}^{1} \int_{-1}^{1} \tilde{U}_{ij}(\mathbf{x}^\beta, \mathbf{x}, s) N_p J^m d\xi d\eta \tag{17}$$

$$\text{for} \quad \beta = 1, 2, ..., B_1$$

and

$$\frac{1}{2}\tilde{t}_j(\mathbf{x}^\beta, s) + n_i(\mathbf{x}^\beta) \sum_{m=1}^{M} \sum_{p=1}^{8} \tilde{u}_k^{mp}(s) \int_{-1}^{1} \int_{-1}^{1} \tilde{T}_{kij}(\mathbf{x}^\beta, \mathbf{x}, s) N_p J^m d\xi d\eta$$

$$= n_i(\mathbf{x}^\beta) \sum_{m=1}^{M} \sum_{p=1}^{8} \tilde{t}_k^{mp}(s) \int_{-1}^{1} \int_{-1}^{1} \tilde{U}_{kij}(\mathbf{x}^\beta, \mathbf{x}, s) N_p J^m d\xi d\eta \tag{18}$$

$$\beta = 1, 2, ..., B_2$$

where the same notations are used as for the time domain. The set of discretized boundary integral equations can be written in matrix form as

$$\tilde{\mathbf{H}}\tilde{\mathbf{u}} = \tilde{\mathbf{G}}\tilde{\mathbf{t}} \tag{19}$$

where the matrices $\tilde{\mathbf{H}}$ and $\tilde{\mathbf{G}}$ depend on integrals of the fundamental solutions in equations (17) and (18). It is clear that when $r = |\mathbf{x}^\beta - \mathbf{x}|$ tends to zero, the fundamental solutions $\tilde{U}_{ij}(\mathbf{x}^\beta, \mathbf{x}, s), \tilde{T}_{ij}(\mathbf{x}^\beta, \mathbf{x}, s)$ and the functions $\tilde{U}_{kij}(\mathbf{x}^\beta, \mathbf{x}, s), \tilde{T}_{kij}(\mathbf{x}^\beta, \mathbf{x}, s)$ have the same form as in the static case, and so an evaluation technique for singular integrals similar to the static case can also be used in the Laplace domain (see Fedelinski et al(1996a) and Wen et al(1998c)).

Putting the unknowns on the left, equation (19) becomes:

$$\tilde{\mathbf{A}}\tilde{\mathbf{x}} = \tilde{\mathbf{y}} \tag{20}$$

The unknown transformed displacements and tractions on the boundary can be obtained from this equation for a particular Laplace parameter s. The time–dependent values of any of the transformed variables must be obtained by an inverse transform. There are many Laplace transformation inversion methods; here, the method put forward by Durbin (1974) is used. For the Heaviside function, this method can be used to obtain accurate inversion results for long durations. The calculation formula used is as follows

$$f(t) \approx 2\frac{e^{at}}{T}\left[-\frac{1}{2}\mathrm{Re}[\tilde{f}(a)]+\right.$$

$$\left. + \sum_{k=0}^{L} \left(\mathrm{Re}[\tilde{f}(a + i\frac{2k\pi}{T})] \cos\frac{2k\pi t}{T} - \mathrm{Im}[\tilde{f}(a + i\frac{2k\pi}{T})] \sin\frac{2k\pi t}{T} \right) \right] \tag{21}$$

where $\tilde{f}(s_k)$ stands for the value in Laplace space at the sample point $s_k = a + 2k\pi i/T, i = \sqrt{-1}$ and the sample points are chosen for $k = 0, 1, ..., L$. Good results have been obtained for $aT = 5$ and $T/t_0 = 20$, where t_0 is unit time. The results of many test examples have shown that the sample number L should be at least 25.

4. Dual Reciprocity Method (DRM)

Considering the acceleration terms in the equilibrium equation as body forces, the boundary integral equations can be expressed in the same way as for the static case:

$$c_{ij}u_j(\mathbf{x}') - \int_\Gamma U_{ij}(\mathbf{x}',\mathbf{x})t_j(\mathbf{x})d\Gamma + \int_\Gamma T_{ij}(\mathbf{x}',\mathbf{x})u_j(\mathbf{x})d\Gamma$$
$$= -\rho \int_\Omega U_{ij}(\mathbf{x}',\mathbf{X})\ddot{u}_j(\mathbf{X},\tau)d\Omega \tag{22}$$

for displacement and

$$\frac{1}{2}t_j(\mathbf{x}') - n_i(\mathbf{x}') \int_\Gamma U_{kij}(\mathbf{x}',\mathbf{x})t_k(\mathbf{x})d\Gamma + n_i(\mathbf{x}') \int_\Gamma T_{kij}(\mathbf{x}',\mathbf{x})u_k(\mathbf{x})d\Gamma$$
$$= -\rho n_i(\mathbf{x}') \int_\Omega U_{kij}(\mathbf{x}',\mathbf{X})\ddot{u}_k(\mathbf{X},\tau)d\Omega \tag{23}$$

for traction on a smooth boundary, where ρ is the mass density; $U_{ij}(\mathbf{x}',\mathbf{x})$, $T_{ij}(\mathbf{x}',\mathbf{x})$, $U_{kij}(\mathbf{x}',\mathbf{x})$ and $T_{kij}(\mathbf{x}',\mathbf{x})$ are fundamental solutions of elastostatics and $\ddot{u}_k$ denotes the acceleration at a domain point $\mathbf{X}$.

In equations (22) and (23), the domain integrals can be transformed into boundary integrals by the dual reciprocity method. The acceleration is approximated as a sum of M coordinate functions $f^n(\mathbf{X})$ multiplied by unknown time–dependent coefficients $\alpha_j^n(t)$:

$$\ddot{u}_m(\mathbf{X},t) = \sum_{n=1}^{M} f^n(\mathbf{X})\alpha_m^n(t), \tag{24}$$

where f^n is a radial basis function chosen to be $1 + r$. The unknowns α_m^n are related to the values of acceleration at M collocation points as

$$\boldsymbol{\alpha} = \mathbf{F}^{-1}\ddot{\mathbf{u}}, \tag{25}$$

where $\mathbf{F}$ is a coefficient matrix from (24). If particular solutions $\hat{u}_{mk}^j$ can be found satisfying following equation

$$G\hat{u}_{mk,ll}^n + \frac{G}{1-2\nu}\hat{u}_{lk,lm}^n = -\delta_{mk}f^n, \tag{26}$$

then the volume integral may be transformed to a boundary integral. Then the displacement boundary integral equation (22) becomes

$$c_{ij}u_j(\mathbf{x}') - \int_\Gamma U_{ij}(\mathbf{x}',\mathbf{x})t_j(\mathbf{x})d\Gamma + \int_\Gamma T_{ij}(\mathbf{x}',\mathbf{x})u_j(\mathbf{x})d\Gamma =$$

$$\rho\sum_{n=1}^{N}\left(c_{ij}\hat{u}^n_{mj}(\mathbf{x}') - \int_\Gamma U_{ij}(\mathbf{x}',\mathbf{x})\hat{t}^n_{mj}(\mathbf{x})d\Gamma + \int_\Gamma T_{ij}(\mathbf{x}',\mathbf{x})\hat{u}^n_{mj}(\mathbf{x})d\Gamma\right)\alpha^n_m \tag{27}$$

and the traction equation (23) becomes

$$t_j(\mathbf{x}') - n_i(\mathbf{x}')\int_\Gamma U_{kij}(\mathbf{x}',\mathbf{x})t_k(\mathbf{x})d\Gamma + n_i(\mathbf{x}')\int_\Gamma T_{kij}(\mathbf{x}',\mathbf{x})u_k(\mathbf{x})d\Gamma$$

$$= \rho\sum_{n=1}^{N}\left(\hat{t}^n_{mk}(\mathbf{x}') - n_i(\mathbf{x}')\int_{\Gamma^e} U_{kij}(\mathbf{x}',\mathbf{x})\hat{t}^n_{mj}(\mathbf{x})d\Gamma\right. \tag{28}$$

$$\left. + n_i(\mathbf{x}')\int_{\Gamma^e} T_{kij}(\mathbf{x}',\mathbf{x})\hat{u}^n_{mk}(\mathbf{x})d\Gamma\right)\alpha^n_m$$

where $\hat{u}^n_{mk}$ and $\hat{t}^n_{mk}$ represent the particular solutions which satisfy equation (26) and are listed in Fedelinski et al(1994) and Wen et al(1999b). Analytically the unknowns of displacement or displacement discontinuity and traction on the boundary can be determined from these two equations. For the collocation point in the domain Ω, equation (22) becomes

$$u_i(\mathbf{X}') - \int_\Gamma U_{ij}(\mathbf{X}',\mathbf{x})t_j(\mathbf{x})d\Gamma + \int_\Gamma T_{ij}(\mathbf{X}',\mathbf{x})u_j(\mathbf{x})d\Gamma =$$

$$\rho\sum_{n=1}^{N}\left(c_{ij}\hat{u}^n_{mj}(\mathbf{X}') - \int_\Gamma U_{ij}(\mathbf{X}',\mathbf{x})\hat{t}^n_{mj}(\mathbf{x})d\Gamma + \int_\Gamma T_{ij}(\mathbf{X}',\mathbf{x})\hat{u}^n_{mj}(\mathbf{x})d\Gamma\right)\alpha^n_m \tag{29}$$

The displacements and the tractions on boundary element n are approximated in terms of quadratic shape functions as

$$u^n_i(\xi,\eta) = \sum_{p=1}^{8} N_p(\xi,\eta)u^{np}_i, \quad \hat{u}^n_{mi}(\xi,\eta) = \sum_{p=1}^{8} N_p(\xi,\eta)\hat{u}^{np}_{mi} \tag{30}$$

$$t^n_i(\xi,\eta) = \sum_{p=1}^{8} N_p(\xi,\eta)t^{np}_i, \quad \hat{t}^n_{mi}(\xi,\eta) = \sum_{p=1}^{8} N_p(\xi,\eta)\hat{t}^{np}_{mi} \tag{31}$$

where $u_i^{np}, \hat{u}_{mi}^{np}(\xi,\eta), t_i^{np}(\xi,\eta)$ and $\hat{t}_{mi}^{np}$ are values of displacement and traction at the node p. Finally the set of discretized boundary integral equations can be written in matrix form as

$$\mathbf{Hu} - \mathbf{Gt} = \rho(\mathbf{H}\hat{\mathbf{U}} - \mathbf{G}\hat{\mathbf{T}})\boldsymbol{\alpha} \tag{32}$$

where $\mathbf{u}$ is the vector of displacement on the boundary Γ or in the domain Ω, $\mathbf{t}$ is the vector of traction on the boundary, matrix $\mathbf{H}$ contains integrals involving static fundamental solutions T_{ij} and T_{kij}, and matrix $\mathbf{G}$ contains integrals involving U_{ij} and U_{kij}. Substitution of equation (25) into the above equation gives

$$\mathbf{Hu} - \mathbf{Gt} - \rho(\mathbf{H}\hat{\mathbf{U}} - \mathbf{G}\hat{\mathbf{T}})\mathbf{F}^{-1}\ddot{\mathbf{u}} = 0 \tag{33}$$

which can be written in the form

$$\mathbf{M}\ddot{\mathbf{u}} + \mathbf{Hu} = \mathbf{Gt} \tag{34}$$

where

$$\mathbf{M} = -\rho(\mathbf{H}\hat{\mathbf{U}} - \mathbf{G}\hat{\mathbf{T}})\mathbf{F}^{-1},$$

$\hat{\mathbf{U}}$ and $\hat{\mathbf{T}}$ are matrices with particular solutions of displacement and traction respectively. Equation (34) can be solved by a direct time integration method for given boundary conditions for $\mathbf{u}$ and $\mathbf{t}$ and initial conditions. Here the Houbolt integration scheme was used; the acceleration is expressed at time step N as

$$\ddot{\mathbf{u}}_{N\Delta t} = \frac{1}{\Delta t^2}[2\mathbf{u}_{N\Delta t} - 5\mathbf{u}_{(N-1)\Delta t} + 4\mathbf{u}_{(N-2)\Delta t} - \mathbf{u}_{(N-3)\Delta t}], \tag{35}$$

and the approximate solution at times $\Delta t, 2\Delta t, 3\Delta t, ...,$ can be calculated successively. The values of the first three steps $\mathbf{u}_0, \mathbf{u}_{-1}$ and $\mathbf{u}_{-2}$ can be evaluated from the boundary initial conditions.

5. Cauchy and Hadamard Principal–Value Integrals

The fundamental solutions $U_{ij}^{N-n+1}, T_{ij}^{N-n+1}, U_{kij}^{N-n+1}$ and T_{kij}^{N-n+1} in equations (10) and (11) in time domain, $\tilde{U}_{ij}, \tilde{T}_{ij}, \tilde{U}_{kij}$ and $\tilde{T}_{kij}$ in equations (17) and (18) in the Laplace transform domain and U_{ij}, T_{ij}, U_{kij} and T_{kij} in the equations (27) and (28) by dual reciprocity method, contain singular terms of the form $O(1/r^n)$ when $r \to 0$, where $n = 2, 3$. The singular terms in the fundamental solutions both in Laplace transform domain and in time domain are of the same form as those for elastostatics. For instance, the fundamental solutions in the Laplace space can be written as

$$\tilde{U}_{ij}(\mathbf{x}',\mathbf{x},s) = U_{ij}(\mathbf{x}',\mathbf{x}) + \tilde{U}^{m}_{ij}(\mathbf{x}',\mathbf{x},s),$$
$$\tilde{T}_{ij}(\mathbf{x}',\mathbf{x},s) = T_{ij}(\mathbf{x}',\mathbf{x}) + \tilde{T}^{m}_{ij}(\mathbf{x}',\mathbf{x},s) \quad (36)$$

and

$$\tilde{U}_{kij}(\mathbf{x}',\mathbf{x},s) = U_{kij}(\mathbf{x}',\mathbf{x}) + \tilde{U}^{m}_{kij}(\mathbf{x}',\mathbf{x},s),$$
$$\tilde{T}_{kij}(\mathbf{x}',\mathbf{x},s) = T_{kij}(\mathbf{x}',\mathbf{x}) + \tilde{T}^{m}_{kij}(\mathbf{x}',\mathbf{x},s) \quad (37)$$

where $\tilde{U}^m, \tilde{T}^m$ contain only weak singularities. Thus the evaluation technique for singular integrals used in static case can also be used for dynamic problems. There are higher singularities in the integrands in the following integrals, which appear in $\mathbf{H}^1$ in equation (12), in $\tilde{\mathbf{H}}$ in (19) and in $\mathbf{H}$ in (32). These are dealt with accurately in papers by Aliabadi and co-workers listed in the reference section.

6. Numerical Examples

6.1. A CENTRAL INCLINED CRACK

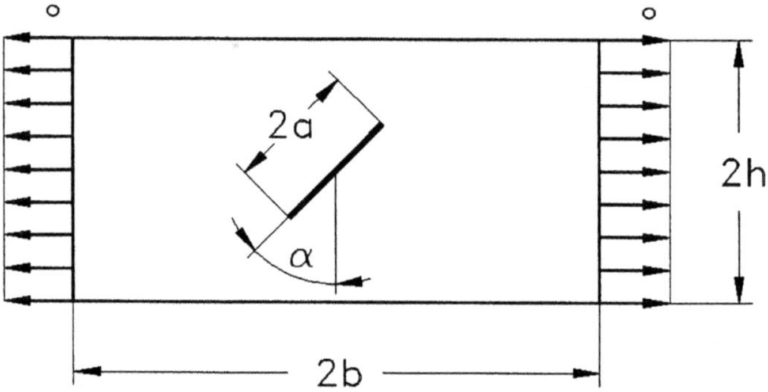

Figure 2. A rectangular plate with an inclined central crack.

A rectangular plate of length $2b = 60mm$ and width $2h = 30mm$ contains a central inclined crack length of $2a = 14.14mm$ slanted at an angle $\alpha = 45^o$, as shown in Figure 2. The material properties are the shear modulus $\mu = 76.92 \times 10^{-9}Pa$; Poisson's ratio $\nu = 0.3$; the density $\rho = 5000kgm^{-3}$. The opposite ends of the plate are loaded by the stress σ_o at $t = 0$. The boundary is divided into 40 boundary elements and 40 additional domain points are used for the DRM. The time step $\Delta\tau = 0.2\mu s$ for the DRM; 25 Laplace parameters are used for the LTM.

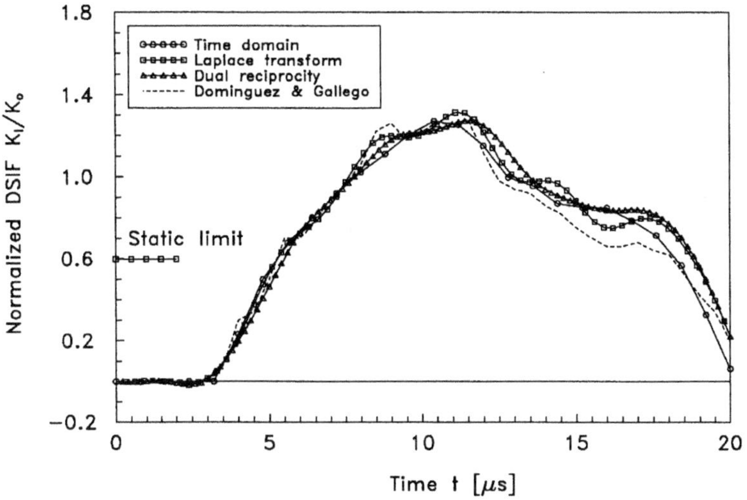

Figure 3. Normalized stress intensity factors K_I/K_o.

The normalized DSIF K_I/K_o and K_{II}/K_o are plotted in Figures 3 and 4, respectively, and compared with those of Dominguez and Gallego(1992), who used the time domain formulation and a subregion technique in the BEM. The solutions obtained by the three methods are similar. The normalized K_I/K_o are bigger and K_{II}/K_o are smaller at later times than those obtained by Dominguez and Gallego(1992).

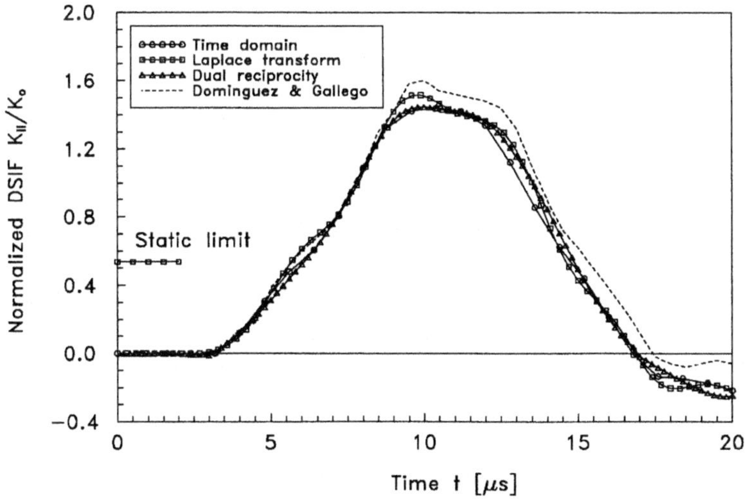

Figure 4. Normalized stress intensity factors K_{II}/K_o.

6.2. ELLIPTICAL CRACK

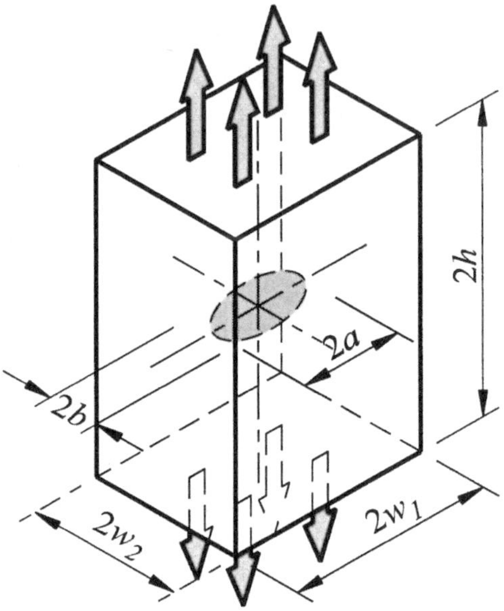

Figure 5. A rectangular bar with an elliptical crack.

Consider a rectangular bar of cross–section $2w_1 \times 2w_2$ (as shown in Figure 5) and height $2h$ containing a centrally located central elliptical crack (pure mode I) subjected to uniform load $\sigma_0 H(t)$ at two ends. The dimensions of the bar are $h = 15\text{cm}, w_1 = 9\text{cm}, w_2 = 6\text{cm}$; the two principal axes of the crack are $a = 3.5\text{cm}$ and $b = 2\text{cm}$ and the material constants are bulk modulus $K = 165\text{GPa}$, shear modulus $G = 77\text{Gpa}$ and density $\rho_0 = 7.9\text{mg/m}^3$. There are 40 quadratic elements on the body and 20 elements on the crack surface. The normalized time increment $\Delta\bar{t}$ is chosen as 0.1 ($\bar{t} = c_1 t/h$) for both the time domain method and the dual reciprocity method and 50 time steps are calculated. For the Laplace transform method, the number of transform parameters L is 25 and unit time $t_0 = c_1 t/h$. The dynamic stress intensity factor at the end of the minor axis is plotted, as $\sqrt{\pi} K_I(t)/\sigma_0$, against time in Figure 6. The numerical results given by Chen and Sih(1977) and Nishioka(1995) are also plotted in this figure for comparison. The stress intensity factors from all methods are close to zero until the dilatational wave from the loaded portion of the boundary arrives at the crack tip.

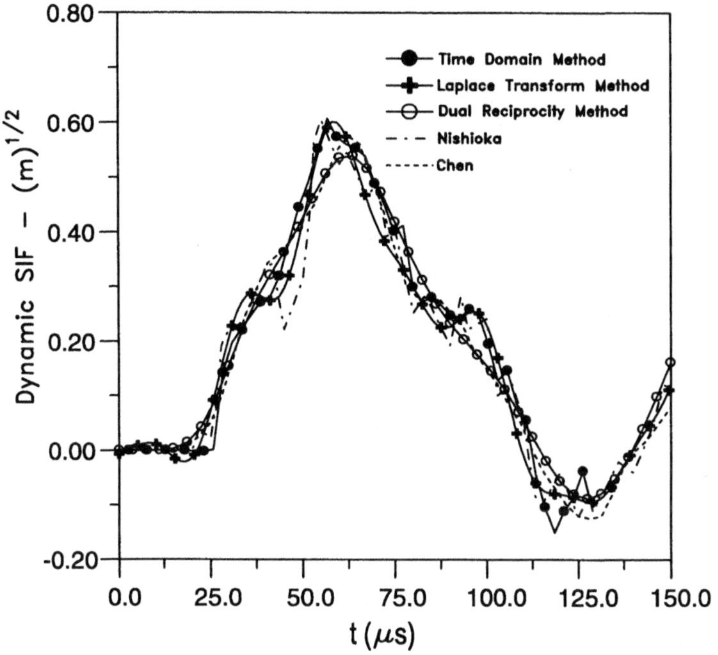

Figure 6. Dynamic stress intensity factors for elliptical crack.

7. Conclusions

Dual boundary element methods for analysing two and three dimensional cracked structures under dynamic loading were considered. The boundary integral equations are given in time domain, Laplace transform domain and by the dual reciprocity technique. For cracked structures, a distinct set of equations is obtained by using the displacement equation on the outer boundary and the traction equation on the crack surface (DBEM). The dynamic stress intensity factors are determined from the crack opening displacement directly in the time domain method and the dual reciprocity method, and with the Durbin inversion technique in the Laplace transform method. The accuracy of the methods have been demonstrated by two examples.

8. References

Achenbach,J.D. and Nuismer,R., Fracture generated by a dilatation wave, *International Journal of Fracture*, **7**, 77-88 (1971).
Aliabadi, M. H., and Rooke, D. P., Numerical Fracture Mechanics, Computational Mechanics Publications and Kluwer Academic Publishers (1991).
Aliabadi,M.H. (Editor), Dynamic Fracture Mechanics, Computational Mechanics Publication, Southampton, (1994).
Baker,B.R., Dynamic stresses created by a moving crack, *Journal of Applied Mechanics*, **39**, 449-458 (1962).
Bathe, K. J. & Wilson, E. L. Numerical Methods in Finite Element Analysis, Prentice-Hall, Inc., New Jersey (1976).
Beskos, D.E., Boundary element methods in dynamic analysis: part II (1986–1996), *Appl. Mech. Rev*, **50**, 149–197 (1997).
Chen, E. P. and Sih, G. C., Transient response of cracks to impact loads, Elastodynamic Crack Problems, Noordhoof (1977).
Dominguez, J. & Gallego, R. Time domain boundary element method for dynamic stress intensity factor computations, *Int. J. Num. Mech. Engng*, **33**, 635–647 (1992).
Dominguez, J., Boundary Elements in Dynamics, Computational Mechanics Publications, Southampton and Boston (1993).
Durbin, F., Numerical inversion of Laplace transforms: an efficient improvement to Dubner and Abate's method, *Comput. J.*, **17**(4), 371–376 (1974).
Fedelinski,P, Aliabadi,M.H. and Rooke,D.P The dual boundary element method in dynamic fracture mechanics, *Engng. Anal.*, **12**, 203-210 (1993).
Fedelinski,P, Aliabadi,M.H. and Rooke,D.P The dual boundary element method: J-integral for dynamic stress intensity factors, *Int.J.Frac.*, **65**, 369-381 (1994).
Fedelinski,P., Aliabadi,M.H. and Rooke,D.P., A single-region time domain BEM for dynamic crack problems. *International Journal of Solids and Structures*, **32**, 3555-3571 (1995).
Fedelinski, P., Aliabadi, M. H. and Rooke, D. P. The Laplace transform DBEM method for mixed-mode dynamic crack analysis, *Computers and Structures*, **59**, 1021-1031 (1996a).
Fedelinski, P., Aliabadi, M. H. and Rooke, D. P., Boundary element formulations for the dynamic analysis of cracked structures, *Engng Anal. Bound. Elem.*, **17** (1), 45–56 (1996b).
Fedelinski,P, Aliabadi,M.H. and D.P.Rooke A time-domain DBEM for rapidly growing cracks. *Int. J. Num.Meth. Eng.*, **40**,1555-1572 (1997).
Hirose, S., Boundary integral equation method for transient analysis of 3–D cavities and inclusions, *Engineering Analysis with Boundary Elements*, **8** (3), 146–154

(1991).

Itou, S., Transient dynamic stresses around a rectangular crack under an impact shear load, *Engng Fract. Mech.*, **39**, 487–492 (1991).

Nishimura, N., Guo, Q.C. and Kobayashi, S., Boundary integral equation methods in elastodynamic crack problems, Boundary Elements IX, 2, Computational Mechanics Publications, Southampton, 279–291 (1987).

Nishioka, T., Recent developments in computational dynamic fracture mechanics, Dynamic Fracture Mechanics, Edited by M. H. Aliabadi, Computational Mechanics Publications, Southampton (1995).

Sladek,J. and Sladek,V. Dynamic stress intensity factors studied by boundary integro-differential equations. *International Journal for Numerical Methods in Engineering*, **23**, 919-928 (1986).

Wen,P.H, Aliabadi,M.H and Rooke,D.P. An indirect boundary element method for three-dimensional dynamic problems, *Eng. Anal.*, **16**, 351-362 (1995a).

Wen,P.H., Aliabadi,M.H. and Rooke,D.P.An approximate analysis of dynamic contact between crack surfaces, *Engng. Anal.* **16**, 41-46 (1995b).

Wen P.H., Aliabadi,M.H. and Rooke,D.P., The influence of elastic waves on dynamic stress intensity factors (three–dimensional problems), *Archive of Applied Mech.*, **66**, 385–394 (1996a).

Wen P.H., Aliabadi,M.H. and Rooke,D.P., The influence of elastic waves on dynamic stress intensity factors (two–dimensional problems), *Archive of Applied Mech.*, **66**, 385–394 (1996b).

Wen,P.H., Aliabadi,M.H. and Rooke,D.P. A variational technique for boundary element analysis of 3D fracture mechanics weight functions: Dynamic, *International Journal for Numerical Methods in Engineering*, **42**, 1425-1439 (1998a).

Wen,P.H., Aliabadi,M.H. and Rooke,D.P. Mixed-mode weight functions in three-dimensional fracture mechnaics: Dynamic, *Engineering Fracture Mechanics,* **59**, 577-587 (1998b).

Wen,P.H., Aliabadi,M.H. and Rooke,D.P. Cracks in three dimensions: a dynamic dual boundary element analysis, *Computer methods in applied mechanics and engineering,* **167**, 139-151 (1998c).

Wen,P.H., Aliabadi,M.H. and Young,A. A time-dependent fromulation of dual boundary element method for 3D dynamic crack problems, *International Journal of Numerical Methods in Engineering,* **45**, 1887-1905 (1999a).

Wen,P.H., Aliabadi,M.H. and Rooke,D.P. A mass-matrix formulation for three-dimensional dynamics fracture mechanics, *Computer Methods in Applied Mechanics and Engineering,* **173**, 365-374 (1999b).

Wen,P.H., Aliabadi,M.H. and Young,A. Dual boundary element methods for three-dimensional dynamic crack problems., *J.Strain Analysis*, **34**, 373-394 (1999c).

Wen,P.H. and Aliabadi,M.H. Approximate dynamic crack frictional contact analysis for 3D structures, *Journal of Chinese Institute of Engineering*, **22**, 785-793 (1999d).
Wen P.H., Dynamic Fracture Mechanics: Displacement Discontinuity Method, Computational Mechanics Publications, Southampton UK and Boston USA (1996).
Zhang, C. H. and Gross, D., A non–hypersingular time–domain BIEM for 3–D transient elastodynamic crack analysis, *Int. J. Num. Meth. Engng*, **36**, 2997–3017 (1993).
textbf

CHAPTER 8. SYMMETRIC GALERKIN BOUNDARY ELEMENT ANALYSIS IN THREE-DIMENSIONAL LINEAR-ELASTIC FRACTURE MECHANICS

A. FRANGI (attilio.frangi@polimi.it) and G. MAIER
Department of Structural Engineering, Politecnico of Milano, P.za L. da Vinci 32, 20133 Milan, Italy

G. NOVATI and R. SPRINGHETTI
Department of Mechanical and Structural Engineering, University of Trento, via Mesiano 77, 38050 Trento, Italy

Abstract. With reference to three-dimensional linear elastic solids susceptible to fracture processes, a symmetric Galerkin boundary element method is developed, based on the regularized version of the weak-form displacement and traction integral equations, and thus involving only kernels $O(1/r)$. When the singularity is active, the numerical evaluation of the double surface integrals is carried out by using special integration schemes which exploit regularizing coordinate transformations. The performance of the method is assessed by solving some example problems involving cracks in unbounded domains and edge cracks in finite bodies.

Key words: variational BEM, fracture mechanics, stress intensity factors, weakly singular integrals

1. Introduction

In the last decade a large number of research contributions have been published concerning the formulation of symmetric Galerkin boundary element methods (SGBEMs) in various contexts. This is well documented by a 1998 review paper (Bonnet *et al.*, 1998) and in more recent literature.

In the linear-elastic fracture mechanics context the SGBEM is very attractive since cracks are modelled as displacement discontinuities on surfaces and, in the absence of body forces, problems can be numerically solved by discretizing only the boundary of the problem domain and the crack surface itself (the extension to the case of multiple cracks being not different).

W.S. Hall and G. Oliveto (eds.), Boundary Element Methods For Soil-Structure Interaction, 315–344.

For two dimensional (2D) linear elasticity problems, a number of implementations of the method into computer codes have been reported in the literature (see, e.g. Balakrishna *et al.*, 1994, Frangi and Novati, 1996, Sirtori *et al.*, 1992), some of which also allow for the presence of cracks. On the contrary few research contributions describing 3D implementations of the Galerkin approach to elastic analyses, with or without cracks, have been published to the authors' knowledge (e.g. Li *et al.*, 1998, Xu and Ortiz, 1993). The double surface integrals involved in the 3D symmetric formulation are not easy to evaluate in the singular cases (that is when the kernel singularity is activated), if generally curved elements are employed. Efficient ad-hoc integration techniques for such singular cases have been developed by applied mathematicians (Andrä and Schnack, 1997 and Sauter and Schwab, 1997) and, although with some delay, are now filtering through to the engineering BE research community.

The present work addresses the application of the SGBEM in the context of 3D linear elastic fracture mechanics using a "regularized" symmetric formulation, which is essentially the same as that expounded in Bonnet (1993), Nishimura and Kobayashi (1989), Frangi (1998); a first fairly general implementation of this approach is documented by Li *et al.* (1998) where, however, few details are given on the integration techniques adopted for the singular cases. On the contrary here the focus is set on the development of efficient algorithms for the crucial singular double surface integrals, according to the new schemes introduced by Andrä and Schnack (1997) and Sauter and Schwab (1997). Some example problems are solved and the relevant stress intensity factors are evaluated, in order to assess and evidence the accuracy of the proposed approach.

2. Formulation

Let Ω denote the volume occupied by a generic body with boundary S in the Cartesian reference system (x_1, x_2, x_3), subject to tractions $p_i(\mathbf{x}) = \bar{p}_i(\mathbf{x})$ given on S_p, and to displacements $u_i(\mathbf{x}) = \bar{u}_i(\mathbf{x})$ enforced on S_u, S_p and S_u being complementary parts of S. Let surface S_c denote a crack inside Ω, conceived as a locus of displacement discontinuity $w(\mathbf{x}) = u(\mathbf{x}^+) - u(\mathbf{x}^-)$, with $\mathbf{x}^+ \in S_c^+$ and $\mathbf{x}^- \in S_c^-$, S_c^+ and S_c^- being the (say upper and lower) faces of the crack. The positive orientation of S_c is associated with the normal unit vector to S_c pointing from S_c^+ to S_c^-. Equal and opposite tractions can be applied to the crack surfaces: $p_i(\mathbf{x}) \equiv p_i(\mathbf{x}^-) = -p_i(\mathbf{x}^+)$ on S_c.

The two-point Kelvin kernel $G_{ik}^{uu}(\mathbf{x} - \tilde{\mathbf{x}})$ expresses the displacement at $\mathbf{x}$ in the kth direction due to a concentrated force acting at $\tilde{\mathbf{x}}$ in the ith direction:

$$G_{ik}^{uu}(\mathbf{x}-\tilde{\mathbf{x}}) = \frac{1}{16\pi\mu(1-\nu)}\frac{1}{r}\left[(3-4\nu)\delta_{ik} + r_{,i}r_{,k}\right] \tag{1}$$

Let kernel $G_{ikj}^{u\sigma}(\mathbf{x}-\tilde{\mathbf{x}})$ denote the kj component of stresses at $\mathbf{x}$ due to the same source, obtained by differentiation of G_{ik}^{uu} and the application of Hooke's law, C_{kjmn} being the elastic stiffness tensor:

$$G_{ikj}^{u\sigma}(\mathbf{x}-\tilde{\mathbf{x}}) = C_{kjmn}G_{im,n}^{uu}(\mathbf{x}-\tilde{\mathbf{x}}). \tag{2}$$

Hence Somigliana equation for a point $\mathbf{x} \in \Omega$ may be written as:

$$u_i(\tilde{\mathbf{x}}) = \int_S \left[G_{ik}^{uu}(\mathbf{x}-\tilde{\mathbf{x}})p_k(\mathbf{x}) - G_{ikj}^{u\sigma}(\mathbf{x}-\tilde{\mathbf{x}})n_j u_k(\mathbf{x})\right] \mathrm{d}S_x + \int_{S_c} G_{ikj}^{u\sigma}(\mathbf{x}-\tilde{\mathbf{x}})n_j w_k(\mathbf{x})\,\mathrm{d}S_x \tag{3}$$

where the unit vector $\mathbf{n}$ defines the outward normal to S at point $\mathbf{x}$.

A second integral equation (traction equation), which turns out to be essential in fracture mechanics problems, can then be obtained for a point $\tilde{\mathbf{x}} \in \Omega$ by differentiation of equation (3):

$$p_i(\tilde{\mathbf{x}}) = \int_S \left[G_{ijk}^{\sigma u}(\mathbf{x}-\tilde{\mathbf{x}})\tilde{n}_j(\tilde{\mathbf{x}})p_k(\mathbf{x}) - G_{ijkp}^{\sigma\sigma}(\mathbf{x}-\tilde{\mathbf{x}})\tilde{n}_j(\tilde{\mathbf{x}})n_p(\mathbf{x})u_k(\mathbf{x})\right] \mathrm{d}S_x + \int_{S_c} G_{ijkp}^{\sigma\sigma}(\mathbf{x}-\tilde{\mathbf{x}})\tilde{n}_j(\tilde{\mathbf{x}})n_p(\mathbf{x})w_k(\mathbf{x})\,\mathrm{d}S_x \tag{4}$$

where

$$G_{ijk}^{\sigma u} = C_{ijab}G_{ak,\tilde{b}}^{uu} \qquad G_{ijkp}^{\sigma\sigma} = C_{ijab}C_{kpcd}G_{ac,b\tilde{d}}^{uu} \tag{5}$$

are the so-called double layer kernels. Unit vector $\tilde{\mathbf{n}}$ defines a reference normal associated with $\tilde{\mathbf{x}}$, while symbol $(\cdot)_{,\tilde{\imath}}$ denotes differentiation with respect to $\tilde{x}_i$.

Alternatively, equation (4) might directly be generated from Betti theorem. In fact $G_{ijk}^{\sigma u}\tilde{n}_j$ and $G_{ijkp}^{\sigma\sigma}\tilde{n}_j n_p$ can be interpreted as the kth components of displacement and traction (relevant to $\mathbf{n}$) in $\mathbf{x}$, respectively, due to a concentrated displacement discontinuity acting at $\tilde{\mathbf{x}}$ in the ith direction across a surface element with outward unit normal vector $\tilde{\mathbf{n}}$ (see Maier *et al.*, 1992).

In customary collocation approaches equations (3) and (4) are enforced pointwise in given points on the boundary with $\tilde{\mathbf{n}}$ in equation (4) being set equal to the actual outward normal in the collocation boundary point. In view of the singular nature of kernels for $\tilde{\mathbf{x}} \to \mathbf{x}$, an infinitesimal domain

Ω_ε (of linear dimension ε) is often excluded from Ω and the limit as $\varepsilon \to 0$ is analysed.

A totally different approach is followed in the symmetric Galerkin method. A detailed explanation of the procedure can be found e.g. in Bonnet *et al.* (1998) and Frangi (1998) and is only briefly summarized hereafter. Let us introduce a surface $\tilde{S}$ representing a fictitious contour *internal* to Ω. We assume the existence of a one to one correspondence between points $\mathbf{x} \in S$ and $\tilde{\mathbf{x}} \in \tilde{S}$: $\tilde{\mathbf{x}} = \boldsymbol{\mathcal{X}}(\mathbf{x}, h)$, where h is a parameter such that $\mathbf{x} = \boldsymbol{\mathcal{X}}(\mathbf{x}, 0)$ and, hence, the two surfaces coincide for $h = 0$. In particular $\tilde{S}$ will consist of portions $\tilde{S}_u$, $\tilde{S}_p$ and, in the presence of an internal crack, also of $\tilde{S}_c^+, \tilde{S}_c^-$ mapped by one-to-one correspondences onto the respective portions of S and of S_c.

The procedure basically consists of two distinct steps: first, equations (3) and (4) are enforced in a weak sense on the auxiliary contour $\tilde{S}$ *distinct* from S (that is with $h \neq 0$) and an analytical regularization procedure is performed via integration by parts removing all higher-order singularities. Secondly, the limit $\tilde{S} \to S$ ($h \to 0$) is performed and the discretization procedure is initiated. Therefore the definition of an auxiliary surface $\tilde{S}$ separated from S is only an artifice which proves useful to the rather difficult purpose of guaranteeing a firm mathematical and computational basis for the evaluation of the singular double integrals involved. However, $\tilde{S}$ does not play any role in the final implementation of the method, since for $h \to 0$, $\tilde{S} \equiv S$.

More specifically, equation (3) is enforced on $\tilde{S}_u$ using as test function the static field $\tilde{p}_i(\tilde{\mathbf{x}})$, while equation (4) is enforced both on $\tilde{S}_p$ and on $\tilde{S}_c$, using as test function the kinematic field $\tilde{u}_i$. By applying the regularization procedure detailed by Frangi (1998) and Bonnet *et al.* (1998), and taking the limit $\tilde{S} \to S$ (hence $h \to 0$), the variational equations listed in the following paragraphs are obtained.

Regularized weak form of the displacement integral equation on S_u.

$$
\begin{aligned}
\frac{1}{2}\int_{S_u} u_i(\mathbf{x})\tilde{p}_i(\mathbf{x})\,\mathrm{d}S_x &= \int_{S_u}\int_S \tilde{p}_i(\tilde{\mathbf{x}})G_{ik}^{uu}(\mathbf{x}-\tilde{\mathbf{x}})p_k(\mathbf{x})\,\mathrm{d}S_x\,\mathrm{d}S_{\tilde{x}} \\
&+ \int_{S_u}\int_S \tilde{p}_i(\tilde{\mathbf{x}})\Big[\frac{1}{4\pi}\frac{1}{r^2}r_{,i}\,n_i(\mathbf{x})u_i(\mathbf{x}) + G_{ikj}^{u\varphi}(\mathbf{x}-\tilde{\mathbf{x}})R_j[u_k](\mathbf{x})\Big]\,\mathrm{d}S_x\,\mathrm{d}S_{\tilde{x}} \\
&- \int_{S_u}\int_{S_c} \tilde{p}_i(\tilde{\mathbf{x}})\Big[\frac{1}{4\pi}\frac{1}{r^2}r_{,i}\,n_i(\mathbf{x})w_i(\mathbf{x}) + G_{ikj}^{u\varphi}(\mathbf{x}-\tilde{\mathbf{x}})R_j[w_k](\mathbf{x})\Big]\,\mathrm{d}S_x\,\mathrm{d}S_{\tilde{x}}
\end{aligned}
\tag{6}
$$

where

$$
G_{ikj}^{u\varphi}(\mathbf{x}-\tilde{\mathbf{x}}) = \frac{1}{8\pi(1-\nu)r}\left[(1-2\nu)\,e_{ikj} - e_{ipj}r_{,p}r_{,k}\right], \tag{7}
$$

In equation (6) R_j denotes the surface rotor operator defined in Section 7. It should be stressed that both G^{uu} and the auxiliary kernel $G^{u\varphi}$ are weakly singular; in fact the additional terms in equation (6) (which coincide with the double layer kernel for potential problems) are weakly singular as well since

$$\frac{1}{r^2} r_{,i}\, n_i \,\mathrm{d}S_x = \mathrm{d}\Theta$$

represents the differential solid angle under which $\mathrm{d}S_x$ is seen from $\tilde{\mathbf{x}}$.

Regularized weak form of the traction integral equation on S_p.

$$\frac{1}{2}\int_{S_p} p_i(\mathbf{x})\tilde{u}_i(\mathbf{x})\,\mathrm{d}S_x = \int_{S_p}\int_{S}\Big[-\frac{1}{4\pi}\tilde{u}_i(\tilde{\mathbf{x}})\frac{1}{r^2}r_{,\tilde{p}}\,\tilde{n}_p(\tilde{\mathbf{x}})p_i(\mathbf{x}) \tag{8}$$

$$-\tilde{R}_j[\tilde{u}_i](\tilde{\mathbf{x}})G^{\varphi u}_{ijk}(\mathbf{x}-\tilde{\mathbf{x}})p_k(\mathbf{x}) - \tilde{R}_j[\tilde{u}_i](\tilde{\mathbf{x}})G^{\varphi\varphi}_{ijkq}(\mathbf{x}-\tilde{\mathbf{x}})R_q[u_k](\mathbf{x})\Big]\,\mathrm{d}S_x\,\mathrm{d}S_{\tilde{x}}$$

$$+\int_{S_p}\int_{S_c}\tilde{R}_j[\tilde{u}_i](\tilde{\mathbf{x}})G^{\varphi\varphi}_{ijkq}(\mathbf{x}-\tilde{\mathbf{x}})R_q[w_k](\mathbf{x})\,\mathrm{d}S_x\,\mathrm{d}S_{\tilde{x}}$$

where:

$$G^{\varphi u}_{ijk} = \frac{1}{8\pi(1-\nu)r}\left[(1-2\nu)\,e_{kij} - e_{kpj}r_{,p}r_{,i}\right] \tag{9}$$

$$G^{\varphi\varphi}_{ijkq} = -\frac{\mu}{8\pi}\frac{1}{r}\Big(\delta_{eg} - r_{,e}\,r_{,g}\Big)e_{iep}e_{jgr}\Big[\frac{2\nu}{1-\nu}\delta_{pk}\delta_{rq} + \delta_{pr}\delta_{kq} + \delta_{pq}\delta_{kr}\Big] \tag{10}$$

It is worth noting in equation (8) that:

$$\frac{1}{r^2}r_{,\tilde{i}}\,\tilde{n}_i\,\mathrm{d}S_{\tilde{x}} = \mathrm{d}\tilde{\Theta}$$

represents the differential solid angle under which $\mathrm{d}S_{\tilde{x}}$ is seen from $\mathbf{x}$. Hence, all the kernels in equation (8) are weakly singular.

Regularized weak form for the traction integral equation on S_c. Let us define the auxiliary displacement discontinuity field $\tilde{w}_i = \tilde{u}_i^+ - \tilde{u}_i^-$ which naturally follows from the $\tilde{u}_i$ field after the limit process $h \to 0$ enforces $\tilde{S}^+$ to coincide with $\tilde{S}^-$.

$$\int_{S_c} p_i(\mathbf{x})\tilde{w}_i(\mathbf{x})\,\mathrm{d}S_x = \int_{S_c}\int_{S}\Big[-\frac{1}{4\pi}\tilde{w}_i(\tilde{\mathbf{x}})\frac{1}{r^2}r_{,\tilde{p}}\,\tilde{n}_p(\tilde{\mathbf{x}})p_i(\mathbf{x})$$

$$-\tilde{R}_j[\tilde{w}_i](\tilde{\mathbf{x}})G^{\varphi u}_{ijk}(\mathbf{x}-\tilde{\mathbf{x}})p_k(\mathbf{x}) - \tilde{R}_j[\tilde{w}_i](\tilde{\mathbf{x}})G^{\varphi\varphi}_{ijkq}(\mathbf{x}-\tilde{\mathbf{x}})R_q[u_k](\mathbf{x})\Big]\,\mathrm{d}S_x\,\mathrm{d}S_{\tilde{x}}$$

$$+\int_{S_c}\int_{S_c}\tilde{R}_j[\tilde{w}_i](\tilde{\mathbf{x}})G^{\varphi\varphi}_{ijkq}(\mathbf{x}-\tilde{\mathbf{x}})R_q[w_k](\mathbf{x})\,\mathrm{d}S_x\,\mathrm{d}S_{\tilde{x}} \tag{11}$$

Continuity requirements. Weak continuity requirements must be enforced on u_k, w_k, $\tilde{u}_k, \tilde{w}_k$ to guarantee the validity of equations (6)-(11), since they must belong to the class of $C_0^0(S)$ continuous functions. The class $C_0^0(S)$ is defined as follows: $C_0^0(S) \equiv C^0(S)$ if S is closed, and $C_0^0(S) = \{f(\mathbf{x}) \in C^0(S) : f|_{\partial S} = 0\}$ if S is open (e.g. admissible kinematic fields for cracks inside bodies are assumed to be in $C_0^0(S)$). On the contrary no special constraints are set on the static fields p_k and $\tilde{p}_k$.

Discretization. By introducing the set of data into the system of the previous equations, a self-adjoint bilinear form is obtained (see Bonnet *et al.*, 1998). At this stage the boundary surface S and the crack S_c are discretized into boundary elements and symmetry is preserved also in the discrete formulation, if the auxiliary and real fields are interpolated over the BEs according to a Galerkin scheme. Further details relevant to the discretization phase and proofs of symmetry properties can be found in Bonnet *et al.* (1998).

3. Numerical Evaluation of Weakly Singular Integrals

Let us assume that the surface S has been partitioned into 9-noded quadrilateral and/or 6-noded triangular isoparametric elements; f and $\tilde{f}$ are given functions of $\mathbf{x}$ and $\tilde{\mathbf{x}}$, respectively, and $B(\mathbf{x}-\tilde{\mathbf{x}})$ is a generic weakly singular kernel. Our purpose is to compute

$$\mathcal{I}_{mn} = \int_{S_m}\int_{S_n} \tilde{f}(\tilde{\mathbf{x}})B(\mathbf{x}-\tilde{\mathbf{x}})f(\mathbf{x})\,\mathrm{d}S_x\,\mathrm{d}S_{\tilde{x}} = \int_{S_m}\int_{S_n} \mathcal{B}(f,\tilde{f},\mathbf{x},\tilde{\mathbf{x}})\,\mathrm{d}S_x\,\mathrm{d}S_{\tilde{x}} \tag{12}$$

where S_m ($\tilde{\mathbf{x}} \in S_m$) and S_n ($\mathbf{x} \in S_n$) represent a generic element pair. Intrinsic parameters η_1, η_2 and $\tilde{\eta}_1, \tilde{\eta}_2$ are introduced on the parent (master) elements, such that $\tilde{\mathbf{x}} = \tilde{\boldsymbol{\mathcal{X}}}(\tilde{\eta}_1, \tilde{\eta}_2)$ on the physical element S_m, and $\mathbf{x} = \boldsymbol{\mathcal{X}}(\eta_1, \eta_2)$ on S_n.

For example, if S_m and S_n are quadrilateral source and field elements, equation (12) becomes:

$$\mathcal{I}_{mn} = \int_0^1\int_0^1\int_0^1\int_0^1 \tilde{\mathcal{B}}[\tilde{f}(\tilde{\boldsymbol{\eta}}), f(\boldsymbol{\eta}), \tilde{\mathbf{x}}(\tilde{\boldsymbol{\eta}}), \mathbf{x}(\boldsymbol{\eta})]\,\mathrm{d}\eta_1\,\mathrm{d}\eta_2\,\mathrm{d}\tilde{\eta}_1\,\mathrm{d}\tilde{\eta}_2 \tag{13}$$

where $\tilde{\mathcal{B}}$ includes also the Jacobians of the transformations. The evaluation of such double surface integrals represents a crucial aspect of the method and is computationally expensive. In this paper the approach described by Erichsen and Sauter (1998) and Sauter and Schwab (1997) is adopted.

Four different situations must be accounted for, in general, according to whether S_m and S_n are: (i) coincident elements, (ii) adjacent elements sharing one edge, (iii) adjacent elements sharing one vertex; (iv) distinct elements (see Figure 1 for the case of quadrilateral-quadrilateral elements).

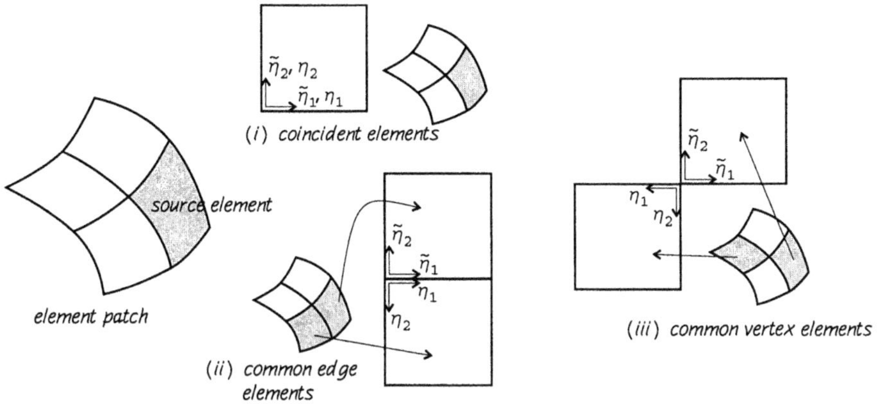

Figure 1. Singular cases for quadrilateral-quadrilateral element-element integration

In the case of distinct elements, standard product Gauss formulae are employed choosing an appropriate number of Gauss points. For the first three cases the procedure can be outlined as follows. (i) The domain of integration is expressed, via suitable coordinate transformations, as the sum of "pyramidal" shaped subdomains in which the singularity is concentrated at one vertex. (ii) For each subdomain a regularizing variable transformation (involving Duffy generalized coordinates) makes the integrand analytic, introducing a Jacobian which cancels the singularity in the kernel. With reference to the case of quadrilateral elements, the following sections give the final ready-to-implement expressions of the regular integrals, to which original integrals of type (12) turn out to reduce.

3.1. COINCIDENT ELEMENTS

A somewhat unorthodox notation will be employed for parameter domains (treated as functions of suitable Cartesian coordinates).

The symbol $\mathcal{D}(\tilde{\eta}_1, \tilde{\eta}_2, \eta_1, \eta_2)$ denotes a four-dimensional polyhedron collecting all the points $\{\tilde{\eta}_1, \tilde{\eta}_2, \eta_1, \eta_2\}$ spanned by the multivariate integral in equation (12).

Let us introduce the relative variables $u_1 \equiv \eta_1 - \tilde{\eta}_1$, $u_2 \equiv \eta_2 - \tilde{\eta}_2$. The singularity in equation (12) is activated whenever $u_1 = u_2 = 0$, while the integrand is regular with respect to $\tilde{\eta}_1, \tilde{\eta}_2$.

The procedure can be outlined as follows. (i) The domain $\mathcal{D}$ is expressed in terms of $\tilde{\eta}_1, \tilde{\eta}_2, u_1, u_2$ and algebraic manipulations are performed in order to make the inequalities concerning u_1, u_2 the outermost ones (so exchanging the integration order) and to provide a partition of $\mathcal{D}$ into subdomains sharing the point $u_1 = u_2 = 0$ as common vertex. (ii) For each subdomain a regularizing variable transformation (involving Duffy generalized coordinates) makes the integrand analytic by introducing a Jacobian which cancels the singularity in the kernel.

Domain partition.

$$\mathcal{D}(\tilde{\eta}_1, \tilde{\eta}_2, \eta_1, \eta_2) \equiv \begin{cases} 0 \leq \tilde{\eta}_1 \leq 1 \\ 0 \leq \tilde{\eta}_2 \leq 1 \\ 0 \leq \eta_1 \leq 1 \\ 0 \leq \eta_2 \leq 1 \end{cases} = \begin{cases} 0 \leq \tilde{\eta}_1 \leq 1 \\ -\tilde{\eta}_1 \leq u_1 \leq 1 - \tilde{\eta}_1 \\ 0 \leq \tilde{\eta}_2 \leq 1 \\ 0 \leq \eta_2 \leq 1 \end{cases}$$

$$\equiv \begin{cases} 0 \leq u_1 \leq 1 \\ 0 \leq \tilde{\eta}_1 \leq 1 - u_1 \\ 0 \leq \tilde{\eta}_2 \leq 1 \\ 0 \leq \eta_2 \leq 1 \end{cases} \cup \begin{cases} 0 \leq -u_1 \leq 1 \\ -u_1 \leq \tilde{\eta}_1 \leq 1 \\ 0 \leq \tilde{\eta}_2 \leq 1 \\ 0 \leq \eta_2 \leq 1 \end{cases} \equiv \mathcal{D}_1^c \cup \mathcal{D}_2^c \quad (14)$$

where Transformation C of Section 8 has been exploited setting $s = \tilde{\eta}_1, t = u_1$. In Section 9 it is shown that $\mathcal{D}_2^c(\tilde{\eta}_1, \tilde{\eta}_2, \eta_1, \eta_2) = \mathcal{D}_1^c(\eta_1, \tilde{\eta}_2, \tilde{\eta}_1, \eta_2)$, that is $\mathcal{D}_2$ coincides with $\mathcal{D}_1$ provided that $\tilde{\eta}_1$ is exchanged with η_1 (and, hence, u_1 with $-u_1$). The above statement of equivalence must be understood as follows: let us imagine in the four-dimensional space two reference systems (with superscripts a and b, respectively) sharing the same origin and oriented such that the $\tilde{\eta}_1^a$ axis coincides with the η_1^b axis, η_1^a with $\tilde{\eta}_1^b$, $\tilde{\eta}_2^a$ with $\tilde{\eta}_2^b$, η_2^a with η_2^b and so on. Hence, domain $\mathcal{D}_1^c$ (thought of as plotted in the reference system a) coincides with $\mathcal{D}_2^c$ (plotted in the reference system b). Only $\mathcal{D}_1^c$ will be considered hereafter and the conclusions immediately extended to $\mathcal{D}_2^c$.

Domain partition. Once expressed in terms of $u_2, \tilde{\eta}_2$, the domain $\mathcal{D}_1^c$ is further partitioned into $\mathcal{D}_{11}^c$ and $\mathcal{D}_{12}^c$ exploiting once again Transformation C in Section 8:

$$\mathcal{D}_1^c = \begin{cases} 0 \leq u_1 \leq 1 \\ 0 \leq \tilde{\eta}_1 \leq 1 - u_1 \\ 0 \leq \tilde{\eta}_2 \leq 1 \\ -\tilde{\eta}_2 \leq u_2 \leq 1 - \tilde{\eta}_2 \end{cases} \quad (15)$$

$$= \begin{cases} 0 \leq u_1 \leq 1 \\ 0 \leq \tilde{\eta}_1 \leq 1 - u_1 \\ 0 \leq u_2 \leq 1 \\ 0 \leq \tilde{\eta}_2 \leq 1 - u_2 \end{cases} \cup \begin{cases} 0 \leq u_1 \leq 1 \\ 0 \leq \tilde{\eta}_1 \leq 1 - u_1 \\ 0 \leq -u_2 \leq 1 \\ -u_2 \leq \tilde{\eta}_2 \leq 1 \end{cases} = \mathcal{D}_{11}^c \cup \mathcal{D}_{12}^c$$

Also in this case it can be verified that $\mathcal{D}_{12}^c(\tilde{\eta}_1, \tilde{\eta}_2, \eta_1, \eta_2) = \mathcal{D}_{11}^c(\tilde{\eta}_1, \eta_2, \eta_1, \tilde{\eta}_2)$, namely the two domains formally coincide if $\tilde{\eta}_2$ and η_2 are exchanged.

Let us focus on $\mathcal{D}^c_{11}(\tilde\eta_1, \tilde\eta_2, \eta_1, \eta_2)$. The singular point $u_1 = u_2 = 0$ is a vertex for the square $0 \le u_1, u_2 \le 1$.

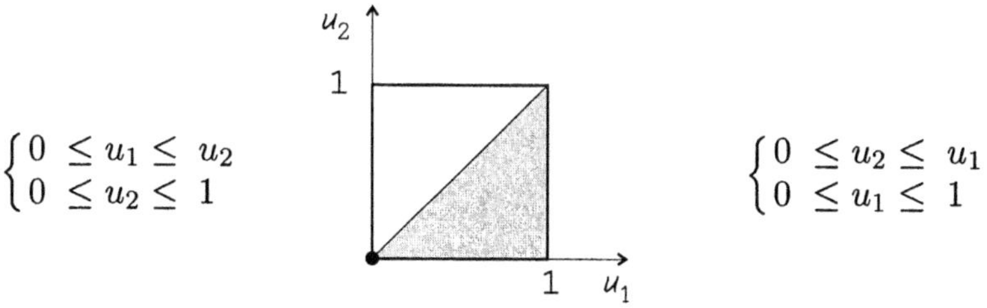

Figure 2. Partition of the square

The square is partitioned into two triangles separated by the diagonal $u_1 = u_2$ as shown in Figure 2:

$$\mathcal{D}^c_{11} = \begin{cases} 0 & \le u_1 \le 1 \\ 0 & \le u_2 \le u_1 \\ 0 & \le \tilde\eta_1 \le 1 - u_1 \\ 0 & \le \tilde\eta_2 \le 1 - u_2 \end{cases} \cup \begin{cases} 0 & \le u_1 \le 1 \\ u_1 & \le u_2 \le 1 \\ 0 & \le \tilde\eta_1 \le 1 - u_1 \\ 0 & \le \tilde\eta_2 \le 1 - u_2 \end{cases} \equiv \mathcal{D}^c_{111} \cup \mathcal{D}^c_{112} \quad (16)$$

Subdomain $\mathcal{D}^c_{112}$ coincides with $\mathcal{D}^c_{111}$ when exchanging $\eta_1, \tilde\eta_1$ with $\eta_2, \tilde\eta_2$. As a conclusion, the original domain $\mathcal{D}$ is obtained as the union of 8 "rotated" subdomains:

$$\begin{aligned} \mathcal{D} \equiv & \mathcal{D}^c_{111}(\tilde\eta_1, \tilde\eta_2, \eta_1, \eta_2) \cup \mathcal{D}^c_{111}(\tilde\eta_1, \eta_2, \eta_1, \tilde\eta_2) \cup \mathcal{D}^c_{111}(\eta_1, \eta_2, \tilde\eta_1, \tilde\eta_2) \cup \\ & \mathcal{D}^c_{111}(\eta_1, \tilde\eta_2, \tilde\eta_1, \eta_2) \cup \mathcal{D}^c_{111}(\tilde\eta_2, \tilde\eta_1, \eta_2, \eta_1) \cup \mathcal{D}^c_{111}(\eta_2, \tilde\eta_1, \tilde\eta_2, \eta_1) \cup \\ & \mathcal{D}^c_{111}(\eta_2, \eta_1, \tilde\eta_2, \tilde\eta_1) \cup \mathcal{D}^c_{111}(\tilde\eta_2, \eta_1, \eta_2, \tilde\eta_1) \end{aligned} \quad (17)$$

Regularizing coordinates and final formula. We seek here a transformation of variables producing a Jacobian which might cancel the weak singularity for $u_1 = u_2 = 0$. At this stage, Duffy coordinates make the integrand regular. Let us define the following variables:

$$\begin{cases} v_1 = \omega \\ v_2 = \xi_1 \omega \\ v_3 = \xi_2(1 - \omega) \\ v_4 = \xi_3(1 - \xi_1 \omega) \end{cases} \quad \text{with} \quad \begin{cases} 0 \le \omega \le 1 \\ 0 \le \xi_1 \le 1 \\ 0 \le \xi_2 \le 1 \\ 0 \le \xi_3 \le 1 \end{cases} \quad (18)$$

For the 8 subdomains in equation (17) the intrinsic variables are expressed as functions of the v_i variables as shown in Table I.

Moreover, the Jacobian of the transformation, $J = \omega(1 - \omega)(1 - \xi_1 \omega)$, cancels the singularity, as required. The rather lengthy procedure outlined herein, however leads to a straightforward implementation formula:

TABLE I. Definition of intrinsic variables: coincident elements

case	1	2	3	4
$\tilde{\eta}_1^i =$	v_3	v_3	$v_1 + v_3$	$v_1 + v_3$
$\tilde{\eta}_2^i =$	v_4	$v_2 + v_4$	$v_2 + v_4$	v_4
$\eta_1^i =$	$v_1 + v_3$	$v_1 + v_3$	v_3	v_3
$\eta_2^i =$	$v_2 + v_4$	v_4	v_4	$v_2 + v_4$

case	5	6	7	8
$\tilde{\eta}_1^i =$	v_4	$v_2 + v_4$	$v_2 + v_4$	v_4
$\tilde{\eta}_2^i =$	v_3	v_3	$v_1 + v_3$	$v_1 + v_3$
$\eta_1^i =$	$v_2 + v_4$	v_4	v_4	$v_2 + v_4$
$\eta_2^i =$	$v_1 + v_3$	$v_1 + v_3$	v_3	v_3

$$\mathcal{I}_{mm} = \int_0^1 \int_0^1 \int_0^1 \int_0^1 \sum_{i=1}^{8} \tilde{\mathcal{B}}[\tilde{f}(\tilde{\boldsymbol{\eta}}^i), f(\boldsymbol{\eta}^i), \tilde{\mathbf{x}}(\tilde{\boldsymbol{\eta}}^i), \mathbf{x}(\boldsymbol{\eta}^i)] \cdot \\ \cdot \omega(1-\omega)(1-\xi_1\omega)\, \mathrm{d}\omega\, \mathrm{d}\xi_1\, \mathrm{d}\xi_2\, \mathrm{d}\xi_3 \tag{19}$$

At this stage equation (19) can be evaluated with standard Gaussian numerical schemes, achieving a high accuracy even for very low numbers of integration points.

3.2. COMMON EDGE

In this case the singularity is activated whenever both the source and field point lie on the common edge, namely when $u_1 = \eta_2 = \tilde{\eta}_2 = 0$.

A procedure similar to the one expounded in the previous Section is devised.

Domain partition. Exploiting once more Transformation C in Section 8, the original intrinsic domain is partitioned as follows:

$$\mathcal{D} \equiv \begin{cases} 0 \leq \tilde{\eta}_1 \leq 1 \\ 0 \leq \eta_1 \leq 1 \\ 0 \leq \tilde{\eta}_2 \leq 1 \\ 0 \leq \eta_2 \leq 1 \end{cases} = \begin{cases} 0 \leq \tilde{\eta}_1 \leq 1 \\ -\tilde{\eta}_1 \leq u_1 \leq 1-\tilde{\eta}_1 \\ 0 \leq \tilde{\eta}_2 \leq 1 \\ 0 \leq \eta_2 \leq 1 \end{cases} \tag{20}$$

$$= \begin{cases} 0 \leq u_1 \leq 1 \\ 0 \leq \tilde{\eta}_1 \leq 1-u_1 \\ 0 \leq \tilde{\eta}_2 \leq 1 \\ 0 \leq \eta_2 \leq 1 \end{cases} \cup \begin{cases} 0 \leq -u_1 \leq 1 \\ -\tilde{\eta}_1 \leq u_1 \leq 0 \\ 0 \leq \tilde{\eta}_2 \leq 1 \\ 0 \leq \eta_2 \leq 1 \end{cases} \equiv \mathcal{D}_1^{ed} \cup \mathcal{D}_2^{ed} \tag{21}$$

As in the coincident element case $\mathcal{D}_2^{ed}(\tilde{\eta}_1, \tilde{\eta}_2, \eta_1, \eta_2) = \mathcal{D}_1^{ed}(\eta_1, \tilde{\eta}_2, \tilde{\eta}_1, \eta_2)$, that is $\mathcal{D}_2^{ed}$ coincides with $\mathcal{D}_1^{ed}$ provided that η_1 is exchanged with $\tilde{\eta}_1$. Hence, only $\mathcal{D}_1^{ed}$ will be considered hereafter.

The inequalities concerning $u_1, \tilde{\eta}_2, \eta_2$ define a cube in the $u_1, \tilde{\eta}_2, \eta_2$ space, which is further partitioned into three different pyramidal subdomains having their vertices in $u_1 = \tilde{\eta}_2 = \eta_2 = 0$, as shown in Figure 3.

$$\mathcal{D}_1^{ed} = \begin{cases} 0 \leq u_1 \leq 1 \\ 0 \leq \tilde{\eta}_2 \leq u_1 \\ 0 \leq \eta_2 \leq u_1 \\ 0 \leq \tilde{\eta}_1 \leq 1-u_1 \end{cases} \cup \begin{cases} 0 \leq \tilde{\eta}_2 \leq 1 \\ 0 \leq u_1 \leq \tilde{\eta}_2 \\ 0 \leq \eta_2 \leq \tilde{\eta}_2 \\ 0 \leq \tilde{\eta}_1 \leq 1-u_1 \end{cases} \cup \begin{cases} 0 \leq \eta_2 \leq 1 \\ 0 \leq u_1 \leq \eta_2 \\ 0 \leq \tilde{\eta}_2 \leq \eta_2 \\ 0 \leq \tilde{\eta}_1 \leq 1-u_1 \end{cases}$$

$$= \mathcal{D}_{11}^{ed} \cup \mathcal{D}_{12}^{ed} \cup \mathcal{D}_{13}^{ed} \tag{22}$$

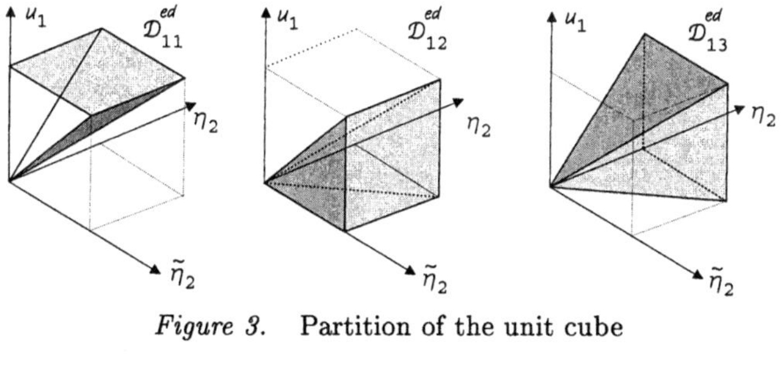

Figure 3. Partition of the unit cube

Hence the original domain $\mathcal{D}$ can be expressed as the union of six subdomains defined by equations (20) and (22):

$$\begin{aligned} \mathcal{D} \equiv & \mathcal{D}_{11}^{ed}(\tilde{\eta}_1, \tilde{\eta}_2, \eta_1, \eta_2) \cup \mathcal{D}_{12}^{ed}(\tilde{\eta}_1, \tilde{\eta}_2, \eta_1, \eta_2) \cup \mathcal{D}_{13}^{ed}(\tilde{\eta}_1, \tilde{\eta}_2, \eta_1, \eta_2) \cup \\ & \mathcal{D}_{11}^{ed}(\eta_1, \tilde{\eta}_2, \tilde{\eta}_1, \eta_2) \cup \mathcal{D}_{12}^{ed}(\eta_1, \tilde{\eta}_2, \tilde{\eta}_1, \eta_2) \cup \mathcal{D}_{13}^{ed}(\eta_1, \tilde{\eta}_2, \tilde{\eta}_1, \eta_2) \end{aligned} \tag{23}$$

Regularizing coordinates and final formula.
Let us define variables $v_1, v_2, v_3, v_4, v_5, \mathcal{J}_1$ and $\mathcal{J}_2$ as follows:

$$\begin{cases} v_1 &= \omega \\ v_2 &= \xi_1\omega \\ v_3 &= \xi_2\omega \\ v_4 &= \xi_3(1-\omega) \\ v_5 &= \xi_3(1-\xi_1\omega) \end{cases} \quad \text{with} \quad \begin{cases} 0 &\le \omega_1 \le 1 \\ 0 &\le \xi_1 \le 1 \\ 0 &\le \xi_2 \le 1 \\ 0 &\le \xi_3 \le 1 \end{cases} \tag{24}$$

and

$$\mathcal{J}_1 = \omega^2(1-\omega) \qquad \mathcal{J}_2 = \omega^2(1-\xi_1\omega)$$

For each of the 6 cases corresponding to subdomains in equation (23), Table II collects the definition of intrinsic variables and Jacobians.

TABLE II. Definition of intrinsic variables and Jacobians: common edge case

case	1	2	3	4	5	6
$\tilde{\eta}_1^i=$	v_4	v_5	v_5	v_1+v_4	v_5+v_2	v_5+v_2
$\tilde{\eta}_2^i=$	v_2	v_1	v_3	v_2	v_1	v_3
$\eta_1^i=$	v_1+v_4	v_5+v_2	v_5+v_2	v_4	v_5	v_5
$\eta_2^i=$	v_3	v_3	v_1	v_3	v_3	v_1
$J_i=$	$\mathcal{J}_1$	$\mathcal{J}_2$	$\mathcal{J}_2$	$\mathcal{J}_1$	$\mathcal{J}_2$	$\mathcal{J}_2$

Finally

$$\mathcal{I}_{mm} = \int_0^1\int_0^1\int_0^1\int_0^1 \sum_{i=1}^{6} \tilde{\mathcal{B}}[\tilde{f}(\tilde{\boldsymbol{\eta}}^i), f(\boldsymbol{\eta}^i), \tilde{\mathbf{x}}(\tilde{\boldsymbol{\eta}}^i), \mathbf{x}(\boldsymbol{\eta}^i)]\, J_i \, \mathrm{d}\omega_1\, \mathrm{d}\omega_2\, \mathrm{d}\xi_1\, \mathrm{d}\xi_2 \tag{25}$$

and the Jacobian of the transformation cancels all singularities.

3.3. COMMON VERTEX

Let us now consider the third case in Figure 1, where the two elements share one vertex. The integrand is singular only if $\tilde{\eta}_1 = \tilde{\eta}_2 = \eta_1 = \eta_2 = 0$.

Domain partition. The domain, a cube in a four-dimensional space, is decomposed in four subdomains on the basis of the partition already devised for the square in $\Re^2$ (Figure 2) and the cube in $\Re^3$ (Figure 3):

$$\mathcal{D} = \begin{cases} 0 \le \tilde{\eta}_1 \le 1 \\ 0 \le \tilde{\eta}_2 \le 1 \\ 0 \le \eta_1 \le 1 \\ 0 \le \eta_2 \le 1 \end{cases} \tag{26}$$

$$= \begin{cases} 0 \le \tilde{\eta}_1 \le 1 \\ 0 \le \tilde{\eta}_2 \le \tilde{\eta}_1 \\ 0 \le \eta_1 \le \tilde{\eta}_1 \\ 0 \le \eta_2 \le \tilde{\eta}_1 \end{cases} \cup \begin{cases} 0 \le \tilde{\eta}_2 \le 1 \\ 0 \le \tilde{\eta}_1 \le \tilde{\eta}_2 \\ 0 \le \eta_1 \le \tilde{\eta}_2 \\ 0 \le \eta_2 \le \tilde{\eta}_2 \end{cases} \cup \begin{cases} 0 \le \eta_1 \le 1 \\ 0 \le \tilde{\eta}_1 \le \eta_1 \\ 0 \le \tilde{\eta}_2 \le \eta_1 \\ 0 \le \eta_2 \le \eta_1 \end{cases} \cup \begin{cases} 0 \le \eta_2 \le 1 \\ 0 \le \tilde{\eta}_1 \le \eta_2 \\ 0 \le \tilde{\eta}_2 \le \eta_2 \\ 0 \le \eta_1 \le \eta_2 \end{cases}$$

$$= \mathcal{D}_1^v \cup \mathcal{D}_2^v \cup \mathcal{D}_3^v \cup \mathcal{D}_4^v$$

Regularizing coordinate transformation and final formula. Duffy coordinates are now introduced on each subdomain. Let us define the following functions

$$\begin{cases} v_1 &= \omega \\ v_2 &= \xi_1 \omega \\ v_3 &= \xi_2 \omega \\ v_4 &= \xi_3 \omega \end{cases} \quad \text{with} \quad \begin{cases} 0 \le \omega \le 1 \\ 0 \le \xi_1 \le 1 \\ 0 \le \xi_2 \le 1 \\ 0 \le \xi_3 \le 1 \end{cases} \tag{27}$$

According to equation (26), the intrinsic variables for the four subdomains are chosen as in Table III. Finally

TABLE III. Definition of variables: common vertex

case	1	2	3	4
$\tilde{\eta}_1^i =$	v_1	v_2	v_2	v_2
$\tilde{\eta}_2^i =$	v_2	v_1	v_3	v_3
$\eta_1^i =$	v_3	v_3	v_1	v_4
$\eta_2^i =$	v_4	v_4	v_4	v_1

$$\mathcal{I}_{mn} = \int_0^1 \int_0^1 \int_0^1 \int_0^1 \sum_{i=1}^{4} \tilde{\mathcal{B}}[\tilde{f}(\tilde{\boldsymbol{\eta}}^i), f(\boldsymbol{\eta}^i), \tilde{\mathbf{x}}(\tilde{\boldsymbol{\eta}}^i), \mathbf{x}(\boldsymbol{\eta}^i)]\, \omega^3 \, \mathrm{d}\omega \, \mathrm{d}\xi_1 \, \mathrm{d}\xi_2 \, \mathrm{d}\xi_3 \tag{28}$$

4. Numerical Examples

Several crack problems have been solved and the results obtained are described here in order to illustrate the accuracy and the effectiveness of the

SGBEM in the fracture mechanics context. The examples concern cracks in the unbounded space and surface breaking cracks in finite bodies. The evaluation of the stress intensity factors is always carried out through extrapolation from nodal values of displacement discontinuities near the crack front utilizing the asymptotic expression for the exact field along lines normal to the crack front. The BEs adopted are isoparametric 6-noded triangles and 9-noded quadrilaterals. The elements employed at the crack front are always quadrilateral 9-noded elements. For the example problems concerning cracks in the unbounded space, the quarter point scheme is adopted for these elements, in order to reproduce the square root behaviour for the displacement discontinuity field. In this case the stress intensity factor (SIF) extraction at a crack front point Q is carried out on the basis of the displacement discontinuity values at the two nodes behind the crack front which lie in the plane through Q normal to the crack edge. Two estimates of the SIFs are obtained from the two nodal values and the final value is computed by linear extrapolation.

The example problems concerning edge cracks are solved using standard elements at the crack front. In this case the SIFs are computed on the basis of the asymptotic formulae and of the nodal value of the displacement discontinuity at the node nearest to the crack front.

The surface meshing for the example problems presented herein has been carried out using a commercial finite element pre-processor.

4.1. FRACTURES IN INFINITE DOMAINS

Here we focus on a general fracture S_c embedded in an infinite isotropic medium with elastic constants ν (Poisson coefficient) and μ (shear modulus) and subjected to remote uniform loading σ^0_{ik}. Three test examples in linear elastic fracture mechanics are solved with the SGBEM: a penny-shaped crack, an elliptical plane crack and a spherical-cap crack.

Penny-shaped crack. Let us consider a penny-shaped crack in the plane x_1, x_2 under remote stresses σ^0_{33} and σ^0_{23} (see Figure 4). Denoting by a the crack radius and by r the distance from the centre, the exact solution reads (Xu and Ortiz, 1993):

$$w_1 = 0 \quad w_2 = \frac{8(1-\nu)}{\pi\mu(2-\nu)}\sqrt{a^2-r^2}\,\sigma^0_{23}, \quad w_3 = \frac{4(1-\nu)}{\pi\mu}\sqrt{a^2-r^2}\,\sigma^0_{33}$$

$$K_I = \frac{2}{\pi}\sqrt{\pi a}\,\sigma^0_{33}, \quad K_{II} = \frac{4}{\pi(2-\nu)}\sqrt{\pi a}\sin\theta\,\sigma^0_{23},$$

$$K_{III} = \frac{4(1-\nu)}{\pi(2-\nu)}\sqrt{\pi a}\cos\theta\,\sigma^0_{23} \tag{29}$$

The problem is solved for $\nu = 0.3$ using the two meshes illustrated in Figure 4, with 12 and 20 elements respectively.

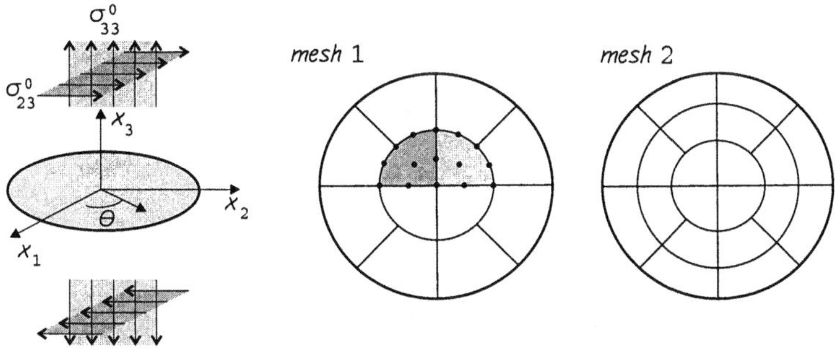

Figure 4. Penny-shaped crack: loading conditions and meshes

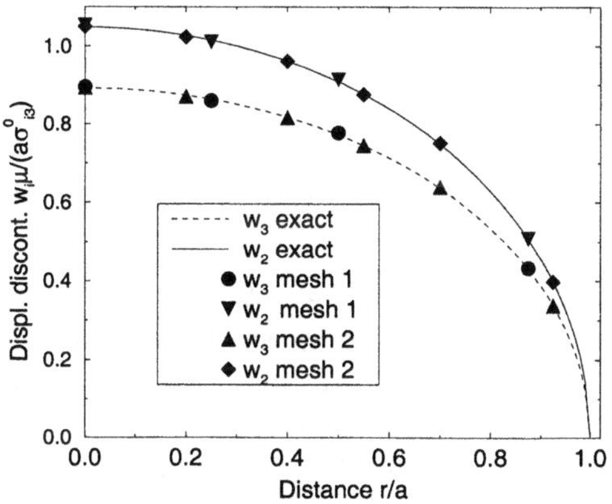

Figure 5. Penny-shaped crack: numerical results for displacement discontinuities $w_i\mu/(a\sigma_{i3})$

Figure 5 shows the nodal values obtained for the opening and sliding displacements and compared with the exact solutions, equation (29).

Table IV gathers the relative errors for the SIFs computed for $\theta = 0$, (that is at $x_1 = a$, $x_2 = 0$). Despite the coarseness of the adopted meshes, the numerical results exhibit very good accuracy.

Elliptical crack. Let us now consider an elliptical crack with major semi-axis a and minor semi-axis b and subjected to the remote stress σ^0_{33}.

TABLE IV. Relative errors for computed SIFs

Mesh	K_I	K_{III}
1	$6.15\,10^{-3}$	$4.62\,10^{-3}$
2	$2.76\,10^{-3}$	$1.15\,10^{-3}$

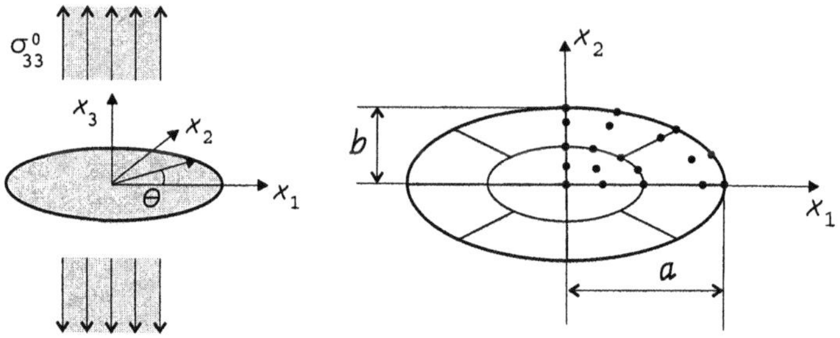

Figure 6. Elliptical crack: applied loading and mesh adopted

The analytical expression of K_I is available (see Kassir and Sih, 1966) and reads:

$$K_I = \frac{\sigma^0_{33}\sqrt{\pi b}}{E(k)}\left(\sin^2\theta + \frac{b^2}{a^2}\cos^2\theta\right)^{1/4} \tag{30}$$

where $E(k)$ is the complete elliptic integral of the second kind and $k = 1 - b^2/a^2$. The problem is numerically solved for two values of the aspect ratio a/b: $a/b = 2, a/b = 3$. In both cases the mesh is obtained by a linear contraction (in the x_2 direction) of Mesh 1 in Figure 4. A comparison between exact and computed normalized stress intensity factors is presented in Figure 7, where $K_0 = (2/\pi)\sqrt{\pi b}\sigma^0_{33}$.

Spherical-cap crack. As an example of a non-planar crack, a spherical-cap crack bounded by a circular front and subjected to surface tractions $t_1 = t_2 = 0$, $t_3 = \pm t^0_3$ is considered (see Figure 8); a is the radius of the spherical surface and 2α is the subtended angle.

For this problem, numerical results in terms of SIFs are given by Xu and Ortiz (1993) for a given range of α. The analysis has been carried out for three values of α ($\alpha = 15°, 30°, 45°$) and $\nu = 0.3$, using three meshes with 40, 112 and 240 elements on the spherical surface. All the elements adjacent to the crack front are quadrilateral quarter-point elements, the remaining elements being quadrilateral for Mesh 1 and triangular for Meshes 2 and 3. Figure 8 gives a planar representation of the actual meshes adopted for

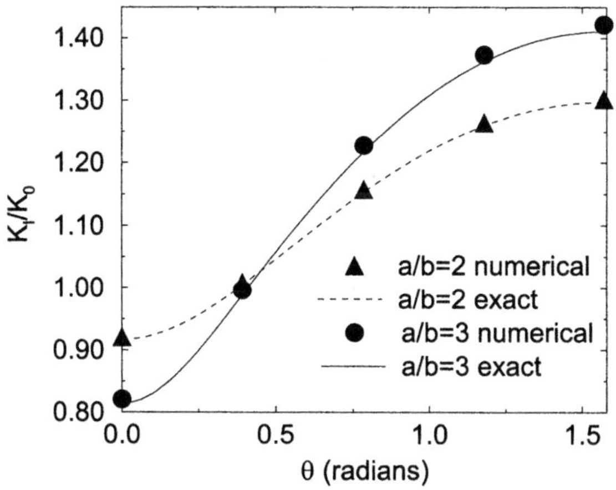

Figure 7. Elliptical crack: normalized stress intensity factor along one quarter of the crack front

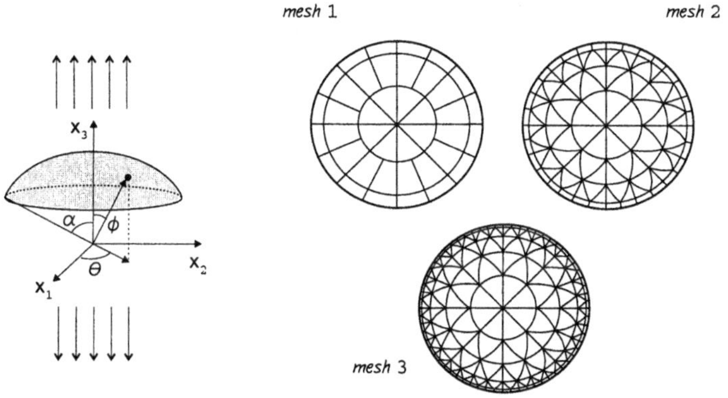

Figure 8. Spherical-cap crack: applied loading and meshes adopted

the spherical-cap crack, obtained by prescribing that the polar coordinate ρ equals $a\phi$. All results (see Table V), normalized by means of the equivalent mode I SIF for a penny-shaped crack ($K^0 = (2t_3^0/\pi)\sqrt{\pi a \sin\alpha}$), compare well with the graphical solution provided by Xu and Ortiz (1993).

4.2. EDGE CRACKED BAR

The problem considered below a single edge crack in a rectangular bar subject to uniform normal tractions applied at the end surfaces.

The bar geometry is shown in Figure 9, where the thickness is denoted by t, the width by w, the height is $2H$ and the crack depth is a. The following parameters are adopted, since for this case existing numerical

TABLE V. Spherical-cap crack: computed SIFs

Mesh	K_I/K^0 (15°)	K_{II}/K^0 (15°)	K_I/K^0 (30°)	K_{II}/K^0 (30°)	K_I/K^0 (45°)	K_{II}/K^0 (45°)
1	0.964	0.263	0.845	0.520	0.655	0.769
2	0.966	0.266	0.849	0.525	0.662	0.774
3	0.966	0.267	0.851	0.527	0.665	0.776

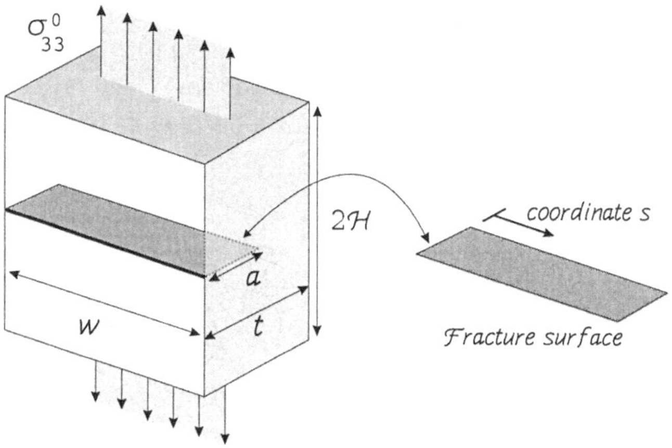

Figure 9. Edge cracked bar: geometry and loading conditions

results are available for comparison: $H/t = 0.875$, $a/t = 0.5$, $w/t = 1.5$ and Poisson's ratio $\nu = 1/3$.

The problem is numerically solved using the four discretizations, Mesh A to Mesh D, shown in Figures 10-11, having 4,6,8 and 14 elements adjacent to the straight crack front, respectively.

The variation of the mode I SIF along the crack front computed with the four meshes is displayed in Figure 12, where the dashed line represents the SIF relevant to the a plane-strain idealization.

In Figure 13 the SIF obtained with the finest mesh is compared to the results due to Li *et al.* (1998), Raju and Newman (1977) and Mi (1996).

In view of the dispersion among different results presented in the literature (see Figure 13), especially with respect to the plane strain idealization, different configurations and boundary conditions are considered for the edge cracked bar in an attempt to approach a plane strain situation.

For this purpose, increasing values of the ratio w/t are taken ($w/t = 1.5; 3; 4.5$), leaving unaltered the ratios $H/t = 0.875$ and $a/t = 0.5$. Vanishing displacements in the direction parallel to the width w are imposed

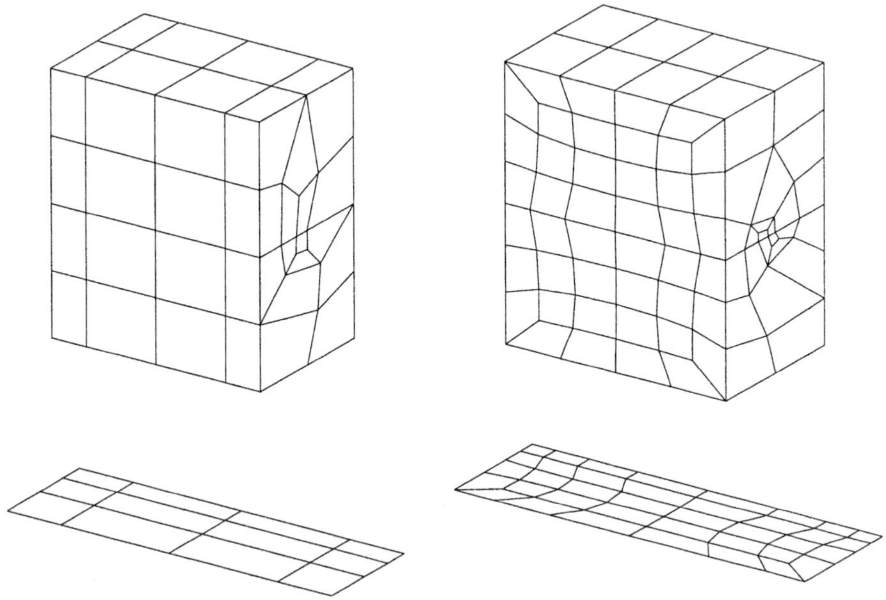

Figure 10. Mesh A Mesh B

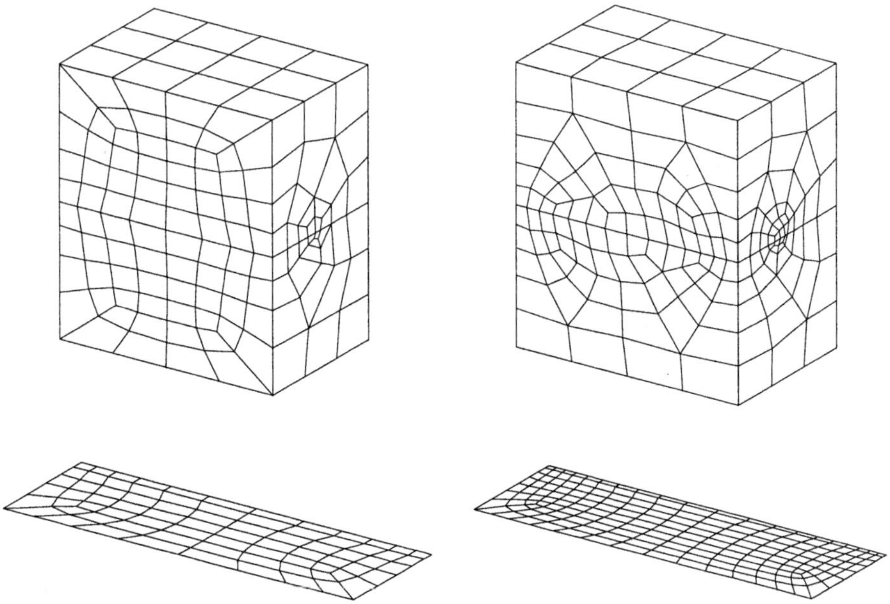

Figure 11. Mesh C Mesh D

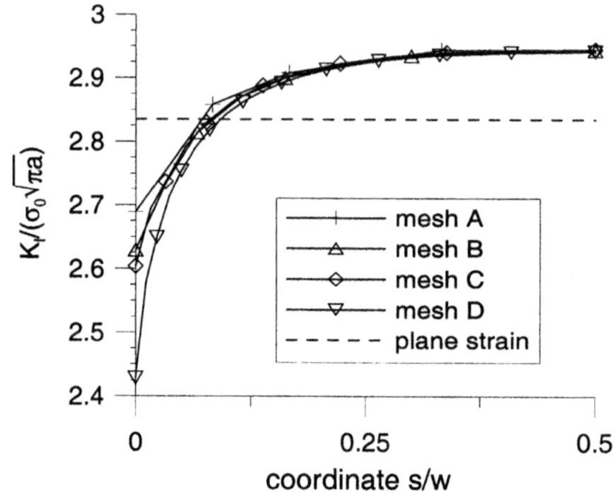

Figure 12. Normalized stress intensity factor for an edge cracked bar as function of position along the crack front; results obtained with the four meshes shown in Figures 10-11

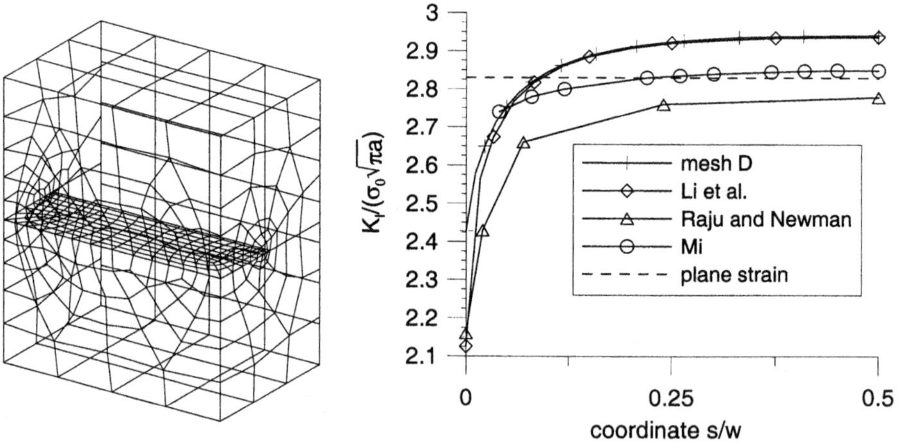

Figure 13. Edge cracked bar: comparison of present approach (Mesh D) with other available results: Raju and Newman (1977), Li *et al.* (1998) and Mi (1996)

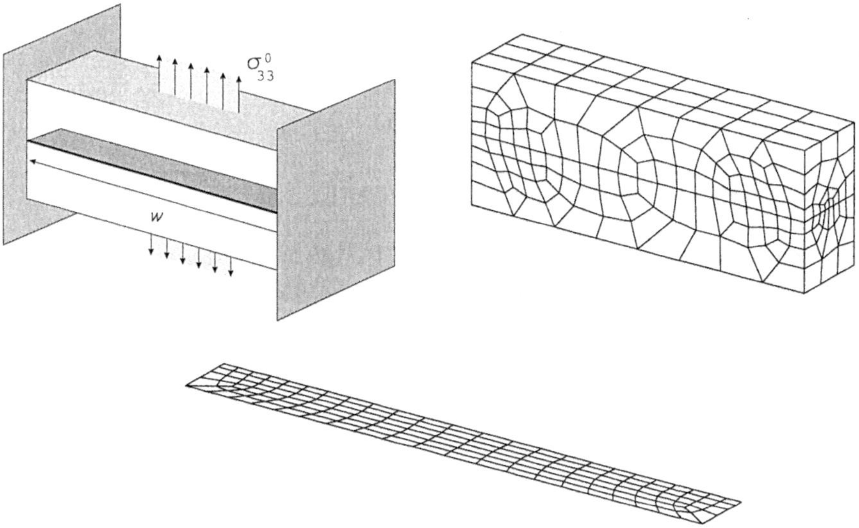

Figure 14. Edge cracked bar with constrained lateral displacements: modified geometry and mesh

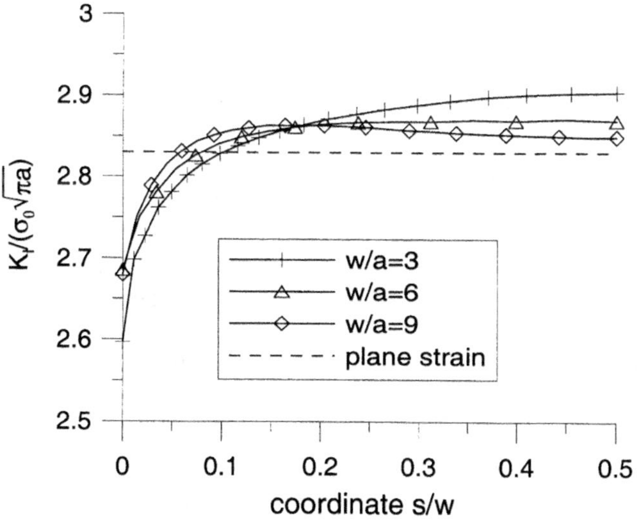

Figure 15. Edge cracked bar: convergence to the plane strain solution

on the lateral faces of the bar, as indicated in Figure 14. The SIF results obtained with the boundary element mesh shown in Figure 14 are plotted in Figure 15. The values of the SIF at the central section, i.e. for $s/w = 0.5$, slowly converge to the plane strain value as the ratio w/t is increased, thus corroborating the validity of the present variational approach.

4.3. CIRCULAR EDGE CRACK IN A PLATE

The geometry of the problem is shown in Figure 16; uniform tensile stresses are applied at two opposite faces of the bar (plate) in the direction perpendicular to the crack; a value of Poisson's ratio $\nu = 0.3$ is adopted.

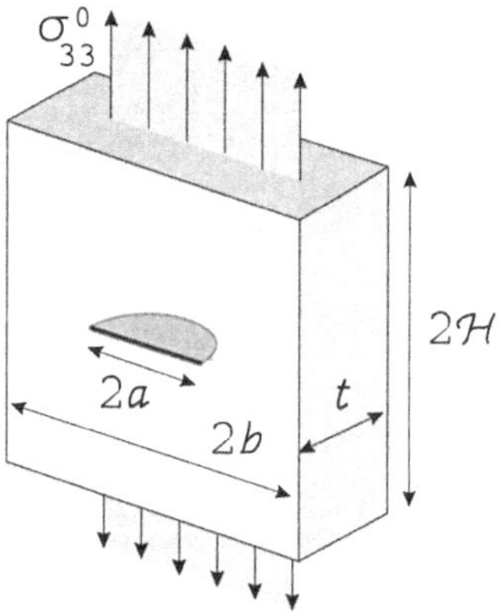

Figure 16. Plate containing a semi-circular surface crack

The configuration considered is characterized by the geometric ratios $H/a = 5$, $b/a = 5$ and $a/t = 0.4$. The values adopted for H/a and b/a are large enough to effectively represent an edge crack in an infinite plate.

The problem is analyzed using the three meshes, Mesh A, Mesh B and Mesh C, depicted in Figure 17, having 12, 24 and 40 elements along the circular crack front, respectively. The results obtained in terms of mode-I SIF are plotted in Figure 18 as a function of the angular parameter $2\phi/\pi$ (with $\phi = 0$ at the free surface), where they are compared to the finite element results of Raju and Newman (1979). It is well known that, if a surface breaking crack intersects the surface itself at a right angle, the SIFs, as defined on the base of the classical Williams-Westergaard asymptotic formulae, tend to zero in a boundary layer the thickness of which depends on the problem's geometry and material properties. From Figure 19 it turns out that the analyses carried out by the SGBEM (in particular with the refined Mesh C) show a much thinner boundary layer than in Raju and Newman (1979). However SIF values far from the external surface are not significantly affected by the accuracy with which the boundary layer effect

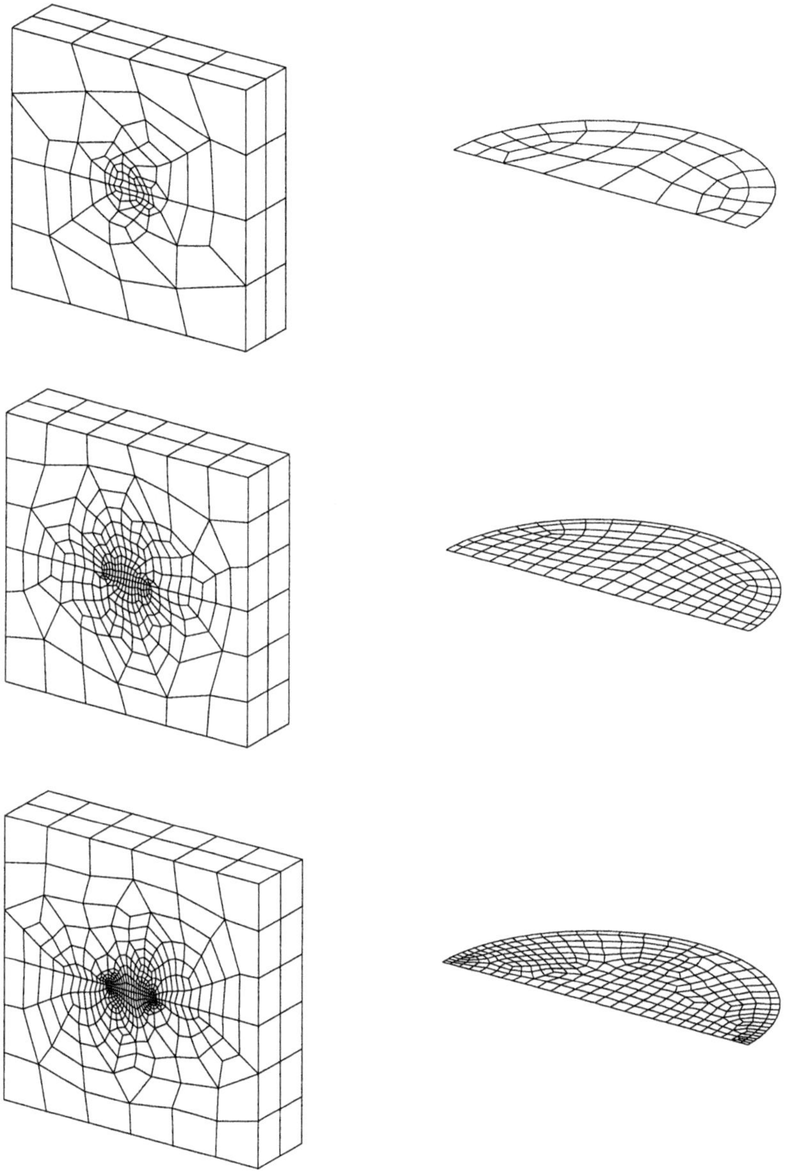

Figure 17. Meshes A, B and C

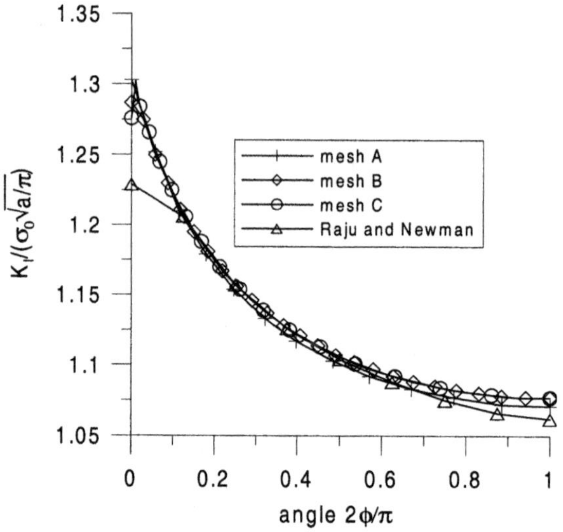

Figure 18. Normalized stress intensity factor for a semi-circular surface crack in a plate as function of the angular position along the crack front; results for the meshes A B and C and earlier FE results of Raju and Newman (1979).

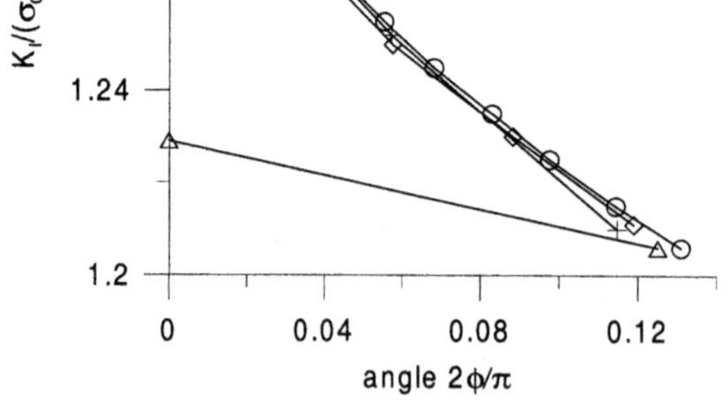

Figure 19. Normalized stress intensity factor for a semi-circular surface crack in a plate: "boundary layer" effect for meshes A B and C and earlier FE results of Raju and Newman (1979).

is accounted for and can be accurately predicted by using even rather coarse meshes.

4.4. QUARTER ELLIPTIC CORNER CRACK IN A PLATE

As a final example we consider the rectangular plate in Figure 20 with a quarter elliptic crack emanating from the inner circular hole. The problem geometry is characterized by the following ratios: $R/t = 2$, $a/t = 0.2$, $a/c = 0.8$, $R/W = 0.2$ and $H/W = 2$. Poisson's ratio is taken as $\nu = 0.3$.

The mesh adopted (only half of the specimen was analyzed due to symmetry) is depicted in Figure 21 with an enlarged view of the fracture-area. The resulting values of the normalized SIF along the crack front are plotted in Figure 22 as a function of the non-dimensional curvilinear coordinate s/ℓ, where ℓ is the quarter-ellipse perimeter length.

5. Concluding Remarks

The 3D regularized formulation illustrated in the present contribution parallels the one developed by Li *et al.* (1998), but differs from it in the treatment of the singular double surface integrals the accurate evaluation of which represents a key ingredient for an efficient implementation of the method. Several example problems concerning cracks in the infinite medium and edge cracks in finite solids have been solved using quadratic elements. The results obtained for the stress intensity factors, extracted through extrapolation from displacement discontinuity nodal values, show good accuracy even in the absence of special elements at the crack front. For crack propagation problems, the latter feature, together with the limited remeshing work required, makes the SGBEM a very attractive tool compared to alternative domain methods. The extension to crack propagation of the present formulation and of the relevant computer code is currently under way. The SGBEM, like all BEM approaches, entails fully populated coefficient matrices. For the viability of the SGBEM in large-scale problems it is mandatory to reduce the computational cost and memory requirements by making recourse to recent algorithms such as fast multipole methods and panel clustering. Several contributions in this line have appeared recently with promising results (e.g. Lage and Schwab, 2000,Yoshida *et al.*, 2001).

Acknowledgement

A research grant from MURST ("Cofinanziamento" on Integrity Assessment of Large Dams, 2000) is acknowledged

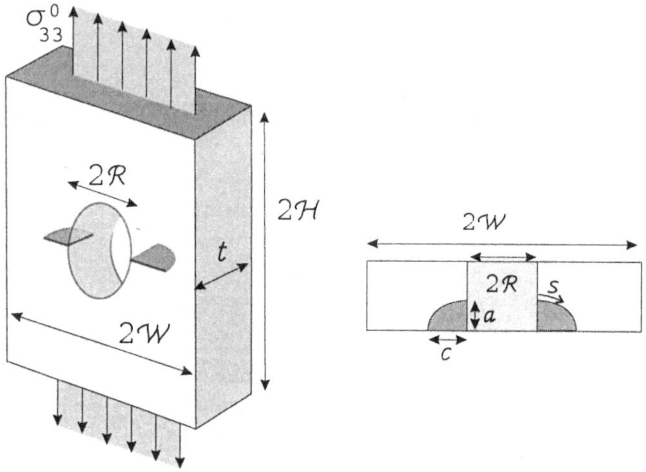

Figure 20. Quarter elliptic corner crack at circular hole in a plate

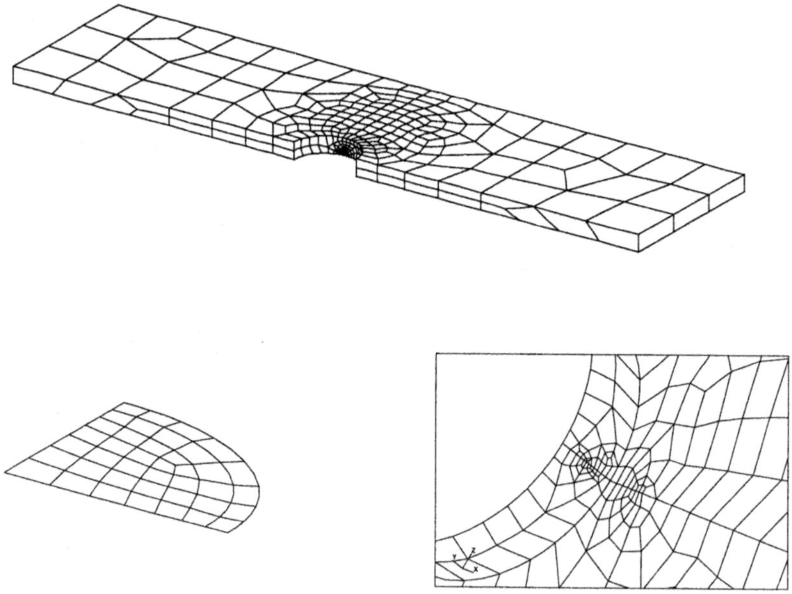

Figure 21. Quarter elliptic corner crack: mesh adopted for the plate and for the crack

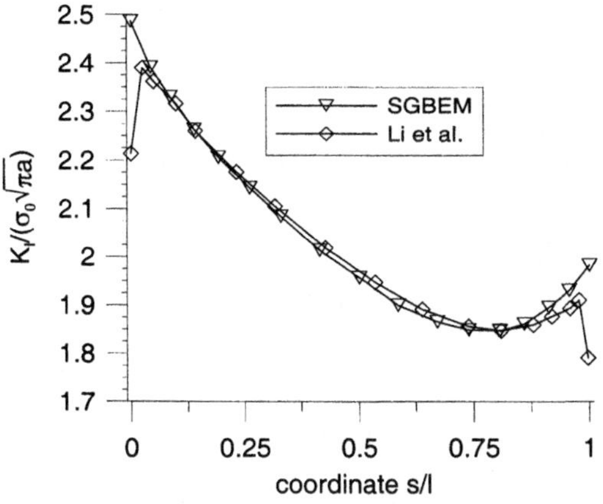

Figure 22. Normalized stress intensity factor for quarter elliptic corner crack: results for the adopted mesh and earlier BEM results of Li *et al.* (1998)

6. References

Andrä, H and Schnack, E. (1997) Integration of singular Galerkin-type boundary element integrals for 3D elasticity, *Numerische Mathematik*, **76**, 143–165.

Balakrishna, C., Gray, L.J. and Kane, J.H. (1994) Efficient analytical integration of symmetric Galerkin boundary integrals over curved elements: elasticity, *Comp. Meth. Appl. Mech. Engng.*, **117**, 157–179.

Bonnet, M., Maier, G. and Polizzotto, C. (1998) Symmetric Galerkin boundary element method, *Appl. Mech. Rev.*, **51**, 669–704.

Bonnet, M. (1993) A regularized Galerkin symmetric BIE formulation for mixed 3D elastic boundary values problems, *Boundary Elements Abstracts & Newsletters*, **4**, 109–113.

Frangi, A. and Novati, G. (1996) Symmetric BE method in two dimensional elasticity: evaluation of double integrals for curved elements, *Computat. Mech.*, **19**, 58–68.

Frangi, A. (1998) Regularization of boundary element formulations by the derivative transfer method, in Sládek, V., Sládek, J. (eds.), *Singular Integrals in Boundary Element Methods, Advances in Boundary Elements, chap. 4*, Computational Mechanics Publications, 125–164.

Erichsen, S. and Sauter, S.A. (1998) Efficient automatic quadrature in 3-D Galerkin BEM, *Comp. Meth. Appl. Mech. Engng.*, **157**, 215–224.

Hartranft, R.J. and Sih, G.C. (1970) An approximate three-dimensional theory of plates with application to crack problems, *Int. J. Engng. Sci.*, **8**, 711–729.

Hills, D.A., Kelly, P.A. (1996) *Solution of Crack Problems*, Kluwer Academic Press, Dortrecht.

Kassir, M.K. and Sih, G.C. (1966) Three dimensional stress distribution around an elliptical crack under arbirary loadings, *J. Applied Mech.* , **33**, 602–615.

Lage, C. and Schwab, C. (2000) Advanced boundary element algorithms, in Whiteman, J.R. (eds.), *Mafelap 1999*, Elsevier, 283–306.

Li, S., Mear, M.E. and Xiao, L. (1998) Symmetric weak-form integral equation method for three-dimensional fracture analysis, *Comp. Meth. Appl. Mech. Engng.*, **151**, 435–459.

Maier, G., Miccoli, S., Novati, G. and Sirtori, S. (1992) A Galerkin symmetric boundary element method in plasticity: formulation and implementation, in Kane, J.H., Maier, G. , Tosaka, N. and Atluri, S.N. (eds.), *Advances in Boundary Elements Techniques*, Springer Verlag, 288–328.

Nishimura, N. and Kobayashi, S. (1989) A regularized boundary integral equation method for elastodynamic crack problems, *Computat. Mech.*, **4**, 319–328.

Mi, Y. (1996) *Three-dimensional Analysis of Crack Growth*, Computational Mechanics Publications, Southampton.

Paulino, G.H. and Gray, L.J. (1999) Galerkin residuals for adaptive symmetric-Galerkin boundary element methods, *Journal of Engineering Mechanics (ASCE)*, **125**, 575–585.

Raju, I.S. and Newman, J.C. (1977) Three dimensional finite-element analysis of finite-thickness fracture specimens, *NASA-TN*, D-8414.

Raju, I.S. and Newman, J.C. (1979) Stress-intensity factors for a wide range of semi-elliptical surface cracks in finite-thickness plates, *Engng. Fracture Mech.*, **11**, 817–829.

Sauter, S.A. and Schwab, C. (1997) Quadrature for hp-Galerkin BEM in 3-d, *Numerische Mathematik*, **78**, 211–258.

Sirtori, S. (1979) General stress analysis method by means of integral equations and boundary elements, *Meccanica*, **14**, 210–218.

Sirtori, S., Maier, G., Novati, G. and Miccoli, S. (1992) A Galerkin symmetric boundary element method in elasticity: formulation and implementation, *Int. J. Num. Meth. Engng.*, **35**, 255–282.

Tada, S. , Paris, P. and Irwin, G. (1985) *The Stress Analysis of Cracks Handbook*, Dell Research Corporation, St. Louis.

Xu, G. and Ortiz, M. (1993) A variational boundary integral method for the analysis of 3-D cracks of arbitrary geometry modelled as continuous distributions of dislocation loops, *Int. J. Num. Meth. Engng.*, **36**, 3675–3701.

Yoshida, K., Nishimura, N. and Kobayashi, S. (2001) Application of fast multipole Galerkin boundary integral equation method to elastostatic crack problems, *Int. J. Num. Meth. Engng.*, **50**, 525–547.

Appendices

7. Surface Rotors

Surface rotors are defined as:

$$\tilde{R}_j[\tilde{w}_i](\tilde{\mathbf{x}}) = e_{bcj}\tilde{n}_b(\tilde{\mathbf{x}})\frac{\partial \tilde{w}_i}{\partial \tilde{x}_c}(\tilde{\mathbf{x}}) \qquad R_s[w_k](\mathbf{x}) = e_{bcs}n_b(\mathbf{x})\frac{\partial w_k}{\partial x_c}(\mathbf{x}) \tag{31}$$

They express the vector product between the gradients of the argument functions and the unit normal vector to the surface

$$\tilde{R}_j[\tilde{w}_i](\tilde{\mathbf{x}}) = (\tilde{\mathbf{n}} \wedge \nabla\tilde{w}_i)\cdot\mathbf{e}_j \quad \text{in } \tilde{\mathbf{x}}, \qquad R_s[w_k](\mathbf{x}) = (\mathbf{n} \wedge \nabla w_k)\cdot\mathbf{e}_s \quad \text{in } \mathbf{x}$$

where $\mathbf{e}_k$ denotes the unit vector along the k_{th} Cartesian coordinate. Now, letting $\mathbf{g}_1$ and $\mathbf{g}_2$ denote the local covariant base vectors associated with intrinsic coordinates η_1 and η_2, we have:

$$R_s[w_k](\mathbf{x}) = e_{bcs} n_b \frac{\partial w_k}{\partial x_c} = e_{bcs} e_{bij} g_{1i} g_{2j} \frac{\partial w_k}{\partial x_c} J^{-1} =$$
$$= (\delta_{ic}\delta_{js} - \delta_{ic}\delta_{js}) g_{1i} g_{2j} \frac{\partial w_k}{\partial x_c} J^{-1} = \left(\frac{\partial w_k}{\partial \eta_1} g_{2s} - \frac{\partial w_k}{\partial \eta_2} g_{1s} \right) J^{-1} \tag{32}$$

Hence, $R_s[w_k](\mathbf{x})$ can be expressed only in terms of the in-plane components of the displacement discontinuity gradient.

8. Transformations and Equivalence of Domains

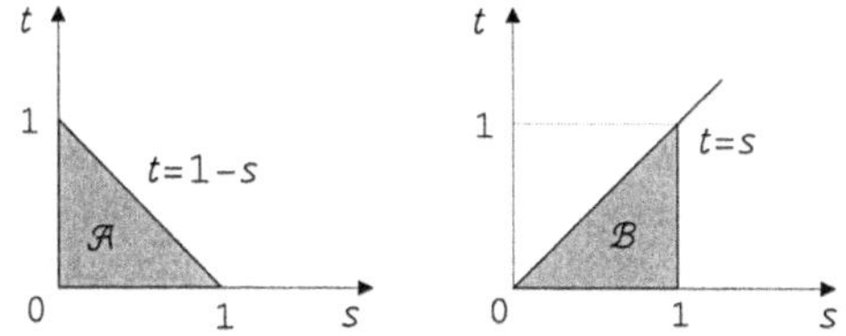

Figure 23. Geometrical representations of domains for Transformations A and B

Transformation A. Let us consider domain $\mathcal{A}$ in Figure 23 in the two dimensional space s, t. The following two sets of inequalities both define $\mathcal{A}$

$$\mathcal{A}(s,t) = \begin{cases} 0 \le s \le 1 \\ 0 \le t \le 1-s \end{cases} = \begin{cases} 0 \le t \le 1 \\ 0 \le s \le 1-t \end{cases} \tag{33}$$

Moreover, the variable transformation $\tilde{s} = s + t$, $\tilde{t} = t$ allows us to write

$$\mathcal{A}(\tilde{s} - \tilde{t}, \tilde{t}) = \begin{cases} 0 \le \tilde{t} \le 1 \\ \tilde{t} \le \tilde{s} \le 1 \end{cases} = \begin{cases} 0 \le \tilde{s} \le 1 \\ 0 \le \tilde{t} \le \tilde{s} \end{cases} = \mathcal{B}(\tilde{s}, \tilde{t}) \tag{34}$$

thus mapping domain $\mathcal{A}$ onto domain $\mathcal{B}$ in Figure 23. This expedient is useful for the exploitation of the Duffy coordinate transformations.

Transformation B. Similar conclusions hold also for domain $\mathcal{B}$ in Figure 23:

$$\mathcal{B}(s,t) = \begin{cases} 0 \le s \le 1 \\ 0 \le t \le s \end{cases} = \begin{cases} 0 \le t \le 1 \\ t \le s \le 1 \end{cases} \tag{35}$$

Transformation C. Let us now focus on the domain $\mathcal{C}$ in Figure 24:

$$\mathcal{C}(s,t) = \begin{cases} 0 & \leq s \leq 1 \\ -s & \leq t \leq 1-s \end{cases}$$

The quadrangle $\mathcal{C}$ can equivalently be expressed as the union of two triangles $\mathcal{C}_1 \cup \mathcal{C}_2$.

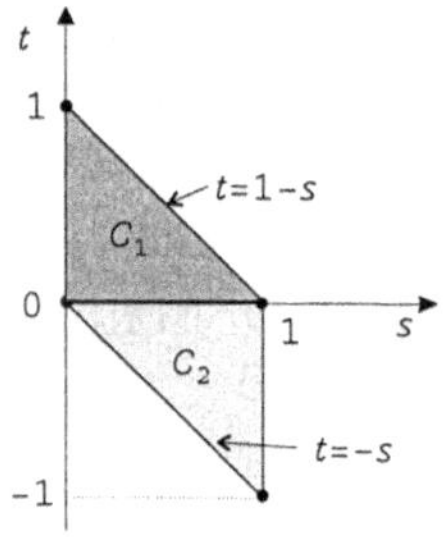

$$\mathcal{C}_1 = \begin{cases} 0 & \leq t \leq 1 \\ 0 & \leq s \leq 1-t \end{cases}$$

$$\mathcal{C}_2 = \begin{cases} 0 & \leq -t \leq 1 \\ -t & \leq s \leq 1 \end{cases}$$

Figure 24. Domain partition for Transformation C

9. Equivalence of $\mathcal{D}_2^c(\tilde{\eta}_1, \tilde{\eta}_2, \eta_1, \eta_2)$ and $\mathcal{D}_1^c(\eta_1, \tilde{\eta}_2, \tilde{\eta}_1, \eta_2)$

Let us consider the two subdomains in equation (14) and focus on the first two inequalities in $\mathcal{D}_2^c$. Denoting $\tilde{u}_1 = \tilde{\eta}_1 - \eta_1 = -u_1$, we can write:

$$\begin{cases} 0 & \leq -u_1 \leq 1 \\ -u_1 & \leq \tilde{\eta}_1 \leq 1 \end{cases} \equiv \begin{cases} 0 & \leq \tilde{u}_1 \leq 1 \\ \tilde{\eta}_1 - \eta_1 & \leq \tilde{\eta}_1 \leq 1 \end{cases} \equiv \begin{cases} 0 & \leq \tilde{u}_1 \leq 1 \\ 0 & \leq \eta_1 \leq 1 - \tilde{u}_1 \end{cases} \tag{36}$$

If u_1 and $\tilde{u}_1$ are exchanged, equation (36) transforms into the first two inequalities defining $\mathcal{D}_1^c$ in equation (14). This completes the proof.

CHAPTER 9.
NUMERICAL SIMULATION OF SEISMIC WAVE SCATTERING AND SITE AMPLIFICATION, WITH APPLICATION TO THE MEXICO CITY VALLEY

L.C. WROBEL[1], E. REINOSO[2] and H. POWER[3]
[1] *Department of Mechanical Engineering, Brunel University, Uxbridge UB8 3PH, UK*
[2] *Instituto de Ingenieria, UNAM, Ciudad Universitaria, Apartado Postal 70-472, Mexico City, D.F. 04510, Mexico*
[3] *Department of Mechanical Engineering, University of Nottingham, University Park, Nottingham NG7 2RD, UK*

Abstract

This chapter presents direct formulations of the boundary element method for the two-dimensional scattering of seismic waves from irregular topographies and buried valleys. The BEM models were formulated with isoparametric, quadratic boundary elements, and were employed to simulate a section of the Mexico City valley. Because the Mexico City valley is relatively flat and shallow, and the contrast of S waves between the clays and the basement rock is very high, it is believed that the one-dimensional theory is sufficient to explain the amplification patterns. Although this is true for many sites, results from accelerometric data suggest that two- and three-dimensional models are needed to explain the amplification behaviour at other sites, particularly near the borders of the valley.

Keywords: Boundary element method, seismic wave propagation, site amplification, Mexico City valley

1. Introduction

For the past thirty years, the seismic effects of subsurface and topographic irregularities have been extensively studied. The damage on human settlements located over alluvial valleys, observed during recent earthquakes, has further encouraged studies of site amplification. Important theoretical and experimental results have been obtained, but the identification of such effects on observed records has not been satisfactorily quantified.

W.S. Hall and G. Oliveto (eds.), Boundary Element Methods For Soil-Structure Interaction, 345–375.

Although the amplification caused by flat layers is well documented through the one-dimensional (1D) theory, in reality all valleys are of finite size where borders and other geometric features play an important role in modifying the 1D response and in generating surface waves. Therefore, two-dimensional (2D) and three-dimensional (3D) wave propagation models are necessary to simulate these problems more realistically. Analytical solutions of 2D seismic wave scattering problems have been obtained for simple geometries, but for more realistic configurations, numerical methods have to be employed.

The simplest way to model 2D seismic wave scattering is by considering SH waves. In this case, the mathematical model results in a scalar problem in which the Helmholtz equation is solved. Different numerical methods have been applied to study the scattering of SH waves. The most common are the Aki-Larner method [1-4], the finite difference method [5], the boundary integral equation method [6-14] and the combination of finite and boundary elements [15-16].

For P, SV and Rayleigh waves, the problem is vectorial and the solution of the equations of elastodynamics is required. Similarly to SH waves, the most common numerical methods to simulate P, SV and Rayleigh wave propagation are the Aki-Larner method [4,17], the finite difference method [18], the boundary integral equation method [19-26] and the finite element method [27].

More accurate predictions of ground motions in topographies and basins require 3D models for several reasons. The response of a 3D topography, whether a mountain or a canyon, is generally very sensitive to the azimuth, angle and type of incident wave. Lateral variations in sediment thickness and velocity could cause site response to be dependent on the azimuth to the earthquake. In closed basins, resonant modes can be set up by multiple waves reflected by the edges of the deposit. In addition, the curvature of the alluvium-basement interface could cause wave focusing for certain locations in the basin.

This article discusses two-dimensional formulations of the direct boundary element method for the scattering of seismic waves. A BEM formulation for the Helmholtz equation is used to model out-of-plane displacements due to incident SH waves [13], while a BEM formulation for the Navier-Cauchy equations of elastodynamics is used to model in-plane displacements due to incident P, SV and Rayleigh waves [24]. A three-dimensional BEM formulation for scattering of seismic waves was also presented by Reinoso *et al.* [28]. More information on BEM formulations for dynamics and wave propagation can be found in the book by Dominguez [29].

The BEM formulations are then used to reproduce observed site amplification due to earthquakes in the Mexico City valley. Owing to the dynamic characteristics of the clay deposits upon which Mexico City rests, the valley is one of the best examples of amplification in alluvial basins. The complexity of the valley response to seismic waves requires reliable numerical models. Because of the shallowness, slow shear-wave velocity of the soil in the valley and the lack of accurate information on the strata and deep structure, detailed 3D modelling is still far from being feasible and realistic. However, important 1D and 2D results have been obtained that may explain some features of the valley amplification.

2. Wave Propagation in a Half-space

The equations of motion for a homogeneous, isotropic and linearly elastic medium are the Navier-Cauchy equations

$$\left(\alpha^2 - \beta^2\right) u_{j,ji} + \beta^2 u_{i,jj} - \frac{\partial^2 u_i}{\partial t^2} = 0 \tag{1}$$

where

$$\alpha = \left(\frac{\lambda + 2\mu}{\rho}\right)^{1/2} \tag{2}$$

$$\beta = \left(\frac{\mu}{\rho}\right)^{1/2} \tag{3}$$

are the longitudinal and transversal wave velocities, with ρ the mass density and λ and μ the two Lamé constants, given by

$$\lambda = \frac{\nu E}{(1+\nu)(1-2\nu)}$$

$$\mu = \frac{E}{2(1+\nu)}$$

in terms of the Young's modulus E and Poisson's ratio ν.

For harmonic problems, with a time dependence e^{iwt}, the equations of motion become

$$\left(\alpha^2 - \beta^2\right) u_{j,ji} + \beta^2 u_{i,jj} + w^2 u_i = 0 \tag{4}$$

with w the circular frequency.

Because equation (4) couples the three displacement components, the most convenient approach to solve it is to express the displacements in terms of derivatives of potentials. These potentials satisfy uncoupled wave equations.

According to the Helmholtz theorem, any vector field can be expressed as the sum of the gradient of a scalar field ϕ, plus the curl of a vector field $\boldsymbol{\psi}$, *i.e.*

$$u_i = \frac{\partial \phi}{\partial x_i} + \epsilon_{ijk} \frac{\partial \psi_k}{\partial x_j} \tag{5}$$

in which ϵ_{ijk} is the alternating tensor, equal to 1 when the indices are in cyclic order, -1 when the indices are in acyclic order, and equal to zero when any two indices are equal; ϕ is the compressional displacement potential and $\boldsymbol{\psi}$ is the distortional displacement vector. The first term on the right-hand side of equation (5) is called the dilatational component of the elastodynamic displacement, and the second term is the rotational component.

The elastodynamic displacement components u_i satisfy the Navier-Cauchy equations (4) if the potentials in equation (5) satisfy the following Helmholtz equations:

$$\nabla^2 \phi + k_\alpha^2 \ \phi = 0 \tag{6}$$

$$\nabla^2 \psi_k + k_\beta^2 \ \psi_k = 0 \tag{7}$$

where $k_\alpha = w/\alpha$ and $k_\beta = w/\beta$ are the longitudinal and transversal wave numbers, respectively. Equations (6) and (7) are uncoupled wave equations.

The completeness theorem guarantees that every solution of the Navier-Cauchy equations is included in the solution of equation (5). It also guarantees the existence of only two types of waves in an unbounded elastic medium. The first are the P waves, also called primary, longitudinal, dilatational or pressure waves, which propagate with velocity α and produce displacements parallel to the direction of propagation. The second type are the S waves, also called secondary, transversal, rotational or shear waves, which propagate with a lower velocity β and produce displacements perpendicular to the direction of propagation. Both are plane waves that satisfy the wave equations (6) and (7) and the Navier-Cauchy equations (4).

Assuming that the elastic plane waves propagate over the $x_1 - x_2$ plane, the derivatives with respect to x_3 will all be zero. The in-plane displacements u_1 and u_2 are then given by the Helmholtz decomposition (5) in the form:

$$u_1 = \frac{\partial \phi}{\partial x_1} + \frac{\partial \psi_3}{\partial x_2} \tag{8}$$

$$u_2 = \frac{\partial \phi}{\partial x_2} - \frac{\partial \psi_3}{\partial x_1} \tag{9}$$

while the out-of-plane displacement is given by

$$u_3 = \frac{\partial \psi_2}{\partial x_1} - \frac{\partial \psi_1}{\partial x_2} \tag{10}$$

For simplicity of notation, the ψ_3–component of the distortional displacement vector will be referred to as ψ, and the out-of-plane displacement u_3 as w.

Because w is a linear combination of ψ_1 and ψ_2, both of which satisfy the wave equation (7), the displacement w also satisfies the same equation, *i.e.*

$$\nabla^2 w + k_\beta^2 \ w = 0 \tag{11}$$

2.1. INCIDENT WAVES

The displacement at any point in the half-space for an incident SH wave is given by the sum of w_I, the incident SH wave, and w_R, the reflected SH wave. Their expressions are as follows:

$$w_I = B_0 \exp\left[-ik_\beta\left(x_1 \sin\theta + x_2 \cos\theta\right)\right]$$

$$w_R = B_0 \exp\left[-ik_\beta\left(x_1 \sin\theta - x_2 \cos\theta\right)\right]$$

with θ the angle of incidence of the SH wave. The total displacement is given by

$$w_0 = w_I + w_R \tag{12}$$

Similarly, the displacement at any point in the half-space for an incident P wave is given by the sum of ϕ_I, the incident P wave, and ϕ_R and ψ_R, the reflected P and SV waves. Their expressions are as follows:

$$\phi_I = A_0 \exp\left[-ik_\alpha\left(x_1 \sin\theta + x_2 \cos\theta\right)\right]$$

$$\phi_R = A_1 \exp\left[-ik_\alpha\left(x_1 \sin\theta - x_2 \cos\theta\right)\right]$$

$$\psi_R = A_2 \exp\left[-ik_\beta\left(x_1 \sin\gamma - x_2 \cos\gamma\right)\right]$$

where the amplitude of the incident wave has been taken as $A_0 = (ik_\alpha)^{-1}$ and the reflection coefficients are defined by

$$\frac{A_1}{A_0} = \frac{\sin(2\theta)\sin(2\gamma) - k^2\cos^2(2\gamma)}{\sin(2\theta)\sin(2\gamma) + k^2\cos^2(2\gamma)}$$

$$\frac{A_2}{A_0} = \frac{-2k\sin(2\theta)\cos(2\gamma)}{\sin(2\theta)\sin(2\gamma) + k^2\cos^2(2\gamma)}$$

with

$$\gamma = \arcsin\left(k^{-1}\sin\theta\right)$$

the angle of the reflected SV wave relative to the x_2-axis, and k the material constant

$$k = \frac{\alpha}{\beta} = \left[\frac{2(1-\nu)}{1-2\nu}\right]^{1/2}$$

The components of the total displacement vector $\mathbf{u}_0$ are given by the equations

$$u_{0_1} = \frac{\partial\phi}{\partial x_1} + \frac{\partial\psi}{\partial x_2} = -ik_\alpha\left(\phi_I + \phi_R\right)\sin\theta + ik_\beta\psi_R\cos\gamma$$

$$u_{0_2} = \frac{\partial\phi}{\partial x_2} - \frac{\partial\psi}{\partial x_1} = -ik_\alpha\left(\phi_I - \phi_R\right)\cos\theta - ik_\beta\psi_R\sin\gamma$$

The displacement at any point in the half-space for an incident SV wave is given by the sum of ψ_I, the incident SV wave, and ψ_R and ϕ_R, the reflected SV and P waves. Only the expressions for angles of incidence smaller than the critical angle are shown. The expressions for the potentials are as follows:

$$\psi_I = B_0 \exp\left[-ik_\beta \left(x_1 \sin\theta + x_2 \cos\theta\right)\right]$$

$$\psi_R = B_1 \exp\left[-ik_\beta \left(x_1 \sin\theta - x_2 \cos\theta\right)\right]$$

$$\phi_R = B_2 \exp\left[-ik_\alpha \left(x_1 \sin\gamma - x_2 \cos\gamma\right)\right]$$

where the amplitude of the incident wave has been taken as $B_0 = (ik_\beta)^{-1}$ and the reflection coefficients are defined by

$$\frac{B_1}{B_0} = \frac{\sin(2\theta)\sin(2\gamma) - k^2\cos^2(2\theta)}{\sin(2\theta)\sin(2\gamma) + k^2\cos^2(2\theta)}$$

$$\frac{B_2}{B_0} = \frac{k\sin(4\theta)}{\sin(2\theta)\sin(2\gamma) + k^2\cos^2(2\theta)}$$

with

$$\gamma = \arcsin(k\sin\theta)$$

the angle of the reflected P wave relative to the x_2-axis. The components of the total displacement vector $\mathbf{u}_0$ are given by the equations

$$u_{0_1} = -ik_\beta\left(\psi_I - \psi_R\right)\cos\theta - ik_\alpha\phi_R\sin\gamma$$

$$u_{0_2} = -ik_\beta\left(\psi_I + \psi_R\right)\sin\theta + ik_\alpha\phi_R\cos\gamma$$

Displacements produced by Rayleigh waves decay exponentially with distance from the free surface. The displacement at any point in the half-space for incident Rayleigh waves is given by

$$u_1 = \exp(-ik_r x_1)\left[\exp(b_\alpha x_2) - \frac{1}{2}\left(2 - \frac{c_r^2}{\beta^2}\right)\exp(b_\beta x_2)\right]$$

$$u_2 = ik_r \exp(-ik_r x_1)\left[\frac{b_\alpha}{k_r^2}\exp(b_\alpha x_2) - \frac{1}{2b_\beta}\left(2 - \frac{c_r^2}{\beta^2}\right)\exp(b_\beta x_2)\right]$$

with

$$b_\alpha = k_r\left(1 - \frac{c_r^2}{\alpha^2}\right)^{1/2}$$

$$b_\beta = k_r\left(1 - \frac{c_r^2}{\beta^2}\right)^{1/2}$$

where c_r and k_r are the Rayleigh wave velocity and wave number, respectively. The Rayleigh wave velocity can be obtained from the equation for the phase velocity of Rayleigh waves

$$\left(2-\frac{c_r^2}{\beta^2}\right)^2=4\left(1-\frac{c_r^2}{\alpha^2}\right)^{1/2}\left(1-\frac{c_r^2}{\beta^2}\right)^{1/2}$$

Rayleigh waves are non-dispersive because the wave number k_r does not appear in the above equation.

3. BEM Formulation for SH Waves

Consider the problem of wave scattering by a canyon, shown in Figure 1. The domain Ω is the unbounded half-space below the infinite traction-free boundary Γ. For the propagation of SH waves in Ω, an elastodynamic problem is considered in which the out-of-plane displacement w is a solution of the Helmholtz equation (11). The boundary condition at the traction-free boundary Γ is given by

$$q=\frac{\partial w}{\partial n}=0 \tag{13}$$

with n the normal vector to Γ.

Applying the principle of superposition, the displacement w and its normal derivative q can be written in the form:

$$w=w_0+w_S \tag{14}$$

$$q=q_0+q_S=\frac{\partial w_0}{\partial n}+\frac{\partial w_S}{\partial n} \tag{15}$$

in which w_0, the free-field displacement given by equation (12), describes an SH wave propagating in a half-space, and w_S is the scattered wave due to the presence of the irregularity. The displacement w_S satisfies the Sommerfeld radiation condition and is defined by the following integral representation formula

$$c(\xi)\,w_S(\xi)=\int_\Gamma w^*(\xi,\chi)\,q_S(\chi)\,d\Gamma(\chi)-\int_\Gamma q^*(\xi,\chi)\,w_S(\chi)\,d\Gamma(\chi) \tag{16}$$

where the free term $c(\xi)$ depends on the internal angle subtended at point ξ. Function w^* is the free-space Green's function for the Helmholtz equation,

$$w^*(\xi,\chi)=\frac{i}{4}H_0^{(1)}(k_\beta\, r_1) \tag{17}$$

and q^* its normal derivative,

$$q^*(\xi,\chi)=\frac{\partial w^*(\xi,\chi)}{\partial n(\chi)}=-\frac{ik_\beta}{4}H_1^{(1)}(k_\beta\, r_1)\frac{\partial r_1}{\partial n} \tag{18}$$

in which r_1 is the distance from the source point ξ to the field point χ, $H_0^{(1)}$ and $H_1^{(1)}$ are the Hankel functions of the first kind of zero and first order, respectively.

Taking into account boundary condition (13) and using the method of images, it is possible to rewrite integral equation (16) for the total displacement w in the form

$$c(\xi)\, w(\xi) + \int_{\Gamma_S} q^*(\xi,\chi)\, w(\chi)\, d\Gamma(\chi) = w_0(\xi) \tag{19}$$

with the fundamental solutions now given by

$$w^*(\xi,\chi) = \frac{i}{4}\left[H_0^{(1)}(k_\beta\, r_1) + H_0^{(1)}(k_\beta\, r_2)\right] \tag{20}$$

$$q^*(\xi,\chi) = -\frac{ik_\beta}{4}\left[H_1^{(1)}(k_\beta\, r_1)\frac{\partial r_1}{\partial n} + H_1^{(1)}(k_\beta\, r_2)\frac{\partial r_2}{\partial n}\right] \tag{21}$$

where r_2 is the distance from point ξ to the image of point χ with respect to the free surface $x_2 = 0$.

Discretizing equation (19) and applying a collocation technique produces the system of equations

$$\mathbf{H}\,\mathbf{w} = \mathbf{w}_0 \tag{22}$$

Consider now the problem of wave scattering by a valley, shown in Figure 2. The domain Ω is now divided into two sub-regions, the half-space Ω_h and the valley Ω_v. Assume that the respective displacements are w_h and w_v. The system of equations obtained for the half-space Ω_h is given by

$$\mathbf{H}_h\,\mathbf{w}_h - \mathbf{G}_h\,\mathbf{q}_h = \mathbf{w}_0 \tag{23}$$

while that for the valley Ω_v is of the form

$$\mathbf{H}_v\,\mathbf{w}_v - \mathbf{G}_v\,\mathbf{q}_v = \mathbf{0} \tag{24}$$

Imposing compatibility and equilibrium conditions at the interface, *i.e.*

$$w = w_h = w_v$$

$$q = \mu_h\, q_h = -\mu_v\, q_v$$

with μ the shear modulus, allows the combination of equations (23) and (24) in the form

$$\begin{bmatrix} \mathbf{H}_h & -\frac{1}{\mu_h}\mathbf{G}_h \\ \mathbf{H}_v & \frac{1}{\mu_v}\mathbf{G}_v \end{bmatrix} \begin{Bmatrix} \mathbf{w} \\ \mathbf{q} \end{Bmatrix} = \begin{Bmatrix} \mathbf{w}_0 \\ \mathbf{0} \end{Bmatrix} \tag{25}$$

Notice that, given the geometry and properties of a canyon or a valley, the solution for as many angles of incidence as required can be easily obtained by

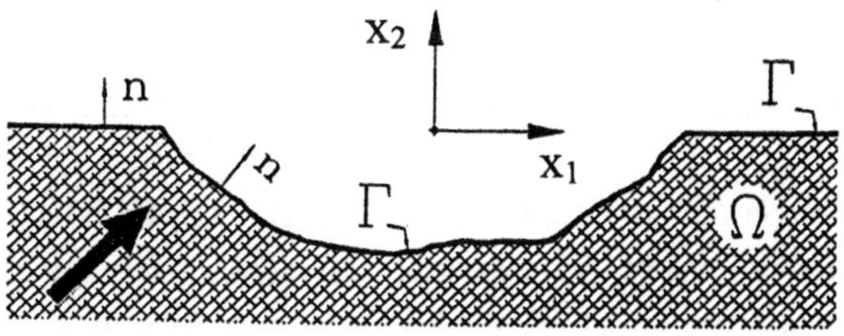

Figure 1: Scattering by canyons: geometrical definitions.

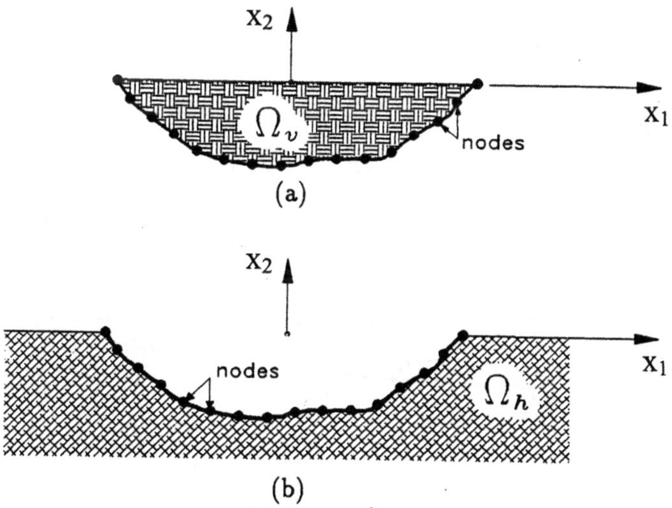

Figure 2: Scattering of SH waves by valleys: (a) Discretization of the valley, (b) Discretization of the half-space.

changing the load vector w_0 in equations (22) or (25). Damping is usually included in the formulation by considering a complex wave number,

$$k'_\beta = k_\beta\,(1 - iD)$$

where D is the hysteretic damping factor.

Figure 3 shows displacements along the surface of a parabolic valley, due to a vertical incidence of SH waves. The characteristics of the valley are the same as described by Sánchez-Sesma *et al.* [9]: the depth is 0.05 times a, the valley half-width, and the mass density and velocity ratios are 2 and 4, respectively. Displacements are shown at nine observation points at the surface (from left to right: $x_1/a = 0, 0.125, 0.25, 0.375, 0.5, 0.625, 0.75, 0.875, 1$). Because of the symmetry of the problem, only half the width is shown. The results of Sánchez-Sesma *et al.* (1993) were obtained with an indirect integral equation method, and the comparison of results presented here is excellent. It can be observed that the amplification obtained is higher than 16, which is the maximum amplitude predicted by the one-dimensional model. The response of this valley is strongly modified by the lateral interferences for frequencies higher than the one which controls the one-dimensional response ($\eta = ak/\pi = 2.5$ for the centre of the valley).

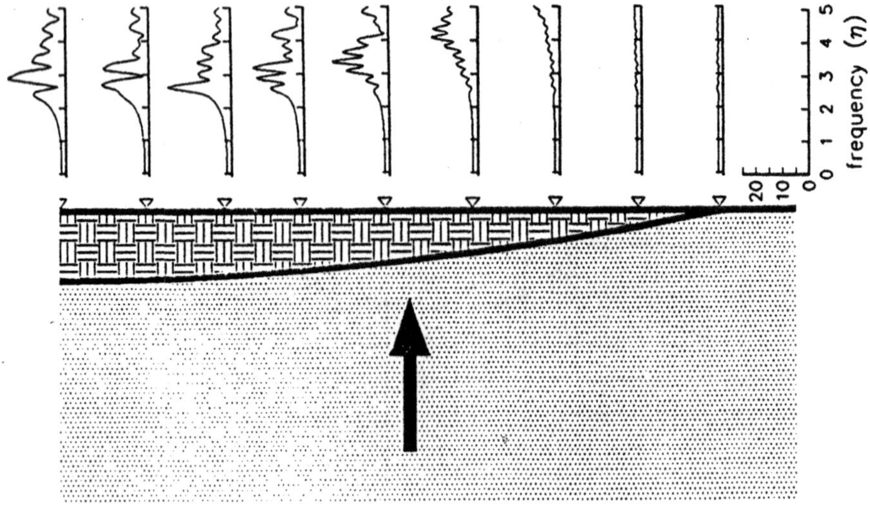

Figure 3: Displacements along the surface of a parabolic valley due to a vertical incidence of SH waves.

Since we are interested in valleys with dimensions similar to the valley of Mexico, the next example shows a very wide and shallow shape. Figure 4 shows a valley 5 km wide and 100 m deep, with a shape defined as type 2 by Bard and Bouchon [3]. The mass density and velocity ratios are 1.15 and 2.65, respectively. Displacements were obtained at different points along the surface of the valley, showing an anomalous type of oscillation in Figure 4a which almost disappear when damping of 2% for the alluvial deposit was included (Figure 4b). The one-dimensional response is also shown by a dotted line in both plots. It is clear that the closer to the edge, the more irregular the response. At the centre of the valley, the response is very close to that computed by the one-dimensional model. More realistic results are obtained when including damping in the model. This means that Love surface waves reflected at the edge of the basin attenuate rapidly and, therefore, do not affect the response at sites far from the borders of the valley.

4. BEM Formulation for P, SV and Rayleigh Waves

For the propagation of P, SV and Rayleigh waves in Ω, an elastodynamic problem is considered in which the in-plane displacement vector $\mathbf{u}$ is a solution of the Navier-Cauchy equation (4). The boundary condition at the traction-free boundary Γ is given by

$$p_i = \sigma_{ij} n_j = 0 \tag{26}$$

with σ_{ij} the components of the stress tensor.

Applying the principle of superposition, the displacements $\mathbf{u}$ and the tractions $\mathbf{p}$ can be written in the form:

$$\mathbf{u} = \mathbf{u}_0 + \mathbf{u}_S \tag{27}$$

$$\mathbf{p} = \mathbf{p}_0 + \mathbf{p}_S \tag{28}$$

The displacement $\mathbf{u}_S$ satisfies the Sommerfeld radiation condition and is defined by the following integral representation formula

$$c_{ij}(\xi)\, u_{S,j}(\xi) = \int_\Gamma u^*_{ij}(\xi,\chi)\, p_{S,j}(\chi)\, d\Gamma(\chi) - \int_\Gamma p^*_{ij}(\xi,\chi)\, u_{S,j}(\chi)\, d\Gamma(\chi) \tag{29}$$

The fundamental solutions to the Navier-Cauchy equations are of the form,

$$u^*_{ij}(\xi,\chi) = \frac{1}{2\pi\rho\beta^2}\left(\Psi\,\delta_{ij} - \Theta\, r_{,i}\, r_{,j}\right) \tag{30}$$

$$p^*_{ij}(\xi,\chi) = \frac{1}{2\pi}\left[\left(\frac{d\Psi}{dr} - \frac{\Theta}{r}\right)\left(\delta_{ij}\frac{\partial r}{\partial n} + r_{,j}\, n_i\right) - \frac{2}{r}\Theta\left(r_{,i}\, n_j - 2\, r_{,i}\, r_{,j}\frac{\partial r}{\partial n}\right) - \right.$$
$$\left. 2\frac{d\Theta}{dr} r_{,i}\, r_{,j}\frac{\partial r}{\partial n} + \left(\frac{\alpha^2}{\beta^2} - 2\right)\left(\frac{d\Psi}{dr} - \frac{d\Theta}{dr} - \frac{\Theta}{r}\right) r_{,i}\, n_j\right] \tag{31}$$

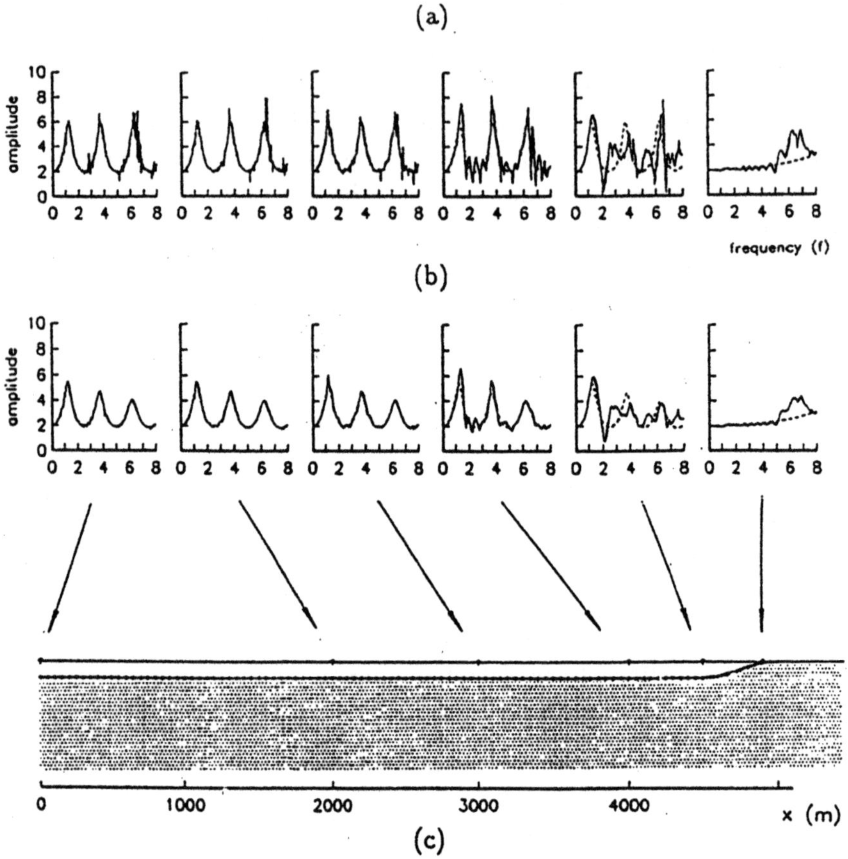

Figure 4: 2-D (solid line) and 1-D (dotted line) displacements along the surface of a shallow valley. (a) Amplification with no damping; (b) Amplification with 2% damping; (c) Geometry of the problem.

with

$$\Psi = K_0\left(i\, k_\beta\, r\right) + \frac{1}{i\, k_\beta\, r}\left[K_1\left(i\, k_\beta\, r\right) - \frac{\beta}{\alpha}K_1\left(i\, k_\alpha\, r\right)\right]$$

$$\Theta = K_2\left(i\, k_\beta\, r\right) - \frac{\beta^2}{\alpha^2}K_2\left(i\, k_\alpha\, r\right)$$

in which K_n is the modified Bessel function of the second kind, with the index n indicating the order. The derivatives of the functions Ψ and Θ are of the form

$$\frac{d\Psi}{dr} = i\, w\, K_1\left(i\, k_\beta\, r\right)\left[\frac{2\beta}{w^2 r^2} - \frac{1}{\beta}\right] + \frac{2}{r^2}\frac{\beta^2}{i\, \alpha^2\, w}K_1\left(i\, k_\alpha\, r\right) -$$

$$\frac{1}{r}\, K_0\left(i\, k_\beta\, r\right) + \frac{1}{r}\frac{\beta^2}{\alpha^2}K_0\left(i\, k_\alpha\, r\right)$$

$$\frac{d\Theta}{dr} = -i\, k_\beta\, K_1\left(i\, k_\beta\, r\right) - \frac{2}{r}K_2\left(i\, k_\beta\, r\right) + i\, k_\alpha\frac{\beta^2}{\alpha^2}K_1\left(i\, k_\alpha\, r\right) + \frac{2}{r}\frac{\beta^2}{\alpha^2}K_2\left(i\, k_\alpha\, r\right)$$

For the case of canyons, discretizing equation (29) and applying a collocation technique produces the system of equations

$$\mathbf{H}\ \mathbf{u}_S = \mathbf{G}\ \mathbf{p}_S \tag{32}$$

By virtue of the traction-free boundary condition on Γ, equation (28) produces $\mathbf{p}_S = -\mathbf{p}_0$, and the above system can be rewritten as

$$\mathbf{H}\ \mathbf{u}_S = -\mathbf{G}\ \mathbf{p}_0 \tag{33}$$

The expressions for the tractions $\mathbf{p}_0$ are obtained from the incident displacements $\mathbf{u}_0$ in the form

$$p_{0_1} = \left[(\lambda + 2\mu)\frac{\partial u_{0_1}}{\partial x_1} + \lambda\frac{\partial u_{0_2}}{\partial x_2}\right] n_1 + \mu\left(\frac{\partial u_{0_1}}{\partial x_2} + \frac{\partial u_{0_2}}{\partial x_1}\right) n_2$$

$$p_{0_2} = \left[(\lambda + 2\mu)\frac{\partial u_{0_2}}{\partial x_2} + \lambda\frac{\partial u_{0_1}}{\partial x_1}\right] n_2 + \mu\left(\frac{\partial u_{0_1}}{\partial x_2} + \frac{\partial u_{0_2}}{\partial x_1}\right) n_1$$

in which the derivatives are evaluated from the expressions for the incident P, SV and Rayleigh waves given in Section 2.

Consider now the problem of wave scattering by a valley, shown in Figure 5. The system of equations obtained for the half-space Ω_h is given by

$$\mathbf{H}_h\ \mathbf{u}_S - \mathbf{G}_h\ \mathbf{p}_S = \mathbf{0} \tag{34}$$

and, according to Figure 5, this equation can be written as

$$\mathbf{H}_h\left[\mathbf{u}_S^a + \mathbf{u}_S^b + \mathbf{u}_S^d\right]\ - \mathbf{G}_h\left[\mathbf{p}_S^a + \mathbf{p}_S^b + \mathbf{p}_S^d\right] = \mathbf{0} \tag{35}$$

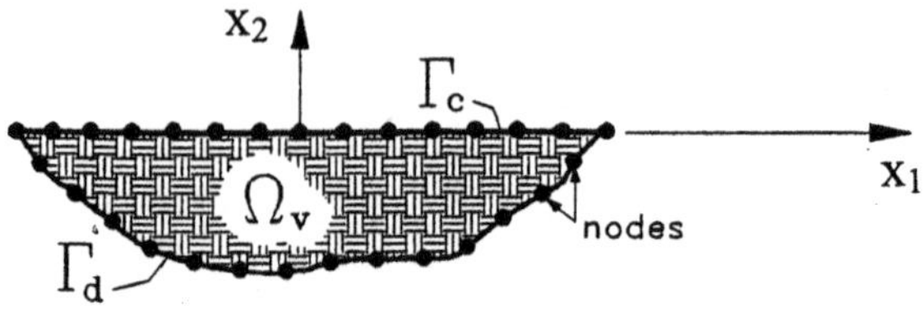

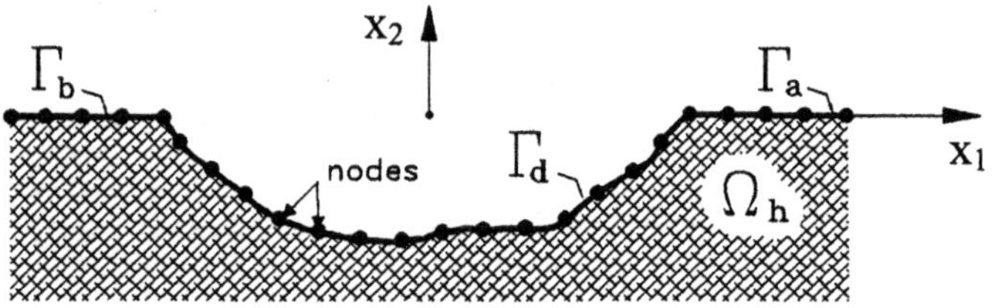

Figure 5: Scattering of P, SV and Rayleigh waves by valleys: (a) Discretization of the valley, (b) Discretization of the half-space.

Because the traction-free condition applies on Γ_a and Γ_b, the above becomes

$$\mathbf{H}_h \left[\mathbf{u}_S^a + \mathbf{u}_S^b + \mathbf{u}_S^d\right] \; - \mathbf{G}_h \; \mathbf{p}_S^d = -\mathbf{G}_h \left[\mathbf{p}_0^a + \mathbf{p}_0^b\right] \tag{36}$$

On the other hand, the system of equations for the valley is given in terms of total displacements, in the form

$$\mathbf{H}_v \; \mathbf{w}_v - \mathbf{G}_v \; \mathbf{p}_v = \mathbf{0} \tag{37}$$

or, according to Figure 5,

$$\mathbf{H}_v \left[\mathbf{u}_v^c + \mathbf{u}_v^d\right] \; - \mathbf{G}_v \left[\mathbf{p}_v^c + \mathbf{p}_v^d\right] = \mathbf{0} \tag{38}$$

where the total tractions $\mathbf{p}_v^c = \mathbf{0}$.

Imposing compatibility and equilibrium conditions at the interface, *i.e.*

$$\mathbf{u}_v^d = \mathbf{u}_h^d$$

$$\mathbf{p}_v^d = -\mathbf{p}_h^d$$

allows the combination of equations (36) and (38) in the form

$$\begin{bmatrix} \mathbf{H}_h & \mathbf{H}_h & \mathbf{H}_h & \mathbf{0} & -\mathbf{G}_h \\ \mathbf{0} & \mathbf{H}_v & \mathbf{0} & \mathbf{H}_v & \mathbf{G}_v \end{bmatrix} \begin{Bmatrix} \mathbf{u}_S^a \\ \mathbf{u}_S^d \\ \mathbf{u}_S^b \\ \mathbf{u}_v^c \\ \mathbf{p}_S^d \end{Bmatrix} = \begin{Bmatrix} -\mathbf{G}_h\, \mathbf{p}_0^a - \mathbf{G}_h\, \mathbf{p}_0^b \\ -\mathbf{H}_v\, \mathbf{u}_0^d - \mathbf{G}_v\, \mathbf{p}_0^d \end{Bmatrix} \quad (39)$$

Similarly to SH waves, given the geometry and properties of a canyon or a valley, the solution for as many angles of incidence as required can bė easily obtained by changing the load vectors $\mathbf{u}_0$ and $\mathbf{p}_0$ in equation (39). Damping can be included by considering complex wave numbers,

$$k'_\alpha = k_\alpha\,(1 - iD)$$

$$k'_\beta = k_\beta\,(1 - iD)$$

$$k'_r = k_r\,(1 - iD)$$

In this way, P, SV and Rayleigh waves decay to infinity as $e^{-Dk_\alpha r}$, $e^{-Dk_\beta r}$ and $e^{-Dk_r r}$, respectively.

Figure 6 shows the amplification obtained at the surface of a semicircular alluvial valley. The transversal wave velocity β of the half-space and the valley are equal to 1 and 1/2, respectively, and their Poisson ratio is 1/3. Figures 6a and 6b show the displacements obtained for incident P and SV waves, respectively; the left plot of both figures shows the response to vertical incidence ($\theta = 0°$), while the right plot shows the response to oblique incidence ($\theta = 60°$ for P waves and $\theta = 30°$ for SV waves). All figures are for a normalized frequency of 0.5. Comparison is shown with the results of Dravinsky and Mossessian (1987), with excellent agreement.

5. Observed Amplification in the Mexico City Valley

Mexico City is located on a valley approximately 110 km long and 80 km wide. The valley is completely surrounded by mountains, some of which reach up to 5,230 m above the sea level. The lowest part of the valley has an altitude of 2,230 m. Figure 7 shows the southwest part of the valley where the city is located. Some reference sites and main streets are indicated, as well as accelerometric stations and the main geotechnical zones: (1) hill zone, localized in the higher parts of the valley, formed by hard soils of high resistance; (2) transition zone, with mixed characteristics of the hill and lake-bed zones; (3) lake-bed zone, consisting of very soft compressive alluvial deposits, where shear wave velocities can be as low as 50 m/s and water content as high as 400%.

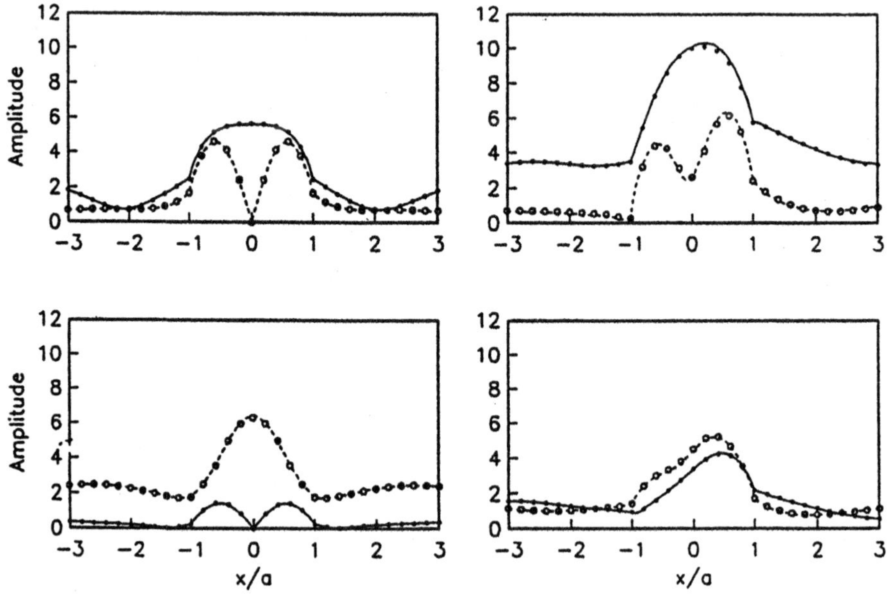

Figure 6: Response of a semicircular canyon for two different incidence angles. (a) P wave; (b) SV wave. Present results are shown by continuous (x_1–direction) and dashed (x_2–direction) lines. Results by Dravinsky and Mossessian are shown by dots and circles.

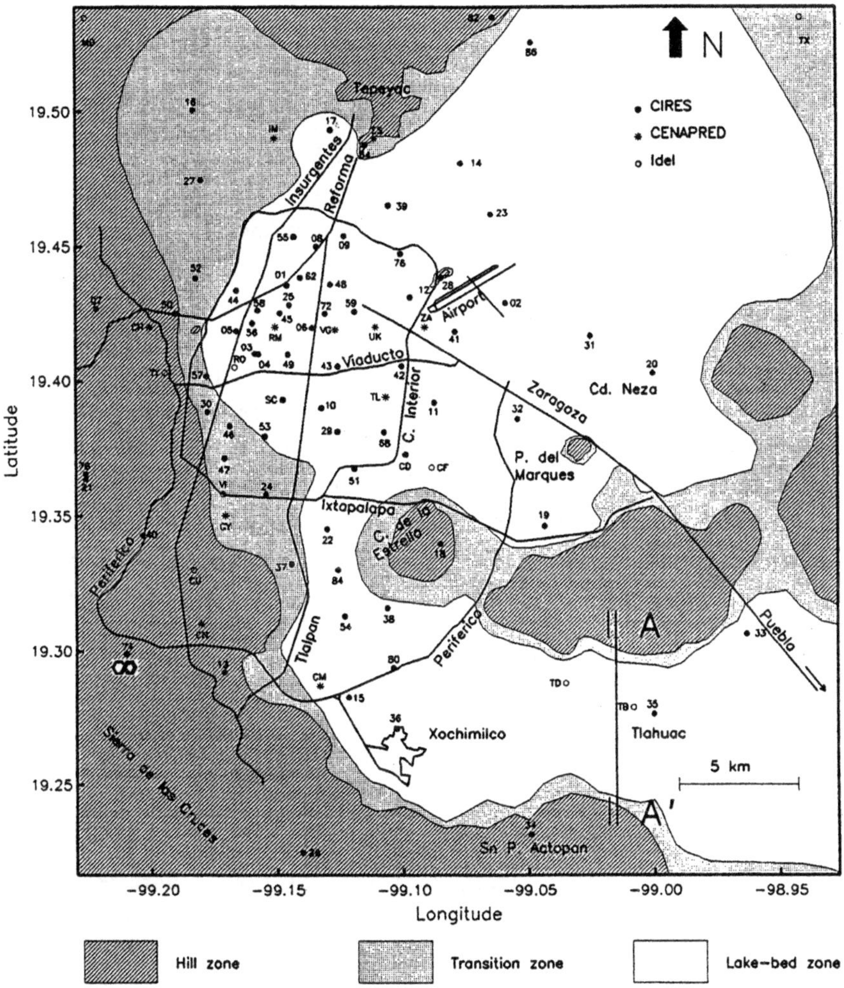

Figure 7: Mexico City valley: accelerometric stations, main streets, reference sites and geotechnical zones.

Since 1965, accelerometric data in the city have been obtained, for a wide range of earthquakes with different magnitudes (M) and epicentral distances (R), at a hill-zone site (CU), south of the city. During the devastating Michoacan earthquake of 1985 (M = 8.1, R = 400 km), data were collected at eleven sites; unfortunately, only site SC recorded data from the damage zone in the city. This record alone provided an important proof of the amplification that can be observed in alluvial valleys. With a larger network of more than 100 digital accelerometers, enormous amounts of data have been collected since 1986 from more than 16 small and moderate subduction earthquakes ($4.5<M<7.5$), all of them with a range of distance from their rupture area to CU between 250 and 500 km.

The motion observed in the Mexico City valley is one of the best known examples of dynamic amplification in alluvial basins. At some frequencies, this amplification could be as large as 500 times relative to the registered motion at near-source stations (Singh *et al.*, 1988).

The dramatic way in which site effects manifest in Mexico City is mainly due to the high contrast between the dynamic characteristics of the alluvial strata and the bed rock. In the frequency domain, the form of the amplification is controlled by the elastic impedances, soil damping, characteristics of the incident field and the geometry of the valley. In the time domain, the response is manifested in harmonic motions and in the larger amplitude and duration of the records.

Singh *et al.* [30] used accelerometric data from subduction earthquakes to compute empirical transfer functions or spectral ratios, in order to obtain the measured amplification at transition and lake-bed sites with respect to CU. However, some differences in the amplification patterns were found that could not be satisfactorily explained.

A later study by Reinoso [31] observed that the amplification obtained at more than 80% of sites was practically the same for all earthquakes, independent of their magnitude, distance or azimuth. This was possible because of the incorporation of data from more recent earthquakes, the selection of data considered reliable (as spectral ratios are very much dependent on the noise-to-signal ratio), and the use of the average Fourier amplitude at hill-zone sites as the reference site. This conclusion was also valid for the earthquake of 1985, in which important non-linear soil behaviour did not seem to be present [32]. The idea of using the average Fourier spectra at hill-zone sites is justified because strong motions there exhibit important differences from one site to another during the same earthquake but, on average, their behaviour in the frequency domain is fairly similar [13].

From the spectral ratio results, it has been shown that most lake and transition site behaviour can be predicted by one-dimensional models and, therefore, are considered as having regular responses. However, at some sites near the edges of the valley and some sites over the thicker alluvial deposits, the ratios show responses suggesting two- and three-dimensional effects, since the ratio is systematically different for both horizontal components. Figure 8, taken from [13], depicts some average ratios (considering all earthquakes) of both horizontal components of motion; they are indicated with an R, if showing a regular response, with a T, if located over a thick alluvial deposit, or with an E, if located near the edges of the lake-bed zone.

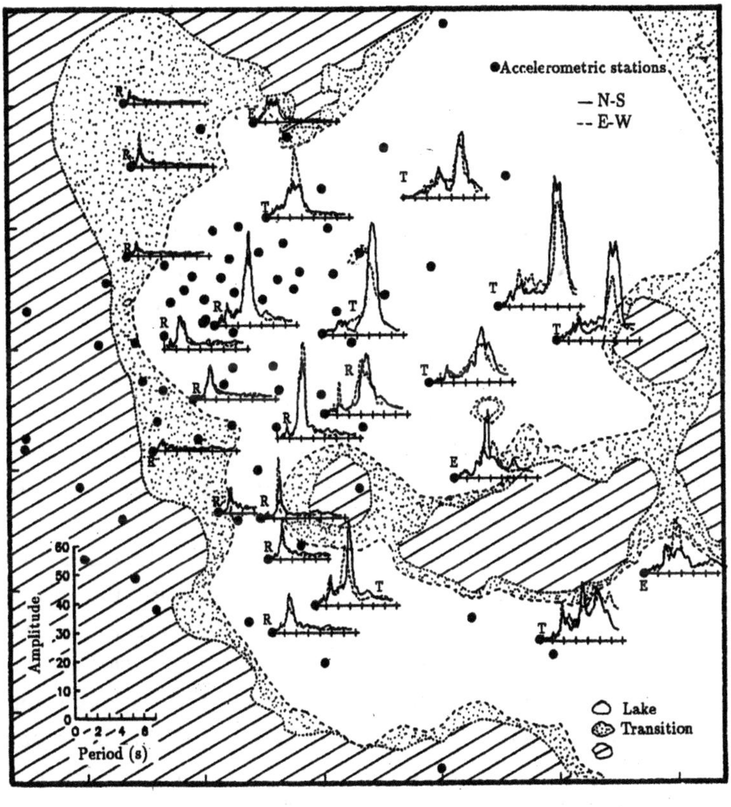

Figure 8: Some average ratios of all earthquakes.

6. One-dimensional Response of the Mexico City Valley

Because the soft alluvial layer in Mexico City is relatively flat and shallow, the one-dimensional model for vertically incident S waves has been the only analytical method actually employed to predict the amplification of seismic motion at lake-zone sites. This was the case of the design spectra included in the 1987 seismic code, which were obtained with one-dimensional results using CU as reference station [33]. Romo and Seed [34] have obtained the response spectra at various locations of the valley using random vibration theory and one-dimensional wave propagation models. Seed *et al.* [35] concluded that one-dimensional models produced a good agreement with observed response spectra for the 1985 earthquake, although Kawase and Aki [21] showed that the model of Seed *et al.* [35] was not accurate for spectral ratios computed for sites SC and CD, for the same 1985 earthquake. Chávez-Garcia and Bard [36] compared the response at one site for the 1989 earthquake with the one-dimensional model, finding a good agreement in the frequency domain but not in the time domain. Ordaz *et al.* [37] analyzed borehole recordings with the one-dimensional model and found an excellent agreement with the observed amplification, while Ordaz and Faccioli [32] used a one-dimensional model to study the non-linear response of two sites of the southern Xochimilco-Chalco lake.

As a typical example of one-dimensional behaviour, we have anayzed site 84 located in an area which is now considered as a high intensity seismic zone [13]. The characteristics of motion at site 84 are very similar to its surrounding stations and, therefore, all the following studies regarding 84 are also valid for the other stations.

With the simple geometry shown in Figure 9 and the Haskell method, we have computed the one-dimensional response at site 84 using data for the north-south (NS) component of motion recorded during the 1989 earthquake ($M = 6.9$, $R = 300$ km), using each of the eleven hill-zone accelerometric stations as reference sites. Figure 10a shows a comparison between these 1D responses and the recorded accelerogram at site 84. The right part of Figure 10b shows the spectral ratio between the respective hill-zone site and site 84; at the top of Figure 10b, the average spectral ratio (dotted line) and the 1D transfer function (solid line) are also shown. Although important differences can be seen for each reference station in both the time and frequency domains, we argue that site 84 exhibits a 1D response because the average ratio is almost identical to the 1D transfer function.

The differences between results computed for each station have reasonable explanations only for sites 18, 28 and 64, which are located over old (Tertiary) features of the valley, and have systematically lower amplitude. The difference for other stations could be attributed to random or noise contributions to the strong

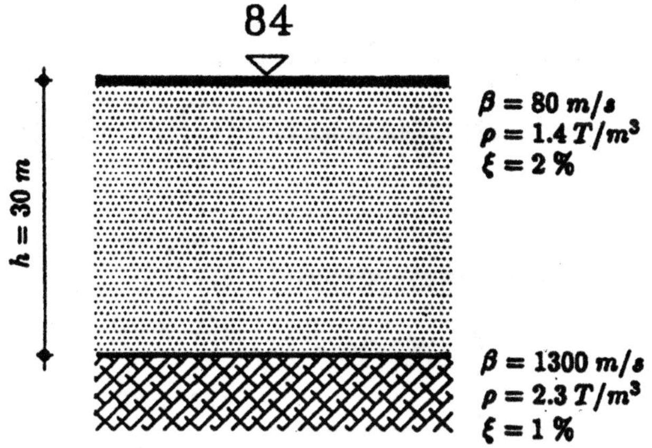

Figure 9: Profile adopted at site 84 to obtain 1D response.

ground motion at hill-zone sites, which disappear when calculating the average motion.

From these results, it is clear that, on average, the 1D model is capable of reproducing the amplification patterns observed at site 84, for a moderate earthquake. But it is also clear that the results can be used as an argument against the efficiency of the 1D theory, by stressing the differences in the time domain (lower or higher amplitudes, shorter durations, different phases and larger harmonic codas) or in the frequency domain (different amplitude and frequency content). These different amplification patterns illustrate the importance of selecting a proper reference site to study the lake-bed zone behaviour, as published by Reinoso *et al.* [13] and later by Ordaz and Faccioli [32].

7. Two-dimensional Modelling Using the BEM

Site TB, located at the centre of the alluvial deposit over strata of more than 100 m of very soft clay, was chosen as an example of possible 2D behaviour because the spectral ratios for this station are very different for both horizontal components and for all earthquakes, so the average ratio presents large standard deviations.

Figure 11 shows the ratios computed for site TB for the 1985 (Figure 11a) and 1989 (Figure 11b) earthquakes with respect to sites CU (solid line), TY (dashed line) and the average Fourier spectra at hill-zone sites (dotted line); also shown are accelerometric records. It can be observed in the figure that results for the two

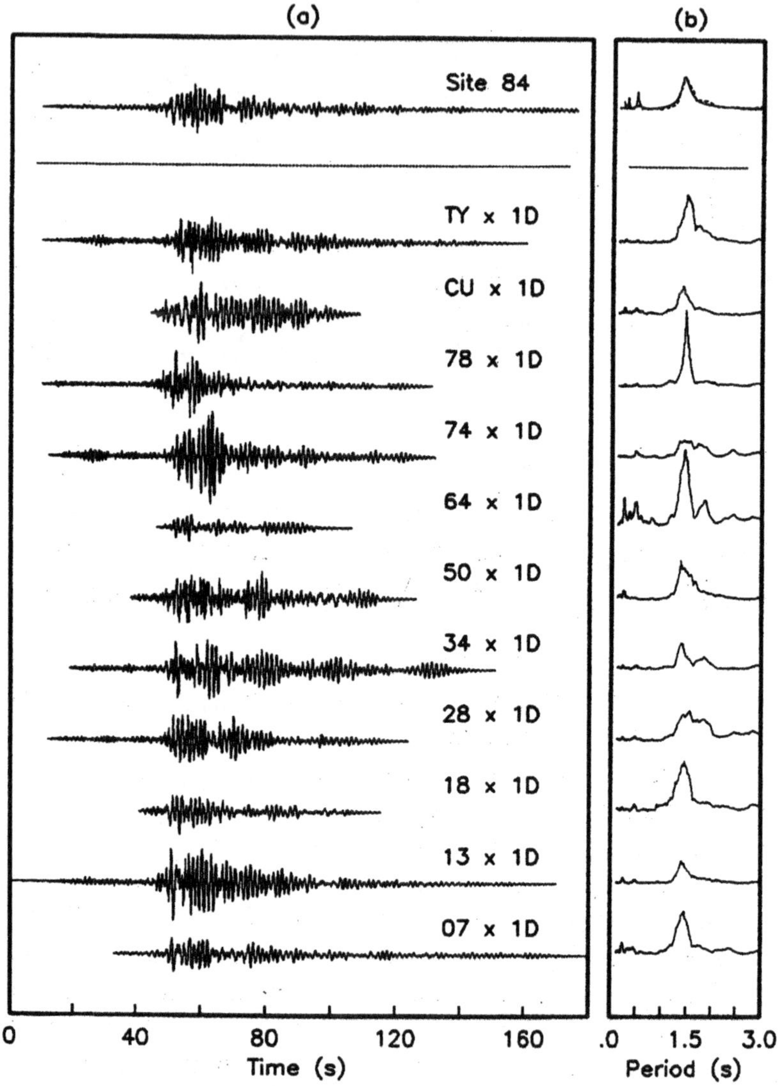

Figure 10: Response at site 84 for the north-south component of motion during the 1989 earthquake: (a) Recorded accelerogram (top) and one-dimensional time responses using all hill zone sites, (b) One-dimensional transfer function and average ratio (top) with solid and dashed lines, respectively, and spectral ratios between each hill-zone site and site 84.

horizontal components of each earthquake differ considerably, a strong evidence of 2D and 3D response. Moreover, the different shape of the ratios for each earthquake suggests that site TB could be very sensitive to the earthquake azimuth: the 1989 earthquake came straight from the south (azimuth = 186°) while the 1985 earthquake came from the southwest (azimuth = 240°). Because of the canal-type geometry of the valley, the latter could have caused 3D amplification patterns, and data from the 1985 earthquake were not used.

Another reason to select site TB in this study is because it was the easiest place to propose a relatively simple (in shape and size) 2D geometry with rough characteristics of the actual valley. The section modelled is section A-A' of Figure 12. The geometry used to represent it was a 5 km wide and 150 m deep parabolic valley (Figure 12). The velocity and mass density ratios between the half-space and the valley were 13 and 2, respectively. The damping and Poisson's ratios of the material inside the valley were 2% and 0.495, and those of the bed rock were 1% and 0.333. The S wave velocity of the bed rock was taken as 1300 m/s. The 1D response of this site was computed using the above properties, by considering a flat valley 145 m deep.

Figure 13 shows the transfer functions at site TB obtained by Reinoso *et al.* (1993) for incident SH waves. Figure 13(a) shows the east-west (EW) component of the 1989 earthquake measured at CU, while Figure 13(b) shows the same for site TB and the spectral ratio between the two records. Figures 13(c), (d) and (e) show the computed time-response using CU as the reference site and the spectral amplification for 0°, 30° and 60° of incidence, respectively, while Figure 13(f) shows the time-response and the transfer function for the 1D case.

From the comparison between the spectral ratio and the transfer functions, it is clear that the 2D results can reproduce some of the amplification patterns not predicted by the 1D model. For instance, the peaks between the first and second modes, specially the one at 0.2 Hz in the transfer function and 0.21 Hz in the spectral ratio. In spite of the short duration of the records at CU, the 2D time-domain results are closer to the observed accelerogram than those of the 1D model.

Figure 14 shows the transfer functions at site TB obtained by Reinoso *et al.* [24] for incident P, SV and Rayleigh waves, for the north-south (NS) component of the 1989 earthquake. Results are shown from top to bottom for angles of incidence of 45°, 30° and 15°, respectively, for P waves, and 30°, 15° and 0°, respectively, for SV waves. Plots on the left, centre and right correspond to results at the surface points x_1 = 100, 400 and 700 m, respectively, measured from the centre of the valley.

The smooth spectral ratio between TB and CU, obtained with the NS com-

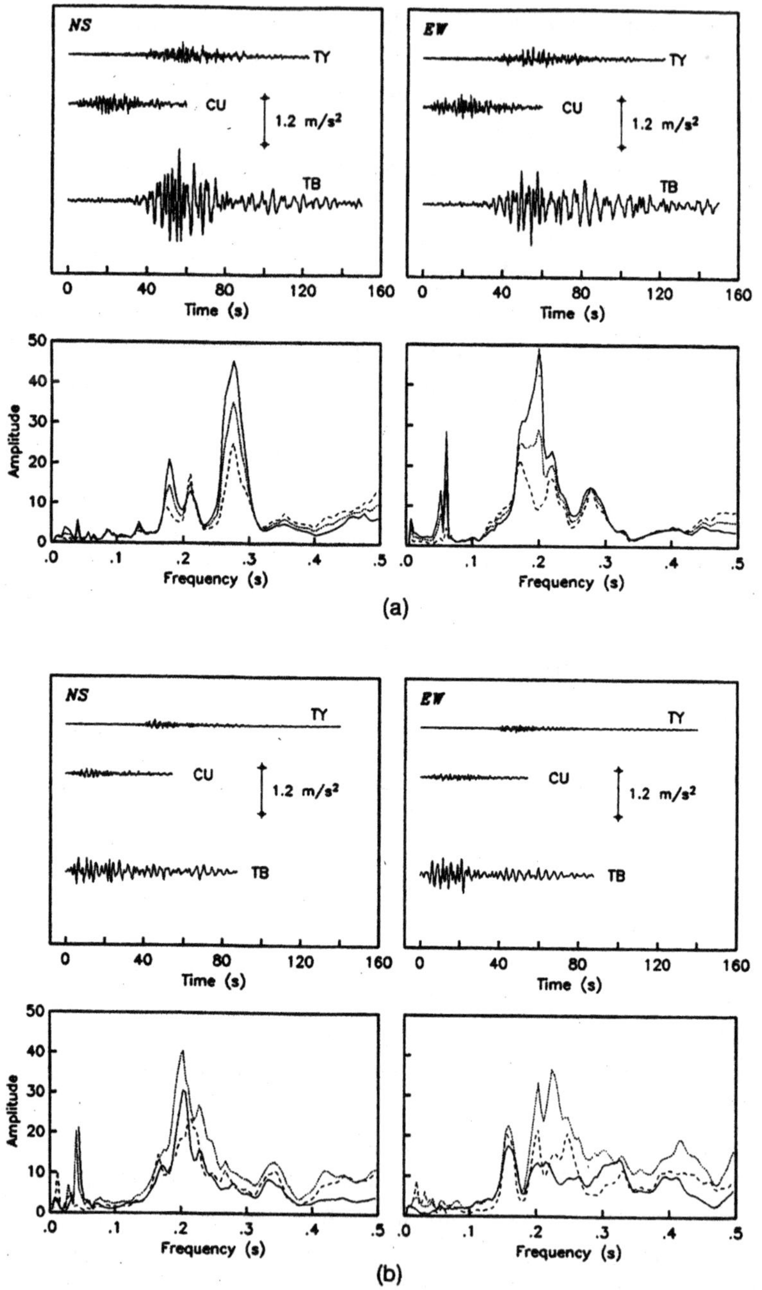

Figure 11: Ratios computed for site TB with respect to CU (solid line) and TY (dashed line); also shown are the transfer functions with respect to the average Fourier spectra at hill-zone sites (dotted line) and the accelerometric records: (a) 19 September 1985 earthquake; (b) 25 April 1989 earthquake.

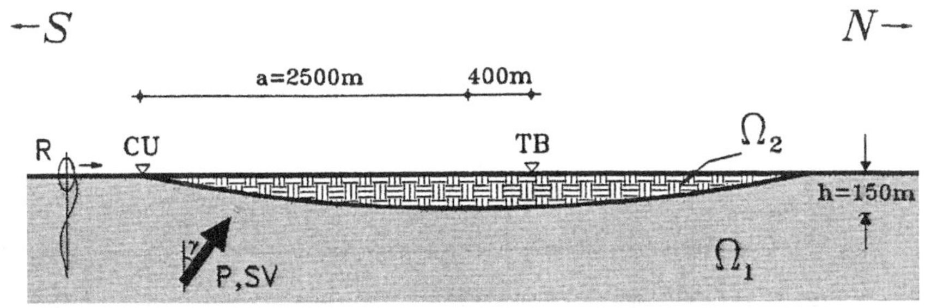

Figure 12: Two-dimensional section used to model the response at site TB.

ponents of the 1989 earthquake, is shown in Figure 15. Figure 16 shows the one-dimensional response for vertically incident SV waves at the site TB ($x_1 = 400$ m) and for the other two sites near TB ($x_1 = 100$ and 700 m), computed using the same properties of the valley at the corresponding depths of 145, 149 and 138 m.

It is clear in the figures that the 2D response for P and SV waves is closer to the spectral ratio at site TB than the 1D response. For non-vertical incidence, some peaks appear that are also observed in the spectral ratio. However, the ratio has a larger amplitude than the 2D responses. For the other two observation points, 2D modelling shows amplification patterns very different from those of the 1D model, particularly at the point $x_1 = 700$ m. The response to an incident Rayleigh wave is very similar to the 1D response.

8. Conclusions

This chapter has reviewed the direct formulation of the boundary element method for the scattering of seismic waves. A BEM formulation for the Helmholtz equation was used to model out-of-plane displacements due to incident SH waves, while a BEM formulation for the Navier-Cauchy equations of elastodynamics was used to model in-plane displacements due to incident P, SV and Rayleigh waves.

The BEM formulations were then used to reproduce observed site amplifications due to earthquakes in Mexico City. An initial comparison between results obtained with a one-dimensional model and the observed amplification at site 84, located in a zone of heavy damage during the 1985 earthquake, was initially presented. The results showed that the 1D model, on average, can explain most of the observed amplification at site 84. This conclusion is also valid for most of the zones where strong structural damage has been observed during recent earthquakes.

Data recorded at site TB, located at the centre of a thick alluvial deposit,

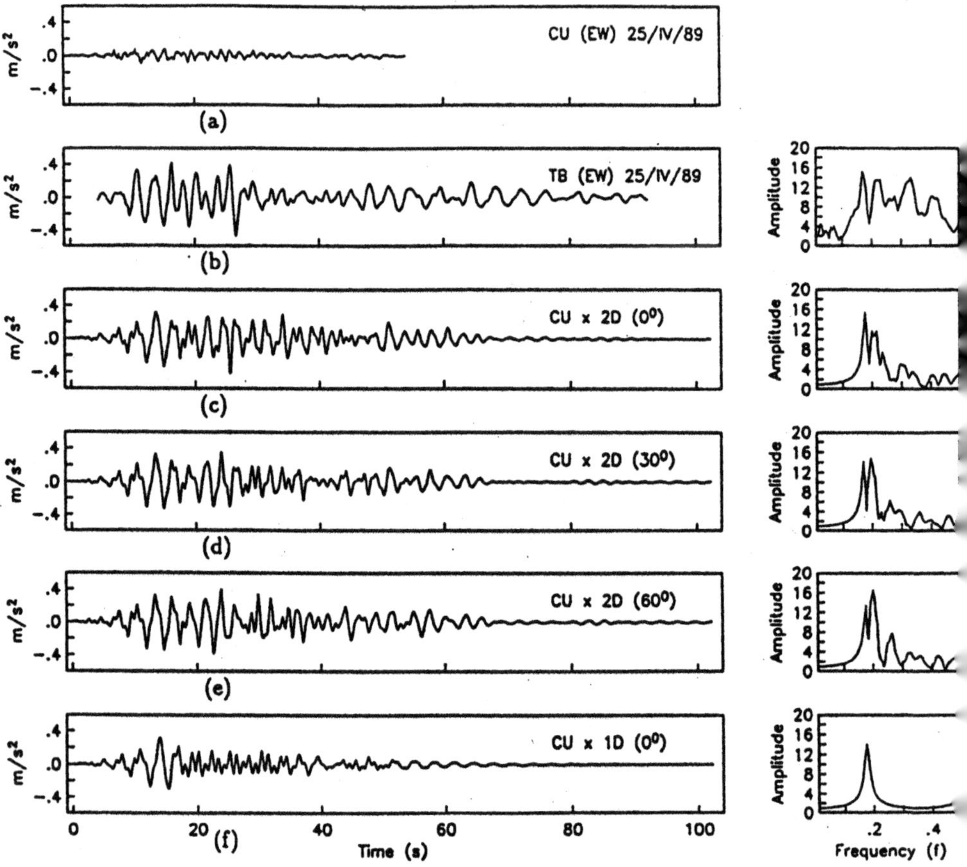

Figure 13: Comparison between observed motion at site TB and results from the 1D and 2D models: (a) reference site CU; (b) observed motion at site TB and spectral ratio with respect to CU; (c), (d), (e) 2D time-response and transfer function at TB for 0°, 30° and 60° of incidence, respectively; (f) 1D time-response and transfer function.

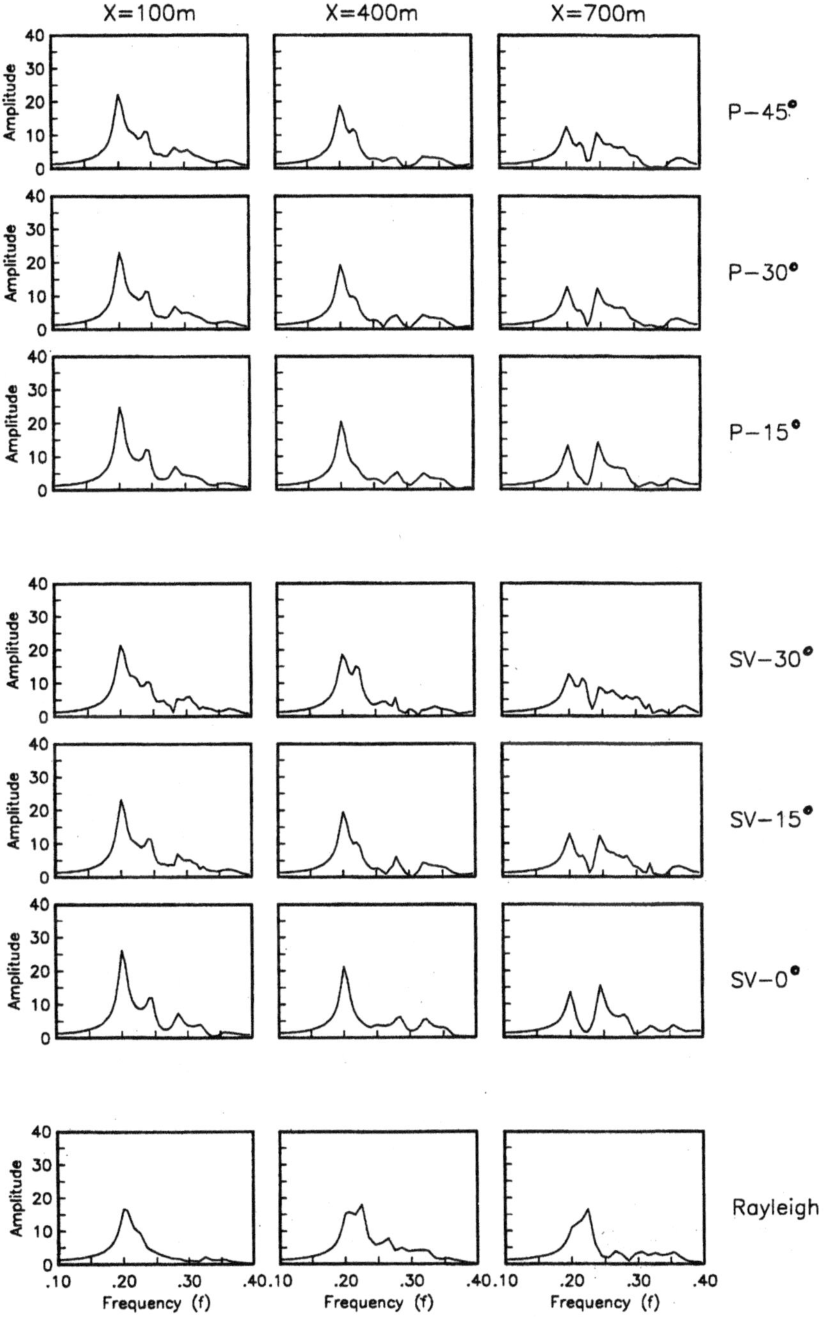

Figure 14: Two-dimensional response at three observation points, due to incident P, SV and Rayleigh waves.

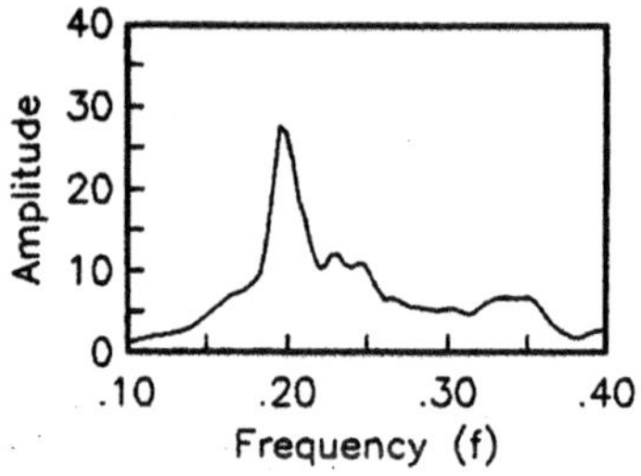

Figure 15: Spectral ratio of the north-south component between CU and TB.

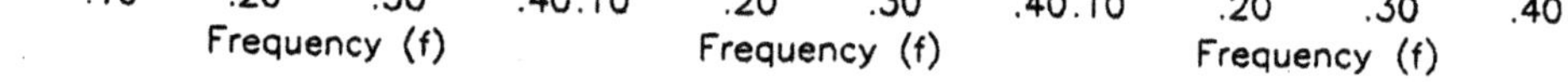

Figure 16: One-dimensional transfer function for vertically incident S waves at $x_1 = 100$, 400 and 700 m.

were used to probe the irregular amplification patterns observed there. Numerical results obtained with two-dimensional boundary element models are closer to the observations than the 1D results suggesting, at least qualitatively, that site TB was probably affected by 2D and 3D effects due to the different azimuth of incoming earthquakes.

9. References

1. Aki, K. and Larner, L. (1970) Surface motion of a layered medium having an irregular interface due to incident plane SH waves, *J. Geophys. Res.*, **75**, 933-954.
2. Bouchon, M. (1973), Effect of topography on surface motion, *Bull. Seismol. Soc. Am.*, **63**, 615-632.
3. Bard, P.-Y. and Bouchon, M. (1980), The seismic response of sediment-filled valleys. Part 1: The case of incident SH waves, *Bull. Seismol. Soc. Am.*, **70**, 1263-1286.
4. Bard, P.-Y. and Bouchon, M. (1985), The seismic resonance of sediment-filled valleys, *Bull. Seismol. Soc. Am.*, **75**, 519-541.
5. Boore, D.M. (1972), A note on the effect of simple topography on seismic SH waves, *Bull. Seismol. Soc. Am.*, **62**, 275-284.
6. Wong, H.L. and Jennings, P.C. (1975), Effects of canyon topography on strong ground motion, *Bull. Seismol. Soc. Am.*, **65**, 1239-1257.
7. Sánchez-Sesma, F.J. and Esquivel, J.A. (1979), Ground motion on alluvial valleys under incident plane SH waves, *Bull. Seismol. Soc. Am.*, **69**, 1107-1120.
8. Sánchez-Sesma, F.J., Herrera, I. and Aviles, J. (1982), A boundary method for elastic wave diffraction: Application to scattering of SH waves by surface irregularities, *Bull. Seismol. Soc. Am.*, **72**, 473-490.
9. Sánchez-Sesma, F.J., Ramos-Martínez, J. and Campillo, M. (1993), An indirect boundary element method applied to simulate the seismic response of alluvial valleys for incident P, S and Rayleigh waves, *Earthquake Engineering and Structural Dynamics*, **22**, 279-295.
10. Dravinski, M. (1982) Influence of interface depth upon strong ground motion, *Bull. Seismol. Soc. Am.*, **72**, 597-614.
11. Dravinski, M. (1983) Ground motion amplification due to elastic inclusions in a half-space, *Earthquake Engineering and Structural Dynamics*, **11**, 313-335.
12. Kawase, H. (1988), Time-domain response of a semi-circular canyon for incident SV, P and Rayleigh waves calculated by the discrete wavenumber boundary element method, *Bull. Seismol. Soc. Am.*, **78**, 1415-1437.
13. Reinoso, E., Wrobel, L.C. and Power, H. (1993), Preliminary results of the modelling of the Mexico City valley with a two-dimensional boundary element method for the scattering of SH waves, *Soil Dynamics and Earthquake Engineering*, **12**, 457-468.
14. Clements, D. and Larsson, A. (1994) A note on surface motion of inhomogeneous alluvial valleys due to incident plane SH waves, *Bull. Seismol. Soc. Am.*, **84**, 192-201.
15. Shah, A.H., Wong, K.C. and Datta, S.K. (1982), Diffraction of plane SH waves in a half-space, *Earthquake Engineering and Structural Dynamics*, **10**, 519-528.

16. Bouden, M., Khair, K.R. and Datta, S.K. (1990) Ground motion amplification by cylindrical valleys embedded in a layered medium, *Earthquake Engineering and Structural Dynamics*, **19**, 497-512.
17. Bard, P.-Y. and Bouchon, M. (1980), The seismic response of sediment-filled valleys. Part 2: The case of incident P and SV waves, *Bull. Seismol. Soc. Am.*, **70**, 1921-1941.
18. Harmsen, S. and Harding, S. (1981) Surface motion over a sedimentary valley for incident plane P and SV waves, *Bull. Seismol. Soc. Am.*, **71**, 655-670.
19. Wong, H.L. (1982), Effect of surface topography on the diffraction of P, SV and Rayleigh waves , *Bull. Seismol. Soc. Am.*, **72**, 1167-1183.
20. Dravinski, M. and Mossessian, T.S. (1987), Scattering of plane harmonic P, SV and Rayleigh waves by dipping layers of arbitrary shape, *Bull. Seismol. Soc. Am.*, **77**, 212-235.
21. Kawase, H. and Aki, K. (1989), A study of the response of a soft basin for incident S, P and Rayleigh waves with special reference to the long duration observed in Mexico City, *Bull. Seismol. Soc. Am.*, **79**, 1361-1382.
22. Sánchez-Sesma, F.J. and Campillo, M. (1991), Diffraction of P, SV and Rayleigh waves by topographical features: A boundary integral formulation, *Bull. Seismol. Soc. Am.*, **81**, 2234-2253.
23. Sánchez-Sesma, F.J. and Campillo, M. (1993), Topographic effects for incident P, SV and Rayleigh waves, *Tectonophysics*, **218**, 113-125.
24. Reinoso, E., Wrobel, L.C. and Power, H. (1997), Two-dimensional scattering of P, SV and Rayleigh waves: Preliminary results for the Valley of Mexico, *Earthquake Engineering and Structural Dynamics*, **26**, 595-616.
25. Papageorgiou, A.S. and Kim, J. (1993), Propagation and amplification of seismic waves in 2D valleys excited by obliquely incident P and SV waves, *Earthquake Engineering and Structural Dynamics*, **22**, 167-182.
26. Chang, W. (1993), Boundary integral equation method for the diffraction of elastic waves using simplified Green's functions, *Earthquake Engineering and Structural Dynamics*, **22**, 773-789.
27. Zhao, C. and Valliapan, S. (1993) Seismic wave scattering effects under different canyon topographic and geological conditions, *Soil Dynamics and Earthquake Engineering*, **13**, 129-144.
28. Reinoso, E., Wrobel, L.C. and Power, H. (1997), Three-dimensional scattering of seismic waves from topographical irregularities, *Soil Dynamics and Earthquake Engineering*, **16**, 41-61.
29. Dominguez, J. (1993), *Boundary Elements in Dynamics*, Computational Mechanics Publications, Southampton.
30. Singh, S.K., Lermo, J., Dominguez, T., Ordaz, M., Espinosa, J.M., Mena, E. and Quaas, R. (1988), A study of amplification of seismic waves in the Valley of Mexico with respect to a hill zone site, *Earthquake Spectra*, **4**, 653-673.
31. Reinoso, E. (1991), Seismic effects in the valley of Mexico: measured amplification at lake-bed zone sites, *Proc. 9th Nat. Conf. on Earthquake Engineering*, Manzanillo, Mexico, **1**, pp. 63-76.
32. Ordaz, M. and Faccioli, E. (1994), Site response analysis in the Valley of Mexico: selection of input motion and extent of non-linear soil behaviour, *Earthquake Engineering and Structural Dynamics*, **23**, 895-908.
33. Rosenblueth, E., Sánchez-Sesma, F.J., Ordaz, M. and Singh, S.K. (1988), Design spectra for Mexico's Federal District, *Earthquake Spectra*, **5**, 273-292.

34. Romo, M.P. and Seed, H.B. (1986), Analytical modeling of dynamic soil response in the Mexico earthquake of September 19, 1985, in *Mexico Earthquakes - 1985*, Amer. Soc. Civil Engineers.
35. Seed, H.B., Romo, M.P., Sun, J.I., Jaime, A. and Lysmer, J. (1988), Relationships between soil conditions and earthquake ground motion, *Earthquake Spectra*, **4**, 687-729.
36. Chávez-Garcia, F.J. and Bard, P.-Y. (1994), Site effects in Mexico City eight years after the September 1985 Michoacan earthquakes, *Soil Dynamics and Earthquake Engineering*, **13**, 229-247.
37. Ordaz, M., Santoyo, M.A., Singh, S.K. and Quaas, R. (1992), Analysis of the borehole recordings obtained in Mexico City during the May 31, 1990 earthquake, *Proc. Int. Symp. Effects of Surface Geology on Seismic Motion*, ESG1992, Odawara, Japan, **1**, pp. 155-160.

INDEX

GPSR Compliance

The European Union's (EU) General Product Safety Regulation (GPSR) is a set of rules that requires consumer products to be safe and our obligations to ensure this.

If you have any concerns about our products, you can contact us on ProductSafety@springernature.com

In case Publisher is established outside the EU, the EU authorized representative is:

Springer Nature Customer Service Center GmbH
Europaplatz 3
69115 Heidelberg, Germany

Batch number: 10584518

Printed by Printforce, the Netherlands